D1190997

Primate Behavior
Developments in Field and Laboratory Research

Volume 1

Contributors

IRWIN S. BERNSTEIN
G. MITCHELL
FRANK E. POIRIER
DUANE M. RUMBAUGH
GENE P. SACKETT
M. W. SORENSON

Primate Behavior

Developments in Field and Laboratory Research

Volume 1

Edited by

Leonard A. Rosenblum

Primate Behavior Laboratory
Department of Psychiatry
Downstate Medical Center
Brooklyn, New York

ACADEMIC PRESS   New York and London 1970

ACADEMIC PRESS, INC.
111 Fifth Avenue, New York, New York 10003

United Kingdom Edition published by
ACADEMIC PRESS, INC. (LONDON) LTD.
Berkeley Square House, London W1X 6BA

LIBRARY OF CONGRESS CATALOG CARD NUMBER: 79-127677

PRINTED IN THE UNITED STATES OF AMERICA

Contents

Behavior of Tree Shrews

M. W. Sorenson

Abnormal Behavior in Primates

G. Mitchell

The Nilgiri Langur (*Presbytis johnii*) of South India

Frank E. Poirier

List of Contributors

IRWIN S. BERNSTEIN, Yerkes Regional Primate Research Center, Emory University, Atlanta, Georgia

G. MITCHELL, Department of Behavioral Biology and Department of Psychology, University of California, Davis, California

FRANK E. POIRIER, Department of Anthropology, Ohio State University, Columbus, Ohio

DUANE M. RUMBAUGH, Yerkes Regional Primate Research Center, Emory University, Atlanta, Georgia and San Diego State College, San Diego, California

GENE P. SACKETT, Regional Primate Research Center and Department of Psychology, University of Wisconsin, Madison, Wisconsin

M. W. SORENSON, Department of Zoology and Space Sciences Research Center, University of Missouri, Columbia, Missouri

Preface

There can be little doubt that we have experienced an explosive increase in research in primate behavior during the last decade. However, despite this expansion of knowledge, the creation of appropriate avenues of communication has not kept abreast of these developments. Although numerous books and monographs have appeared intermittently, the need exists for a publication series which can provide a continuing arena of discourse for all those scientists of varying disciplines concerned with the behavior of primates. *Primate Behavior: Developments in Field and Laboratory Research* has been created in an effort to fulfill this need. It is expected that the participants in this new serial publication and those who will find interest and value in the material it contains will be drawn from a wide array of scientific disciplines, including psychology, anthropology, zoology, psychiatry, physiology, pharmacology, veterinary medicine, and space technology.

As reflected in the contents of Volume 1, it is anticipated that each of the volumes of this series will contain a diverse collection of papers, including reviews of the literature in a given area, the integration of a man's own program of research over a period of years, conceptual or theoretical interpretations, and monographic presentations of extensive single research efforts, in particular those involving major field studies. Thus, in the current volume, Rumbaugh reviews the far-ranging literature on learning in the several species of anthropoids; this review extends back over many years, and includes as well the most recent material, including Rumbaugh's own extensive work in this area. Similarly, Mitchell provides a comprehensive review of current knowledge regarding the various dimensions of abnormal behavior in primates. Both Bernstein and Sorenson provide extensive comparative data gathered in their respective laboratories as these bear upon conceptual or theoretical issues of great significance to the field of primatology. Bernstein provides an incisive empirical analysis of the multidimensional concept of dominance and illuminates clearly the massive ambiguity of the term as it traditionally has been employed. Sorenson details the diversity and communality of behavior patterns in a number of tree shrew species and thereby sheds further light on the complex issues involved in evaluating the taxonomic status of these animals. Likewise, in an integrated review of a number of his own and related studies covering the past several years, Sackett develops a conceptual approach toward and the empirical foundations of the study of social attachments in monkeys. As an excellent illustration of the type of comprehensive presentation of major field studies which the series will present, the current volume contains Poirier's monograph covering his extensive field study on the Nilgiri langur in South India.

The editor was assisted in the preparation of this volume through the support of the National Institute of Mental Health, Research Scientist Development Award—Type II, #5 KO 2 MH23685.

This series is dedicated with warm respect and affection to Margaret and Harry Harlow, each of whom I am proud to call teacher and friend.

Primate Behavior
Developments in Field and Laboratory Research

Volume 1

Learning Skills of Anthropoids

DUANE M. RUMBAUGH[*]

Yerkes Regional Primate Research Center
Emory University, Atlanta, Georgia
and
San Diego State College, San Diego, California

[*] Present address: Yerkes Regional Primate Research Center, Emory University, Atlanta, Georgia.

The main purpose of the study of the animal mind is to learn the development of mental life down through the phylum, to trace in particular the origin of human faculty. In relation to this chief purpose of comparative psychology the associative processes assume a role predominant over that of sense-powers or instinct, for in a study of the associative processes lies the solution of the problem. Sense-powers and instincts have changed by addition and supersedence, but the cognitive side of consciousness has changed not only in quantity but also in quality.

E. L. Thorndike, 1898

I. INTRODUCTION

Reasons for achieving a thorough understanding of the learning skills of anthropoid apes are as varied in profundity as they are legion. The near-universal interest in apes could alone justify exhaustive, scientific study of their behavior and learning skills were it not the case that from certain evolutionary perspectives such study is absolutely compelling.

Of the 11 families and their 52 genera that comprise the Primate order (Napier and Napier, 1967, tree shrews excluded), only 2 families and 5 genera encompass man's nearest living relatives: the apes (lesser apes—Hylobatidae, comprised of *Hylobates* and *Symphalangus;* great apes—Pongidae, comprised of *Pongo, Pan,* and *Gorilla*). Along with man's family, Hominidae (with its single genus, *Homo*), the two families of apes are grouped by Fiedler (1956) to form the superfamily Hominoidea. Apart from Hominoidea, only two other superfamilies are required to embrace all the true monkeys, apes, and man—Cercopithecoidea (Old World monkeys) and Ceboidea (New World monkeys). The long-recognized relationship between man and the apes has presented a compelling reason in its own right for a thorough study of not only their learning skills but of their detailed characteristics as well. Moreover, this relationship is a mere introduction to an even more basic system of reasons for such study.

Just as it is apparent that the great apes (orangutan, gorilla, and chimpanzee) approximate man more closely than do the lesser apes (gibbon and siamang), *all* of the apes approximate man more closely than, in descending order, do the monkeys, tarsiers, lemurs, and the equivocal tree shrews—the latter have many primate characteristics but generally are not now classified as true primates. This graded series of living animal forms from tree shrew to man, as proposed by Le Gros Clark (1959), is not of course to be taken as a linear evolutionary series but rather as one of *approximations* from rather primitive, small primates to man. Such a chain, according to Le Gros Clark, also typifies the basic evolutionary trends that distinguish primates from other mammals. Mason (1968) was the first to use this series of approximations to man to buttress the evolutionary-comparative approach to the study of primate behavior in general and to the understanding of man's behavior in particular. A succinct summary of the most basic implications of Le Gros Clark's and Mason's formulations would be as follows. The non-human primates can be arranged into an ascending series which at least in a gross sense increasingly approximates man; through detailed comparative studies of the major groups that comprise this series, behavioral trends and relationships become apparent; and, possession of knowledge regarding the essence of these trends and relationships is of great worth in understanding not only the behavioral propensities and capacities of all primate forms, but particularly those of man.

Inasmuch as paleontological evidence indicates that primate evolution had its roots in a now-extinct form that bore marked resemblance to the contemporary tree shrew (Tupaioidea) and that the contemporary primate forms reflect most of the important evolutionary developments from the most primitive tupaiids to the most highly evolved apes and man, a rare opportunity for behaviorally oriented evolutionary-comparative studies is at hand. Furthermore, with all primates having a common genetic root that can be traced back at least 75 million years to the Cretaceous period, in a very literal sense it is reasonable to view them in their divergent forms and levels of evolution as alternatives to one another. Many alternatives obviously have not survived, becoming extinct by reason of failing to meet all critical exigencies of the environments of their times. Even so, those alternatives represented by the still-living primate forms constitute a rich array of evolutionary end products which may be studied properly either as alternatives to one another, with man being but one alternative, or as approximations of the evolutionary course that has produced, in its highest form of expression, man.

Such considerations are important, for it is within the framework of theory that research programs are critically formed and shaped. Such considerations should argue for certain kinds of studies of the anthropoid apes in preference to others. For example, though it is perfectly proper in studies in which issues *vital* to human welfare are being pursued to use apes as substitutes for man when risks are extraordinary (e.g., Rohles *et al.*, 1963) or where procedures are unacceptable for man himself to serve as subject material, such should not be the *primary* reasons for the use of apes as subjects. Further, though certain kinds of research, behavioral as well as biomedical, might be conducted as readily with ape subjects as with other animal forms, the use of apes should *not* be condoned if their usage defines even a remote threat to their health and welfare. Just as a tool should be appropriately selected for the mechanical task at hand, so should the animal form be carefully selected for any given study.

Most if not all of the contemporary primate forms, perhaps even man, are not just evolutionary end products—they are literally the end of evolution as determined by forces that have been truly natural, at least to the degree that these forces have not included extraordinary competition with man. Such is particularly the case with the apes. It is unlikely that any of them will survive much longer in a state similar to their natural field conditions. Though all of them are likely to survive in either totally captive or at best in quasinatural reserves, the basis for their selective reproduction will necessarily be radically different from what it has been for them during past millenia. Accordingly, the apes as we now know them will be replaced with ones who will likely be as different from what they now are as *Homo sapiens* is different from *Homo neanderthalensis*—even so, probably a conservative comparison.

With the prospect that in a few generations the genotypes of ape populations will be radically altered as they come to be maintained and sustained in only captive situations, it behooves us to make intensive study of those apes available today, apes which genotypically are products of an evolution in which man's pervasive influences were miniscule by present and future standards.

A. Anthropoid Brains

Diamond and Hall (1969) specify the association cortex as the particular subdivision of the neocortex within which the prime evolutionary advancements in the brains of mammals were made. Though the learning skills of the primates are not to be accounted for solely in terms of volume of the neocortex relative to total brain volume, it is of significance that in the primate the neocortex is proportionately greater (46–58%) (Harman, 1957) than it is in carnivores (40–46%) and rodents (about 30%). It is more likely that the complex cognitive processes and advanced learning skills are accommodated by the increased fissuration of the cortex, the increased numbers of cortical units in the fine structure of the cortex, and the refinement of subcortical structures which interrelate the thalamus and the cortex (Noback and Moskowitz, 1963). As there are advances in these measures from monkeys to the apes, particularly the pongids, it is reasonable to expect apes to be superior to monkeys in complex learning tasks. There is reason to believe that the brain of Hylobatidae is either equal to or slightly superior to the brain of the most advanced Cercopithecidae (Connolly, 1950), the volume in cubic centimeters (cm³) for the gibbon being about 95 and for the siamang being about 125 cm³ on the average (Tobias, 1968). It is also reasonable to conclude that the brains of the great apes are advanced beyond the development that characterizes the brains of the lesser apes (Connolly, 1950). The cranial capacities of the pongids are on the average 394 cm³ for *Pan,* 411 cm³ for *Pongo,* and 506 cm³ for *Gorilla,* with variability among gorilla measurements being more than twice that for chimpanzee and orangutan, but as clarified by Schultz: ". . . there are no true phylogenetic differences in relative cranial capacity among the great apes, besides those due entirely to differences in age and body weight" (Schultz, 1941, p. 282). Though a simple positive relationship between cranial capacity and intelligence is unwarranted, it is of unquestioned significance that between the ages of 3 and 11 months the human child's brain volume increases from the range of the great apes' brains to a twofold gain of about 850 cm³—and language appears. With the average *Homo sapiens* brain being about 1350 cm³, it is clear that at best the Pongidae are compromised so far as at least gross brain capacity is involved in learning and the manifestation of other intelligent behaviors. Harlow (1958) has suggested that even a relatively small intellectual gain by many beyond that of the apes might in fact allow for the great behavioral advancements of man—his language and his culture in particular.

B. Assessment of Learning in the Field and Laboratory

There are among primatologists those who believe that the only valid study of primate intelligence is that made by the skilled observer in the field. The argument is based on the assumption that it is only in the animal's interaction with the presses of its environment that its own idiosyncratic form of intelligence is manifest and that within the confines of the laboratory setting with its anthropocentrically conceived tests no valid measurement of intelligence can be made. It is maintained with equal insistence by others (particularly those in the laboratories who conceive of and administer those tests) that captive animals *do* bring to the laboratory certain behavioral and cognitive capacities just as they also bring to the laboratory their senses, skeleton, musculature, physiology, and so on. This latter group maintains that it is only within the formal laboratory context that the controls necessary for scientific understanding of cognitive functions can be implemented and that the limits of cognitive capacity can be assessed only through the assistance of scientific technology. In support of the latter viewpoint, it is difficult to conceive how assessments could be made of primates' skills for concepts of middleness or numbers and the like except in the laboratory. Further, though study of tool usage in feral animals in the field is totally warranted as well as completely fascinating as a subject, it is surely through either the confines of the laboratory or in the utilization of laboratory technology in the field that the complete understanding of tool-using behaviors of primates will be achieved. Without question the few extremists on these positions of the present day will be replaced by the moderates of tomorrow who will enjoy rich commerce between the field and laboratory, perhaps to the point that the field will become just another type of laboratory in which the rigors of new technology can prevail.

Regardless, it is clearly the case that both field and laboratory assessments of learning and intelligence of primates will continue, with field observations providing great stimulus for laboratory study. Such observations provide definition for behaviors of a kind that otherwise might never be brought to the attention of the laboratory scientist. And it should not be supposed that laboratory study is any less problematic than field study, or vice versa. Problems of observer bias, sampling of behavioral events, and verification of findings characterize both settings, but these matters potentially are under better control in laboratory work. Nonetheless, the laboratory has its unique constellation of difficulties which impede good cross-species study of primate behavior.

In his classic work, *Behavior Mechanisms in Monkeys*, Klüver (1933) discussed some very basic considerations in attempts to measure animal intelligence. It is not sufficient to present certain stimulus situations to different species and to determine simply whether problems are solved; it must first be determined what the situation demands of the animal being studied. The varied sensory and motor functions of the species being compared must not differentially prejudice their performance, unless of course such differences are the subject of study. In addition to these points made by Klüver, there are many others that plague the laboratory researcher–obtaining representative samples of sufficient size to justify cross-species comparisons, maintaining them in an excellent state of health and condition, avoiding performance artifacts and low-performance measurements, and so on.

C. CONDITIONING

Voronin (1962) of the USSR Academy of Sciences in Moscow has reported perhaps the most comprehensive study ever made in which the purpose was to relate certain phenomena associated with the conditioning of reflexes to phylogeny. Though not different from other animal forms so far as learning to approach a feeding place upon a signal, chimpanzees did differ from other animals tested with regard to speed with which either extinction or a reversal of conditioning of approach and avoidance reflexes occurred. Reversals of conditioning were achieved with facility only with monkeys and chimpanzees, particularly with the latter. The stabilizing, synthesizing and chaining of motor patterns, and retention were also marked for the primate subjects, particularly those that were chimpanzees. The highly developed primate capacity to synthesize "complex systems of external stimuli" (Voronin, 1962, p. 187) was attributed to their highly developed neural capacities. Such capacities were assessed as reflecting quantitative, not qualitative, development over other animal forms, with those advanced capacities providing for the physiological precondition for thought in early man. Such capacities presumably also followed for enhanced transfer of experience. On the basis of conditioning phenomena, Voronin arrived at conclusions not too unlike those arrived at in the United States through comparative studies of learning set skills in diverse primate forms (Harlow, 1949).

D. CUE UTILIZATION

As will be seen clearly in Sections O, P, great apes are not necessarily efficient learners. On occasion they have been known to require several hundred trials to learn a single, apparently simple visual discrimination. Jarvik (1953) was intrigued with this inefficient learning and conducted a study in which the discriminations to be learned were initially between pieces of food colored either red or green with vegetable dyes. Color could be used to discriminate between pieces of bread which were either flavored with saccharine, embittered with extract of red pepper, bile, or quinine, or left unaltered. Such discriminations were made extremely rapidly and with essentially no errors by both chimpanzee and monkey subjects. Subsequent discriminations in which either red and green metal pieces or even transparent celluloids of red and green covered foodwells baited with the above kinds of bread were clearly more difficult for the monkeys serving as subjects. As long as the breads were either colored *or* had the appropriate colors pasted to them through the use of colored squares, however, discriminations were essentially errorless. Jarvik concluded that both spatial and temporal proximity of cue to reward probably play critical roles in the determination of the speed with which visual discrimination can be made.

E. AMBIGUOUS-CUE DISCRIMINATION PROBLEMS

In a discrimination task in which certain stimuli are reliably associated with reward and others are not, it is of interest to determine the degree to which the subject uses the two types of cues in the learning of the discrimination. Thompson (1954) investigated this question with 14 chimpanzees ranging in age from 5 to 29 years. In a black-white discrimination task in which only the positive or negative stimulus was present on any given trial, paired with an empty frame or support used to hold the stimuli in front of either the left or right food cup, the animals were significantly better on those trials in which the positive stimulus was present. The same finding was observed in a second problem in which the stimuli were vertical and horizontal stripes. In a third type of problem in which a "neutral" yellow plaque was paired with either the positive or negative stimulus, the results were in the same direction as those from the first two types of problems, but the difference was not statistically significant. Thompson concluded that since the chimpanzees learned an approach habit significantly more rapidly than an avoidance habit, reinforcement theorists could conclude that reward contributes more to learning of discrimination problems than does nonreward. The same findings permit the perceptually oriented theorists to conclude that the positive stimulus served as a more effective cue than did the negative stimulus for the determination of the discrimination response.

Leary (1958) continued the implication of Thompson's findings in a study in which rhesus monkeys served as subjects. Two multidimensional objects were present during every trial of each of eight problems. On each trial there was either an approach object, which was always rewarded, or an avoidance object which was never rewarded. Paired with one of these two stimuli was a second stimulus which was ambiguous in cue value in that it was to be approached or avoided according to whether it was paired with the approach object or the avoidance object. When it was paired with the approach object, the ambiguous-cue object was to be avoided since selection of it was unrewarded (−); when the ambiguous object was paired with the avoidance object, it was to be selected for reward (+) to be obtained.

In each of three experiments, performance was better when the reliably incorrect or unrewarded stimulus was paired with the ambiguous stimulus, these results with rhesus monkeys being exactly the reverse of what Thompson observed with chimpanzee subjects. That there was a variety of differences in the experimental procedures employed in Thompson's and Leary's experiments makes it impossible to attribute the discrepant findings to the species variable. For example, Thompson's chimpanzees took approximately seven times as many trials to achieve problem solution as did Leary's monkeys even though all animals of both experiments were experienced in discrimination problems. Nonetheless, this difference in number of trials to reach criterion, when the criteria employed were very similar, leads one to suspect that at the very least the tasks as structured in the two experiments were quite different in terms of intrinsic difficulty.

Fletcher and Bordow (1965) pursued study of performance in ambiguous-cue problems with rhesus monkey subjects. Results clearly indicated that performance was superior on trials in which a reliably correct stimulus was paired with an ambiguous cue (−) as opposed to those in which a reliably negative stimulus was paired with an ambiguous cue (+). The observations were taken as confirmation of Thompson's observation with chimpanzees. The apparatus used by Fletcher and Bordow was more similar to Thompson's apparatus than was Leary's in that there was discontiguity between the stimulus and the response. Whereas Leary's apparatus required direct response to the multidimensional objects used as stimuli, both Thompson's and Fletcher and Bordow's apparatuses allowed the animals to respond to identical manipulanda which were slightly discontiguous with the stimuli to be discriminated. Further, as was the case in Thompson's first two problem situations, Fletcher and Bordow used a blank plaque as the ambiguous cue. This procedure of course is quite different from that used by Leary in which a variety of three multidimensional objects served as ambiguous cues. Fletcher and Bordow believe that these factors were critical in their obtaining results contrary to Leary's but consistent with Thompson's experiment. Thus, though it is not warranted to conclude that chimpanzees benefit more from the presence of a reliably correct stimulus than a reliably incorrect stimulus, with macaques benefiting from the presence of these cues in just the reverse manner, it is clear that there are conditions that profoundly influence the apparent usefulness of positive and negative cues. These conditions very likely include the use of a variety of ambiguous stimuli as well as discontiguity between the discriminanda and where the choice is made.

Boyer *et al.* (1966) confirmed that when plaques were used as discriminanda macaques did better when the reliably positive stimulus was paired with the ambiguous cue than when the reliably negative stimulus was paired with the ambiguous cue, with the exact reverse holding true when stereometric objects were used as discriminanda. Even more recently, Fletcher *et al.* (1968) again confirmed with monkey, human retardate, and normal children as subjects that when multidimensional test objects were used in problems in which the negative stimulus was paired with the ambiguous one performance was superior to those in which the reliably positive stimulus was paired with the ambiguous stimulus. In basic accord with an interfering-cue model suggested first by Leary (1958), they concluded that performance was superior on the trials pairing the negative and ambiguous cues by reason of avoidance of the former cue complementing the approach tendency to the ambiguous cue (by reason of its being reinforced whenever paired with the reliably negative one). Performance accuracy was relatively poor on the trials pairing the positive and ambiguous cues by reason of competition between approach tendencies of both cues, though strength of approach was relatively stronger to the reliably positive cue. By contrast, the non-reward of responses to the reliably negative cue was not critical. They further concluded that learning was not restricted to the objects chosen but that instead both human and nonhuman subjects respond either directly by choice or inferentially (covertly) to both stimuli of a given trial in this type of problem. They did not detect any qualitative difference in the performances of their human and nonhuman subjects on the ambiguous-cue problems, nor did they conclude that even human primates used anything as complex as a conditional rule, such as choosing the positive stimulus when it was paired with the ambiguous one and shifting choice to the ambiguous stimulus when it was paired with a reliably negative stimulus.

F. The Question of Insight

It is probably impossible to overestimate the significance and value of the experiments conducted and reported by Wolfgang Köhler (1925). His studies were conducted with seven of the chimpanzees of the anthropoid station maintained by the Prussian Academy of Science in Tenerife from 1913 to 1917. Most of his data-gathering occurred in the spring of 1916. In the postscript to his book, he recounts that upon finishing the book he received from a Mr. R. M. Yerkes of Harvard University a monograph entitled, *The Mental Life of Monkeys and Apes; a Study in Ideational Behaviour* (Behaviour Monographs, III, i, 1916). In that monograph Yerkes reported experiments of the same general type that Köhler had conducted, the subject of Yerkes' study being an orangutan. Köhler judged the materials to be in agreement with his. There is probably little doubt that Köhler was delighted to note that Mr. Yerkes concluded that "insight" must have been utilized by the animal in the test situations, for Köhler had also concluded from his own studies that it was surely necessary to appeal to some insightlike process to account for the behaviors observed.

Köhler had many reasons for conducting his studies. In recognition of the close similarity between the chimpanzee and man, he was interested in determining the kind of problem-solving skills that might be provided for by the chimpanzee's advanced brain structure. Also, Köhler was convinced that it was totally reasonable to call certain highly adaptive, plastic-type behaviors insightful and intelligent and to label other behaviors unintelligent by reason of their lacking these characteristics and being more of an association type. Thus it is clear that at least in part Köhler conducted his studies to demonstrate that Thorndikian (1911) association theory could not adequately account for all behaviors, particularly those that he termed "intelligent."

Köhler was an ingenious experimenter who, while bold in his methods and convinced as to the validity of his theoretical position, was nonetheless sensitive to such vital matters as individual differences in the native intellectual endowment of his subjects and their distinctive personalities. He was also extremely sensitive to the implementation of one of his basic experimental principles, that the chimpanzees should be tested in tasks of graded difficulty. He recognized that if animals were confronted with problems too difficult to solve they became very agitated and reluctant to work reliably. With such confused and misleading observations, the experimenter would be likely to draw fallacious conclusions.

Though Köhler probably failed to emphasize, to the degree that he should have, experience as a determining factor in the chimpanzees' abilities to solve problems as structured, there is no question that he recognized its importance. Such allowance is clearly implicit in his tactic of presenting subjects with a series of problems graded according to difficulty. And Köhler cannot properly be criticized for using his more-or-less subjective assessment of what the steps of difficulty really were— he had little of meaning to draw upon from the works of others. Careful reading of his book reveals that he did allow for the possibility that various experiences the animals might have had both prior to his work

with them and also while being maintained at Tenerife might have profoundly influenced their behaviors in the test situations. He recognized
that the animals were highly imitative. Accordingly, he realized that
they possibly might have benefited even from observing attendants of
the station using tools and cleaning implements in the performance of
their duties.

Köhler was highly critical of Thorndike's work, particularly objecting
to structuring tasks so as to predispose the animals to solutions that
would appear to be highly stereotyped and of the type that could be
accounted for readily in terms of stimulus-response associations. Köhler
was a strong believer in availing to subjects a comprehensive visual
survey of the task to be solved. Such survey was critical for the identification of the elements in the situation and the organization of the responses to them in such a sequence as to allow for achievement of the
goal or incentive. Thus it was quite common for Köhler's tasks to entail
such elements as a basket of food suspended by a rope which could be
released by the displacement of a loop or rung from a nail. Or, they
might entail the subject being physically positioned with regard to an
attractive incentive but with a barrier interposed. Still other examples
of tasks were ones in which sticks could be used for reaching some incentive beyond the reach of the animal's extended arm, or boxes which if
stacked one on top of the other would permit the animal to reach a
banana suspended high overhead. In all these situations the subjects had
perceptual access to all the important elements of the situation.

Achievement of the incentives was frequently described as having an
initial stage during which time the animal oriented both visually and
physically to the incentive while from time to time attempting to obtain
it by various approaches. The incentive was selected so as to be attractive enough to sustain attention, a critical factor according to Köhler.
In these situations the animals were at times observed to pause as though
to reflect upon or reassess the problem. Immediately following such
pauses, new responses or attempted solutions would be observed and at
times these responses were successful. Köhler discussed at length the
possibility that chance factors could account for what was observed in
these kinds of settings but rejected as being literally inconceivable the
idea that such intelligent behavior could be innovatively shaped and
"run off" when the correct solution was finally made. This was not to
deny totally the contribution that chance happenings might make to
the animals' attempted solutions, however. Nor did Köhler's reservations
regarding the adequacy of association theory for accounting for these
behaviors rule out his recognition of habits being formed with some facilitating and others interfering with the appearance of insightful,
intelligent behavior.

Many of Köhler's observations were as dramatic as the best of those that have been reported from field studies of recent years. For example, he reported that one chimpanzee, while chasing another in play fight, saw a stone, scratched and pulled at it until it loosened, then immediately resumed chase and threw the stone at its adversary. On other occasions he reported that Mr. Tueber, his predecessor at the Tenerife station, saw a chimpanzee work at a shoe scraper until it succeeded in extracting one of the iron bars. The animal then ran immediately to a far side of its cage and reached out with the bar and drew to within reach incentives that otherwise would have remained beyond access. One of his chimpanzees, Sultan, which was particularly adept at manufacturing implements, was especially noted for attempting to join together only those tubes that would in fact fit together and make a stick of adequate length for the task at hand. He never tried to join together two large tubes of equal size. When he had access to a multiunit stick so long that it hindered pulling in objects, he disconnected a length. He also always held a multiunit stick by the heavier end, leaving the lighter end for the "delicate fishing." This same animal could splinter a wooden board and using his teeth would shape selected splinters into implements with which to probe about in key holes. Another chimpanzee, Grende, would bite a piece off a board to obtain the splinter required to poke someone on the opposite side of her barred cage.

One of the most interesting examples of insight, as evaluated by Köhler, was observed in the box-stacking problem situations. It was observed that the chimpanzees were quite adept at placing smaller boxes on top of larger ones, the result being that a more stable platform on which to climb to an overhead incentive was obtained. In contrast, the stacking of three or four boxes was rarely as precise as it might have been, the result being that the animal had to compensate for the wobble of the stack through movement of its body.

Köhler relegated to a footnote an observation that was to this author one of the most exciting ideas of his entire book: "There seems to be in apes a high positive correlation between intelligence and dexterity" (Köhler, 1925, p. 176). Lack of dexterity was believed to interfere profoundly with the execution of the fine, delicate motions required for successful performance in the task. Recently, Parker (1969) has studied the manipulative manual skills of great ape subjects and concluded that this was indeed a basic parameter of effective tool-using in tasks that involved the dipping of a rope into a vial so as to extract the fruit mash deposited within and also in a task that entailed the use of a tethered hoe to obtain food otherwise beyond arm reach.

Brainard (1930) utilized Köhler's test procedures in a modified form on his daughter who was 2½ year old at the initiation of the testing. At age three her IQ was 141 as indicated by the Stanford–Binet scale. Other children were used for purposes of comparison, but the main work was done with Ruth. It was Brainard's opinion that children from 2 to 4 years of age resembled apes so far as attitudes, skills, and methods were concerned in attempting problem solution. As would be expected, the older children performed somewhat better than the younger ones, though their solutions were not qualitatively distinct from those of the younger children. Brainard's daughter was assessed as having the same difficulties with the problems as did Köhler's chimpanzees, and her emotional responsiveness ranging from discouragement to anger was not unlike that of the chimpanzees. Just as Köhler emphasized that the incentive must be attractive enough to sustain a high degree of attention in the task, Brainard reported that achievement of the incentive was apparently a very satisfying affair. Though Brainard's study was very restrictive in scope, it nonetheless suggested interesting parallels between the behaviors of chimpanzees and children in similar test situations.

The greatest challenge to Köhler's insistence that ideational and insightful processes must be appealed to in order to account for the behaviors of his chimpanzees came from Spence (1938). A strong, unyielding advocate of the stimulus-response reinforcement approach to conditioning and learning of all forms, Spence was one of very few who though actively working with chimpanzees remained convinced that nothing beyond stimulus-response associationistic processes had to be appealed to in order to account satisfactorily for their problem-solving skills. In the late 1930's, Spence conducted extensive studies at what was then called the Yale Laboratories of Primate Biology. Nowhere did he draw the line of distinction more clearly between Köhler's perceptual-insight formulations and his own position of stimulus-response association than in the above-referenced article. He insisted that as Köhler and the other Gestalt psychologists tried to account for learning in terms of insight they were in fact merely restating the problem in new terms (Spence, 1938, p. 214). He also confessed that association theorists needed to give special attention to the kind of phenomenon Köhler pointed to as reflecting ideational processes, and this he set out to do.

Twelve chimpanzees were first trained on each of two discrimination problems involving four stimuli designated A, B, C, and D. Within each pair, for example, A and B as opposed to C and D, one stimulus was correct and netted food reward, whereas the other was incorrect and netted the animal nothing. Following mastery of these problems, the animals were tested on new combinations of the four stimuli. The prediction from association discrimination learning theory was that choice of stimuli in the newly formed pairs would be relative to the excitatory strength of the two as determined in the initial learning. The greater the number of reinforcements in the initial training, the greater the excitatory strength; the greater the number of nonreinforcements, the less the excitatory strength and indeed the greater the inhibitory strength of the stimulus. The net strength of excitation and inhibition was to determine the probability of each stimulus being chosen when paired with the other three stimulus members of the initial discrimination training. The results were "in substantial agreement with the theoretical expectations" (Spence, 1938, p. 217).

Spence further analyzed the performance on the test combinations in terms of performance increments from the 20 trials preceding to the 20 trials within which the criterion of learning (90%) was achieved. Gains of 40% or more were taken as being functionally equivalent to Köhler's "insightful" learning. In light of the high rank order correlation (−0.79) between the error scores of the learning and the estimated relative excitatory strengths of the stimuli at the beginning of learning (the estimate being made on the basis of the number of reinforcements and nonreinforcements to stimuli), Spence concluded that there was no need to introduce a qualitatively different theoretical notion to account for the apparent suddenness of learning. With such findings, Spence challenged the Gestalt psychologists to predict from *their* concepts of structural reorganization of the elements of the task to be solved (and presumed insightful processes associated therewith) the antecedents of sudden solution as satisfactorily as he had done in his study. Though the line was clearly drawn and vital issues were at stake, the challenge of the field was essentially abandoned by all for reasons not clear to this author. Perhaps it can be attributed to the disruption created by World War II, though it is more likely that abandonment of this vital controversy reflected the conviction of Gestalt psychologists that they were in fact talking about something different, something that could not be accounted for within the simple context of a two-choice discrimination situation. As had been so frequently the case in the history of psychology, empirical results failed to alter loyalty to the flags of the theories in confrontation. Thus this classic issue still lives, and its study must be resumed with determination.

G. Transposition

Köhler maintained that the transposition phenomenon constituted strong evidence for the operation of relational processes. In the demonstration of transposition, subjects are trained to discriminate between two stimuli which differ primarily along a single dimension. For example, an animal might learn that the larger of two squares is correct and the choice of it nets food reward. On the test trials for transposition, the larger, initially correct square is paired with another square which is still larger. Choice of this new square, as opposed to the choice of the originally correct square which is now the smaller of two presented, is taken to mean that the animal learned *not* a specific choice response to a specific stimulus but that it learned *a relation*—that it was the larger of the two that netted reward. Spence's studies (i.e., 1941, 1942) either failed to produce the transposition phenomenon or provided an alternative stimulus-response association accounting for it.

The evidence *par excellence* for existence of relational learning in chimpanzees came from a transposition study by Gonzales *et al.* (1954). Initially, the animals were trained to a criterion on a problem involving three boxes in which the middle size box was correct and in which the ratio of areas among the three boxes was 1:1.75. They were trained next in a similar manner on a second problem in which the middle box again was not only the same as in the first problem but also remained correct. In the second problem, however, the ratio of areas of the three boxes was 1:1.32. On the test phase of the experiment, the animals were presented with two new sets of three boxes each, the ratio areas of which were the same, 1:1.32. Now, however, the originally correct box was deleted, but two of the three boxes in each of the two new sets of stimuli were immediately adjacent to the originally correct box so far as area ratio was concerned. One of the boxes was 1.15 the size of the initially correct and middle box; the other box was of a size that made the originally correct and middle box 1.15 its size. As these two boxes were paired with still a third box (in one set the third box was proportionally smaller and in the second set it was proportionally larger than either of these two just defined), a critical test of the relational process was made. If the learning in the initial experimental phase had been one that entailed response to a *specific* box of a specific size, then in the two test problems choice should have been made randomly, on the basis of generalization, of the two boxes that were equidistant in size from the initially correct box. If relational learning had been involved, however, selection of these two boxes should be associated with the one that when combined in presentation with the third box became the middle-sized box of the array. The results were convincingly in support of the relational learning hypothesis, though some evidence for an absolute preference also appeared. The results were considered to be unmistakably in support of the operation of a relational process in the chimpanzee as opposed to just the operation of excitatory and inhibitory valences among an aggregation of stimulus components.

H. LEARNING SET

Clarification regarding the contributions of experience to the manifestation of rapid "insightful" learning came with Harlow's (1949) formulations on the learning set (LS) process. As a function of well-defined experience in working on a long series of discrimination tasks, in which each problem consisted of a pair of stereometric objects and in which one stimulus was correct and the other incorrect, Harlow's monkeys were demonstrated to be transformed essentially from Thorndikian trial-and-error learners into one-trial insightful learners. The task was in fact a two-choice test situation quite *unlike* the tasks presented by Köhler to his chimpanzees. Nonetheless, the importance of sustained encounter with problems of a given type was identified convincingly as the prime determinant of a change of shape in the intraproblem learning curve from one that was initially continuous to one that was discontinuous, with all learning taking place essentially on a single trial, the first. It was clear that *experience* was critical for the appearance of so-called insightful behaviors. The reader's attention is drawn once again to Köhler's prime principle of testing his chimpanzees, namely, that the problems be of graded difficulty. Surely, Köhler recognized the contribution of experience to problem solving in the formulation of his principle, but what he failed to do was to emphasize it adequately in his writing. Undoubtedly, by reason of being so impressed with the chimpanzee as a subject, he failed to consider adequately the critical role of experience in their problem-solving skills.

A very interesting point of contrast as to just how experience contributed to insightful problem solving warrants particular attention, however. If LS reflects the accumulation of positive transfer across a series of problems on which animals work, it seems that various quanta of experience (trials per problem) should be a powerful determinant of LS. Empirical research did not support this proposition, however, and it was Harlow's final conclusion that number of trials per problem was *not* a critical parameter of LS (see Levine *et al.*, 1959). Such a conclusion was certainly in contrast to a clear implication of Köhler's grading of problems according to difficulty. Such grading could be effective only to the degree that successful experience (learning) in one task fosters learning in subsequent tasks.

Rumbaugh (1968) summarized a variety of experiments that relate to this matter and concluded that criterional training LS methods were superior to fixed-trial methods only when, according to any of a variety of factors, the training situation was in effect a very difficult one. Criterional methods require that learning be manifest on each problem as a requisite to encountering the next. "Whenever ability is restricted either by the characteristics of the animal form studied, or by age level . . . , or wherever identification of the relevant cues might be difficult (as when they are encased or mounted on panels), or when the discriminations themselves are difficult to make given the perceptual capacities of the species being studied, criterional training procedures are likely to yield performance levels, assessed on an interproblem measurement, superior to fixed-trial procedures" (Rumbaugh, 1968, p. 273). The author concluded that the apparent effectiveness of criterional LS training, where evidenced, was attributable to the fact that it not only fostered attention to the relevant cues of the problem but also assured that there was some learning to be transferred in a cumulative manner from problem to problem to provide for the eventual appearance of what Harlow termed LS.

One of the early studies of LS in chimpanzees (K. J. Hayes *et al.*, 1953a) suggested that initial training with an extremely small number of trials per problem was not conducive to the development of LS and that training to criterion was the only practicable way to run all problems for an optimal period of time, so as to conserve trials and yet facilitate the formation of LS. This same study reported essentially perfect trial-2 LS for at least one chimpanzee, great individual differences among chimpanzee subjects in LS skills, and a systematic relationship between both the chronological age and IQ's in children so far as LS facility is concerned. Of particular interest was their notation that the language skills of their human subjects did not necessarily provide an advantage in learning. One subject who was particularly talkative was observed to continue making errors even though able to name the objects.

Fischer (1962) reported LS for two young gorillas (*Gorilla gorilla gorilla*) 17 and 21 months of age. Through the course of 232 pattern discrimination problems, one animal eventually achieved trial-2 LS of slightly better than 75% while the other achieved slightly better than 85%. Qualified comparisons by Fischer suggested that at comparable developmental levels her gorilla subjects developed LS with somewhat greater efficiency than did rhesus monkeys (Harlow *et al.*, 1960).

Theorizing regarding the basis of LS typically has been in terms of stimulus-response associationism without invoking some special process (e.g., *insight*) that was both qualitatively distinct and perceptually based. Levine (1959), however, accounted for LS in terms of the learner's testing of various strategies, though no awareness of them was implied. One of the strategies, known as the "win-stay, lose-shift" strategy with regard to object, accounts for LS very elegantly. Throughout his training the learner could be reliably reinforced for *staying* on subsequent trials with any object with which it had "won" by being rewarded on the previous trial and, conversely, shifting from any object from which it had lost.

Schusterman (1962) used chimpanzees to make a critical test of Levine's notion. He hypothesized that one group's training on successive criterional reversals with a limited number of stimuli (he used three) should serve to reinforce the LS strategy—"win-stay, lose-shift." In contrast, a second group of chimpanzees was trained on object alternation which served to reinforce an antithetical strategem, that of "win-shift, lose-stay." Both groups were subsequently tested on 180 discrimination problems. It was observed that the first of the two groups immediately manifested a high-level trial-2 LS, being essentially 85–90% correct, whereas the second group, trained in object alternation, only gradually acquired LS skills. Without employing the hundreds of problems normally used in the establishment of LS in nonhuman primates, Schusterman had found it feasible to train high-level LS skills with specific strategy training on but a few stimulus pairs. In 1964 he demonstrated that training with only *one* stimulus pair where successive criterional reversals were mastered sufficed to establish proficient trial-2 LS (80% correct) with chimpanzee subjects.

The *Schusterman effect* was a major contribution to our understanding of the LS phenomenon. That it was done with great apes proved later not to be the critical point, as both Warren (1966) and Schrier (1966) demonstrated that macaque monkeys also developed multiple-problem LS skills through successive criterional reversal training even with a limited number of different stimuli. Schrier concluded that the transfer from such training was probably *greater* in chimpanzee subjects than in macaques, an interesting comparative notation. Warren also noted an extremely important fact, namely, whereas rhesus monkeys benefited from initial successive reversal training, cats did *not* so far as subsequent test of LS skills in a multiple-problem test situation was concerned.

In 1963 Schusterman reported a comparative study with chimpanzee and human subjects, the latter ranging from 3½ to 11 years. In a two-choice task Schusterman (1963) observed that chimpanzees and 3-year-old children tended to choose according to the position where reward was last hidden. In contrast, the 5- and 10-year-olds were more likely to make choices based on the consequences of *a span* of recent trials, not just the last one.

Rumbaugh and McCormack (1967) concluded that the LS skills of the three great ape genera were probably equivalent and further that they were grossly superior to those of the gibbon and squirrel monkeys tested in the same training program (Fig. 1). The asymptotic LS skills

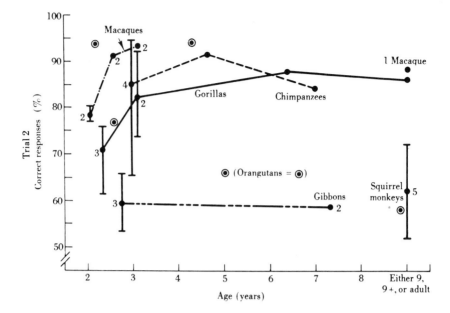

Fig. 1. LS performances for various primate groups and age levels as obtained from phase-III LS testing (after Rumbaugh and McCormack, 1967).

of the great apes were assessed as being equivalent to those of 3- to 4-year-old macaque monkeys.

Changes in problem-solving skills of aging chimpanzees have been comprehensively studied by Riopelle and Rogers (1965). With the exception of a slight though nonsignificant improvement on LS problems with age, there was characteristically either no change (as with serial discrimination learning) or profound drops in performance with age (as with oddity and delayed-response skills). Performance decrements were evaluated as being consistent with a decline "in the functional integrity of the prefrontal-lobe system" (Riopelle and Rogers, 1965, p. 461) with increased age.

Bernstein (1961a) reported no difference in experiments designed to assess adaptability, or variability of response, and proficiency in conditional discrimination and discrimination reversal tasks with chimpanzees of two age groups (11–19 and 28–40 years).

I. Oddity

Bernstein (1961b) studied dimension-abstracted oddity in primates, a particular form of oddity training that requires the subject to respond to a specific relevant cue shared by all "odd" objects. For example, if size were the relevant cue, all negative objects would be of essentially the same size whereas the correct stimulus would be of a different size. Though of approximately equal size, the negative stimuli would vary according to a number of other dimensions, such as brightness and form. Macaque, orangutan, and chimpanzee subjects were able to learn this kind of oddity problem, and it is of interest that there were no discernible differences between the orangutans and chimpanzees, a point that has been borne out in other discrimination studies of more recent years (Rumbaugh and McCormack, 1967).

The excellence of chimpanzees on oddity problems as compared with adolescent rhesus monkeys, adult raccoons, and cats was reported by Strong and Hedges (1966). The training situation was a modified oddity situation in that a total of 9 objects were combined in 72 possible configurations to form oddity problems in which the odd object was never placed in the middle position. The subjects were sustained in training until they either achieved the criterion of 90% correct within the given test session of 48 trials or until they had had a total of 100 sessions. In contrast to the cat and raccoons, none of which achieved the stated criterion, monkeys and chimpanzees did quite well, the monkeys achieving slightly better than 70% and the chimpanzees achieving slightly better than 84% correct responses in the test sessions 41–48. Whereas none of the cats or raccoons achieved criterion, all of the monkeys and chimpanzees did. On a subsequent test series, with the addition of the three new stimuli, two of the monkeys and all of the chimpanzees transferred efficiently. There was an overlap between the monkeys and chimpanzees which serves to temper conclusions that are otherwise tempting. Strong (1967) subsequently demonstrated that both chimpanzees and children were more adept at transferring their oddity skills between qualitatively different test units and in this regard seem to be superior to other primate subjects.

Rumbaugh and McCormack (1967) assessed the oddity concept skills of gorilla, chimpanzee, orangutan, gibbon, and macaque species in a program that entailed: first, training to an intraproblem criterion on each of a series of oddity problems comprised of trials in which three of the four members of two identical stimulus pairs were employed and, second, terminal testing on a series of 100, 1-trial oddity problems unfamiliar to the subjects. All three positions were used for placement of the odd object (Fig. 2). Great apes performed between the 65 and 90% correct levels, there being great overlap between the three genera. The macaque subjects tended to do somewhat better (71–88%) on the average than the great apes, though their performances were within the great ape range. Only one gibbon subject gave any justification for oddity

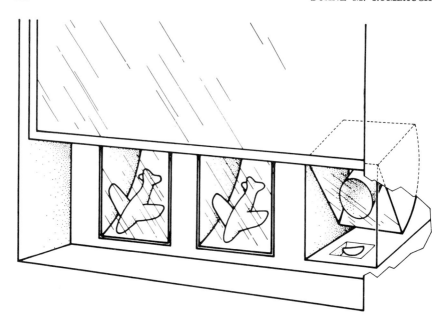

Fig. 2. Schematic of apparatus and materials of the kind used by Rumbaugh and McCormack (1967) for the study of oddity-concept learning in apes and monkeys. The response of pushing on the transparent front of the bin that contained the object chosen on a given trial resulted in the baring of a foodwell, which contained food reward if the choice was correct. An opaque screen was raised to obscure the subject's view while a trial was being arranged. See text for further details of testing procedures.

training, and in that training it evidenced only marginal evidence of an oddity concept, being only about 40% correct on terminal testing.

J. Use of Multiple Cues, Picture Memory, and Object Identification

Nissen (1942) suggested that one of the basic dimensions of adaptive behavior was the ability to take into appropriate account a large number of elements or independent cues in the determination of a correct choice. In 1951 he reported a study with a single young adult male chimpanzee whose training consisted of 401 sessions for a total of almost 18,000 trials. The animal was trained concurrently on 61, two-choice discrimination problems drawn from a total of 32 plaques. These plaques were paired and interrelated so as to require the animal to take into account *simultaneously* five distinct cues for a correct choice to be made on any one trial. The five cues that had to be taken into account in their various combinations were plaque size, color, shape, and presence or absence of a margin or peg, the latter when present positioned in front of the plaques. The animal succeeded in this training and Nissen suspected that its ultimate capacity for this type of problem exceeded the demands of the particular training situation. Though Nissen recognized that the chimpanzee might have learned each of the stimulus configurations independently through the course of the extensive training, he concluded that this was improbable. In 1967, however, Farrer presented evidence indicating that this might have been the case.

Farrer (1967) trained three chimpanzees to a high level of mastery in a four-choice match-to-sample problem situation. They were then tested on the 24 problems mastered *without* the samples present. On each of these problems, the animals chose one of four projected images by depressing one of four levers associated with those images. The 24

PROBLEM NO.	PICTURES				CORRECT POSITION
	LEVER-1	LEVER-2	LEVER-3	LEVER-4	
1	+	(g)	l	X	4
2	△	—	□	◎	2
3	(g)	(w)	+	X	3
4	(b)	—	(r)	□	1
5	(g)	△	—	X	2
6	+	X	(r)	(b)	4
7	l	(g)	(w)	△	3
8	l	(g)	—	◎	1
9	+	◎	(or)	X	3
10	△	□	l	X	2
11	(w)	◎	(b)	l	4
12	◎	(or)	□	—	1
13	(or)	(w)	—	(g)	1
14	X	□	○	(r)	2
15	(r)	—	+	□	3
16	△	(g)	l	(r)	4
17	X	□	◎	(b)	3
18	(b)	l	X	—	4
19	(b)	+	□	—	1
20	+	(g)	l	(b)	2
21	X	(w)	—	(b)	2
22	△	X	(g)	(r)	1
23	□	(b)	X	l	3
24	△	—	□	(r)	4

FIG. 3. The 24 problems of the picture-memory test (from Farrer, 1967).

problems (Fig. 3) shared many common stimuli provided by digital display units; thus it seemingly is the case that without the sample presented prior to the presentation of the four stimuli, better than chance responding is impossible. Nonetheless, it was the case that the animals maintained their high performance accuracies (better than 90% correct) *in the absence of the sample,* demonstrating that they had in fact learned 24 *discrete* stimulus configurations. It was also demonstrated that the animals were able to do well on incomplete stimulus configurations when one of the four stimulus members of a given problem was excluded. When the animals encountered mirror-image sequencing of the stimuli of each problem, however, their recovery in terms of percent correct responses was prolonged enough to warrant the conclusion that they were responding to them as essentially new problems.

Farrer's study does not necessarily dispute Nissen's view that the chimpanzee can make choices contingent upon the processing of several independent cues, but it does bring into question whether Nissen's experiment demonstrated what he thought it did. At the very least, these two experiments serve to demonstrate the remarkable capacity of at least some chimpanzees to recognize considerable numbers of stimulus configurations even though they share many elements or stimulus characteristics. Such insight aids us in understanding the processes whereby the home-raised chimpanzee of the Hayes', Viki, was able to respond to pictures and line drawings so remarkably. Hayes and Hayes (1953) reported that Viki was able to choose with better than 75% accuracy one of two familiar objects portrayed in a picture. She also accurately imitated actions portrayed in still and motion pictures. The animal apparently enjoyed this kind of activity for she would at times "run" herself in this kind of testing by turning on a projector. In a formal testing of picture comprehension, Viki was able to make appropriate selection of pictured objects (matching to sample) even where the choices were between black and white line drawings, one of which resembled the real object presented as the sample to be matched. Interestingly, she performed better when the pictures were colored than when just black and white and performed better with either of these than she did with simple line drawings.

Though impressed with her performance, Hayes and Hayes were interested in her many mistakes and what caused them. Many of them were assessed as being the result of careless observation and quick choices on Viki's part. This judgment frequently recurs with chimpanzee subjects. Once again, it is recalled that Köhler also concluded that attention to the task was frequently problematic to the chimpanzee and unless it was ensured by such a factor as a highly attractive incentive problem solving was either of a low grade or absent.

K. Concurrent Discrimination Learning

K. J. Hayes, Thompson, and Hayes (1953b) studied the concurrent discrimination skills of six chimpanzees, all with extensive histories of discrimination learning. Training was given *concurrently* on lists of either 5, 10, or 20 object-quality discrimination problems; the chimpanzees were given one trial of *each* problem on their lists before receiving the second trial of each problem, and so on. This procedure provides for both proactive and retroactive inhibition to interfere with the mastery of the discriminations. As such, the researchers considered it a technique of probable value for determining the extent to which large numbers of separate associations could be managed concurrently. Performance accuracy was as high as 80% on the third presentation of the 20-pair problem list and higher still on the other two lists, particularly the list that had only 5 problems. Considerable capacity for management of relatively large numbers of separate and potentially conflicting associations was indicated for at least some chimpanzees.

L. Observational Learning

It is quite generally acknowledged that in some manner nonhuman primates can benefit from observing the responses of others. One of the more revealing experiments conducted to determine certain parameters of observational learning was reported by Crawford and Spence (1939). Their apparatus was such as to make it possible for one chimpanzee to observe the choices of the other in a two-choice situation. On the demonstration trials three basic techniques were utilized so as to control systematically the delivery of food to the demonstrator and imitator on those trials in which the demonstrator made correct responses. In one technique, correct responses by the demonstrator resulted in food cups appearing near the correct stimulus, these cups being baited and positioned so that both chimpanzees might take food. In a second technique, again two baited cups appeared when the demonstrator was correct, but in this instance the imitator's cup was a centrally located one, as close to the incorrect as to the correct stimulus. In the third technique, no cup was available to the imitator, but the demonstrator was able to pick food from a cup near the correct stimulus.

Only the first of these three techniques entailed food being delivered to the imitator in a manner that would directly facilitate its orientation to the correct stimulus, and test performances by the imitator did indicate learning on those problems in which it had so oriented. Learning also occurred on those problems in which the second technique was employed, but only when the imitator did in fact orient to the correct stimulus as selected by the demonstrator. Only partial learning was indicated when the third technique was employed, the one that employed no delivery of food to the imitator during the demonstration trials. The latter partial learning was attributed to the possible mechanisms of either social functions or that of *substitute reward* possibly provided by the imitator's observing the demonstrator eat its reward after having chosen the correct stimulus. Orientation to the correct stimulus by the imitator was assessed as a critical act for observational learning to take place.

Hayes and Hayes (1952) reported a study of imitation with their chimpanzee, Viki, and made related observations with another chimpanzee which was caged and still other observations with children. At about 2½ years of age Viki was tested on six problems that could be solved by the experimenter's demonstrations. Prior to the demonstrations Viki was given an opportunity to attempt solution by either insight or trial and error. On one of the tasks Viki solved the problem spontaneously, whereas none of the four children did. Hayes and Hayes concluded on the basis of this and other observations that Viki did not differ appreciably in her imitative skills from human children of comparable age. By contrast, the cage-raised chimpanzee was inept at learning by imitation. The disparate past experiences of the animals surely accounted for this fact. Chimpanzees must have certain kinds of interactions with the environment in order to develop whatever imitative abilities their heredity might allow. These observations reported by Hayes and Hayes are of considerable value and place in proper perspective the frequent failure to obtain imitation with animals raised and tested in rather barren environs.

Hall (1963) presents an insightful review of the work done by a number of investigators on phenomena related to observational learning. The period of mother dependence is examined by Hall as a critical phase during which some of the fundamentals of observational learning are acquired by the infant, such as learning to eat and not eat various objects and to manipulate selectively only parts of the environment. Observational learning quite likely contributes in major proportion to the broader problem of social learning. Also, the affinity between the demonstrator and observer is likely to be a determining factor in whether or not the observer learns from and imitates the behaviors of the demonstrator. Such was surely the case in the remarkable studies by Crawford (1937) in which only by cooperative work, learned through the course of test sessions, were young chimpanzees able to work synchronously with either one another or the experimenter to succeed in pulling within reach a container of food.

Using young chimpanzees as subjects, Crawford devised three problem situations that required cooperative effort by two chimpanzees for food incentive to be obtained. The first of these problems entailed box-pulling. Individually, the chimpanzees were trained first to pull on a rope to bring a box within reach so as to obtain food from it. The box was then made so heavy that it could be pulled only by coordinated effort of two animals, each having a rope on which to pull. Pairs of animals so trained were not successful initially in coordinating their efforts, but with additional training to pull upon signal (the word "pull") cooperative behavior began to emerge. Its emergence was first suggested by the animals observing one another as pulling behavior became imminent. The signal for concurrent pulling provided by the experimenter was then no longer necessary. Once synchrony of pulling so emerged, solicitation of one animal by another was observed in some pairs. Interestingly, the least dominant were the ones that solicited cooperative pulling of their more dominant pair-members. That the least dominant ones were also judged to be the most intelligent animals suggests that probably a con-stellation of characteristics determines soliciting behaviors. Solicitations entailed use of gestures and vocalizations apparently intended to stimu-late the other pair-member ". . . to some activity, oriented with respect to the solicitor, but not immediately with the rope-pulling" (Crawford, 1937, p. 67). The animals were subsequently tested in tasks requiring cooperative cord pulling and lever pushing. Surprisingly, there was little to suggest that in any significant manner the principle of coordination had transferred from the first task of rope pulling. Cooperation had been elegantly manifest, but it had required individual training of the animals of a pair; the cooperation became solicited, though there was no appar-ent "instructional message" or tutoring provided by the solicitor; and the principle of cooperation did not appear to be readily transferred to new tasks.

M. MATCHING BEHAVIOR

Just as nonhuman primates have within limits an ability to imitate the responses of another, they have the capacity to select stimuli that are identical to others, that is, the ability to match a sample. Nissen *et al.* (1948) used what might be called a modified oddity testing situation for the study of matching behavior in chimpanzees. Problems were con-structed through use of three members of two identical pairs of objects. For example, in a problem that entailed basically two cups and two boxes, use of three of the elements would make one of them odd, hence incorrect in a matching situation. In a sequence of box-cup-cup and with the sample always being presented in the middle position, the cup on the right would be correct. (If the "one of a kind" had been required for a correct response, the experiment would have resembled an oddity task.)

The chimpanzees worked on settings of these kinds until a criterion was achieved, and subsequent generalization tests to other objects were made. It was concluded that the animals to varying degrees had learned more than just specific responses to specific objects as presented, that they had in fact learned to match whatever was presented in the middle of a sample.

Among the possible mechanisms that might account for this behavior, the researchers considered that the animals might be responding to the single stimulus member of the trio as "figure" as opposed to the two identical members which side by side might have been perceived as "ground." Another of the more intriguing possibilities considered was that the subject responded to whichever portion of the sequence was homogeneous as opposed to that part which when combined with the middle stimulus was heterogeneous. These possibilities were presented primarily in the interest of guiding additional research rather than accounting for the results of the study of immediate concern. The latter of these possibilities, however, seemingly advises that experimenters *not* use this form of testing if they desire to test for oddity, that is, the single or "odd" element should sometimes be in the middle position, for failure to so place it allows for the subject to be correct by avoidance of homogeneity as opposed to selection on the basis of pure oddity. (This probably accounts for the most errors being made in oddity problems when the odd stimulus is in the center position.)

Robinson (1955) provided additional evidence of chimpanzees' abilities to differentiate two pairs of stimuli according to whether the pairs were comprised of identical or different stimulus elements, that is, homogeneous or heterogeneous. Later (Robinson, 1960) he ascertained that subsequent to extensive sameness-difference training in which choices of the homogeneous pair were reinforced, chimpanzees preferred those variants that had the greater number of identical elements (multiplicity) in contrast to those that had the fewest number of different elements (simplicity). The variants were comprised of the same elements employed in the initial sameness-difference training. He favored a relational interpretation for a comprehensive conceptualization of his data. The relational view would posit that the chimpanzees were responding to either or both dimensions of multiplicity and simplicity. Such an interpretation allows for the observed preference for the variant AAB (where A and B represent different stimuli) when paired with A, but avoidance of AAB when paired with AAA.

Nissen and his associates (1949) continued study of matching behavior of chimpanzees but in a conditional context, one in which the cues to be matched were contingent upon still other cues and events. Again, they were interested in determining the degree to which the chimpanzees could process several units of information in combination for the determination of correct responses. Progress of the chimpanzees was slow, as the tasks proved difficult for them. Their slowness was attributed to persistent responsiveness to too many aspects of the situation. They were not as selective in their attention to the relevant stimuli in combination as they might have been. The authors recognized this capacity as being at a maximum in man, who probably benefits in both selective responses and management of complex relationships between events and responses through use of his language skills.

N. Food Token Experiments

Wolfe (1936) and subsequently Cowles (1937) investigated the conditions under which disks or poker chips or other forms of food tokens acquire reward for the acquisition and maintenance of a variety of behaviors.

These studies clearly demonstrated that such objects could acquire reward value if they were exchangeable for food or other incentives, such as return of the subject to its home cage. Chimpanzees mastered position discriminations, visual size discrimination problems, color-pattern discriminations, and delayed-response tasks when the only rewards were tokens that could be exchanged for food. The animals readily learned the exchange value of tokens, differentiating tokens according to whether they would or would not be exchangeable for food and, also, how much food various kinds of tokens were worth. They would work and accumulate as many as 30 tokens which would be exchangeable as a group though each token had been individually obtained.

In some tasks the animals did about as well when the reward consisted of food tokens as when food itself was used as the reinforcer. In paired preference tests, however, there was a clear-cut preference for food as opposed to food tokens and, as expected, food tokens were preferred to nonfood tokens. Cowles observed that the animals endured as long as 1-hour delays of exchange of food tokens for food.

Kelleher (1957) had chimpanzees press a lever on a 5-minute fixed-interval schedule with poker chips as reinforcers. The chips were exchangeable for food at the end of each hour. Low performance rates and eventual termination of all responding was observed. In an additional study, Kelleher observed that if the animals were required to accumulate numbers of chips for food exchange the rates of responding related directly to the approach of the time when the exchange for food could be made.

O. DELAYED RESPONSE

In the most familiar form of the delayed-response problem, the subject observes the placement of some incentive, such as food, and then after a predetermined amount of time has elapsed is given opportunity to obtain that incentive. For the subject to obtain the incentive, it must in some manner be able to identify the exact location where the incentive was last seen or be able to identify which of various stimuli identify its location. As reviewed by Fletcher (1965), the delayed-response problem has been an attractive one to psychologists for it is sensitive to a very broad spectrum of variables, including those of phylogeny, ontogeny, brain damage, and a host of situational and response modes of behavior in which the subject might engage during the delay interval. Successful performance is known to depend upon a variety of performance variables, hence it is not a pure learning or retention task.

Human capacity for performance on this kind of task is so remarkable that it hardly warrants comment. Undoubtedly, man's language is of great value in handling the demands of such tasks. Performance of non-verbal animals is accordingly of great interest, and it is intriguing to define the maximum delays through which they can in some manner retain information vital to eventual performance of the correct choice.

Yerkes and Yerkes (1928) observed that the chimpanzee had little difficulty in choosing correctly from among four food boxes, one of which had been baited in the animal's view some 3 hours earlier. Performance accuracy dropped off rapidly so that their limit for this particular type of delayed-response task was believed to be about 48 hours. It was demonstrated that the animals were in fact using the exact position or location cues as opposed to the physical color of the food-containing box for execution of the correct response. When color of the food-containing box was the only reliable cue the animal could use, with both the absolute and relative positions of all boxes being unreliable cues from trial to trial, the animals tended to err by going to the place where the box had been when baited or by choosing the box that contained food on the previous trial. With experience, however, the animals were able to utilize color as the only identifying cue for access to food for delays of at least 30 minutes. Of great interest is the fact that as the animals developed such skills they became clearly more careful in their observations, frequently hesitating and vacillating in choice before finally committing themselves to a single box. Errors came to elicit emotional responses in the chimpanzees which were described as expressions of disappointment, mystification, anger, and the like.

Nissen *et al.* (1936) conducted a series of studies to determine the control of certain stimulus and response variables upon the delayed-response performances of chimpanzee subjects. They determined that the distance between food containers (one of which was baited) was critical, with accuracy of choice increasing as distance increased. Distance between the ends of the two cords, attached to the containers for pulling them within reach, was by comparison of no consequence. Latencies were shorter for correct than for incorrect choices. Behavior during the delay interval varied markedly, with two subjects sitting quietly and another two moving about continuously and irregularly. Such differences were not apparently related to accuracy of choice. Requiring the chimpanzees to assume a position in the restraint cage on the side on which the food container with food was to appear (i.e., at the beginning of a given trial) facilitated correct choices.

Somewhat later, Nissen *et al.* (1938) observed that with preadolescent chimpanzees performance was superior when spatial as opposed to visual cues were to be used by the subjects in a delayed-response task. Among the visual cues studied, color or brightness differences were more successfully used by the chimpanzee than were size or pattern differences. Of considerable interest were their findings when animals were trained in successive discrimination reversal problems. One animal became a one-trial learner. Progressive improvement was observed in discrimination reversal training with a single stimulus pair even when the cues were reversed at the end of every 10 or 6 trials. One animal increased in accuracy from 52 to 93% from the start to end of training. Such data are totally consistent with what is now known as the LS phenomenon. Attempts to determine a relationship between delayed-response and discrimination learning skills with the same stimuli employed in both tasks was not successful.

Cowles and Nissen (1937) observed that size of the food incentive played a determining role in performance accuracy when on certain trials "large" and on others "small" incentives were used. Accuracy was greater when the larger incentive was used. The effect was attributed to what was called the expectancy value of the incentive. Such observations appear related to what is now known as the *Crespi effect* (1942).

Riesen and Nissen (1942) concluded that the use of a matching technique did not particularly facilitate the performance of chimpanzees on nonspatial delayed-response problems. Even with delays as brief as 15 seconds, performance based on visual cues dropped as much as 25%, which put one of their chimpanzees at only the 60% level. It was concluded that very extensive training is necessary for high-level performance when only visual cues, as opposed to spatial cues, are to be used by the subject.

Hayes and Thompson (1953) demonstrated that performance decrement was about as marked from the first to the second trials of discrimination LS problems, with delay intervals between trials extending from 10 to 60 seconds, as with similar delays in a delayed-response task. Thus a common factor, possibly that of *forgetting*, was implicated by Hayes and Thompson as the one that accounts for poor performance in visual delayed-response tasks when chimpanzees are used as subjects. In addition to the factor of forgetting, however, a number of other variables should be considered, for example, activity, orientation, and distractions during the delay.

Davenport and Rogers (1968) have recently compared chimpanzees that were raised for the first 2 years of life in a very restricted environment with other chimpanzees that were wild-born and reared with their mothers through the first years of life before being brought into a captive, laboratory colony. The learning of techniques of attention was marked in both groups, though initially the restricted chimpanzees were less task-oriented than were the wild-born chimpanzees. Though the wild-born chimpanzees were better than the restriction-reared chimpanzees on 0-, 5-, and 10-second delays, the restricted animals improved markedly throughout the testing to the point that in many instances their performances rivaled the wild-born group. Two of the restriction-reared subjects had to be eliminated from this study altogether by reason of their nontest behaviors. Thus it is suggested that as long as the effects of an early restricted environment are not such as to preclude behaviors conducive to accurate performance in a delayed-response task the effects of that restrictive environment can be compensated for through a course of prolonged training.

Though much less work has been done on delayed-response with the gorilla and orangutan, Yerkes and Yerkes (1929) did present some very early information which was of the same order as that on the chimpanzee. Berkson (1962) determined that gibbons (*Hylobates lar*) were able to perform at about a 75% correct level with delays as long as 20 seconds when a highly preferred incentive, raisin, was used as the food incentive. With less preferred incentives, celery and sweet potato, performances were not as accurate.

P. DELAYED REWARD

Riesen (1940) studied the learning of color discrimination tasks as influenced by delay of reward and by prior experience using chimpanzees as subjects. Prior to his study it had not been generally recognized that nonspatial tasks of this kind were characteristically highly refractory to learning and accurate performance. Chimpanzees without benefit of prior experience specific to color discrimination tasks were markedly handicapped when reward for correct choices was delayed even 1 or 2 seconds; delays of 4–8 seconds precluded all learning. By contrast, chimpanzees with prior training in a different situation and task which nonetheless entailed experience with color discrimination were not handicapped with delays even to 20 seconds. Apparently, the possession of a preestablished discriminative response to the relevant dimension, that is, color, provided for efficient performance not otherwise obtained in this kind of task when rewards are delayed.

Q. PATTERNED-STRING PROBLEMS

Finch (1941) studied the performances of eight chimpanzees on 11 patterned-string problems. In contrast to the performances of Harlow and Settlage's (1934) monkeys, the chimpanzees improved their performances with practice. The problems are presented in Fig. 4. The chimpanzees were able to solve both problems 10 and 11, whereas Harlow and Settlage's monkeys were not able to do so. One chimpanzee immediately solved problems 4 and 11, patterns that are extraordinarily difficult if not impossible for monkeys. Riesen et al. (1953) presented the same kinds of problems to young gorillas and concluded that they did almost as well as Finch's adult chimpanzees. It was further concluded that their data indicated either essentially equal ability for chimpanzees and gorillas on patterned-string problems, or possibly a slight superiority for the gorilla. Fischer and Kitchener (1965) found that the performances of their orangutan and gorilla subjects were comparable to those of Riesen's gorillas.

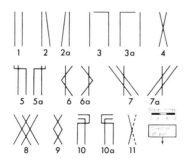

FIG. 4. Patterned-string problems used by Harlow and Settlage (1934) and again by Finch (1941).

Beck (1967) studied the string-pulling skills of four gibbons on a set of problems different from those used by the researchers discussed immediately above. The problems were nevertheless of the basic type used to study the perceptual and presumed insightful processes of primates. In one problem in which a reasonably direct comparison of performances could be made, Beck concluded that his gibbons did better than Köhler's (1925) chimpanzees had done. The task was a rather simple one, however, requiring the subject to displace a string toward the midline of a test board, thereby sweeping a morsel of food to a point within arm reach. In the most difficult of the three problems studied, the gibbon initially pulled on the incorrect string and only with practice shifted to the string that when pulled would bring the food to the subject by a route that went behind a peg, which was further from the subject than the food as initially placed.

Gibbons must be tested on more difficult string problems for us to conclude that their skills are truly consistent with those observed in the chimpanzee and gorilla. Beck's elevation of the strings to facilitate the gibbon's grasping of them is well advised for those who study the gibbon's performance in this kind of task. In view of the gibbon's restricted ability on LS, discrimination reversal, and oddity skills (Rumbaugh and McCormack, 1967), and also for accurate performance in single and double alternation tasks (Schusterman and Bernstein, 1962), it seems unlikely that the gibbon's string-pulling skills could equal those of the great apes.

R. Tool Using

Tool using, defined for the purpose of this chapter as the use of some implement (typically inanimate) for either facilitating some behavior or enhancing the likelihood of incentive achievement, is quite generally believed to reflect, at least in primates, some extraordinary capacity. This belief undoubtedly rests upon evidence obtained in Köhler's early studies as well as from the reports of such activities in feral animals as observed by investigators in the field.

Field reports indicate that all of the great apes, and gibbons as well, either drop branches from trees onto observers below or hurl branches from ground level at either human or animal targets through the course of manifesting displays, this being the specific case for both gorillas and chimpanzees. Only orangutans and chimpanzees have been reported as using twigs, sticks, or branches for investigating strange objects or procuring food in the field. Only the chimpanzee (Kortlandt, 1965) uses clubs and rocks either to intimidate its aggressors or to facilitate its aggressions or defenses against predators, however (see Fig. 5). Other well-known instances of tool-using behaviors have been reported recently in detail by van Lawick-Goodall (1968). A long-term study of the chimpanzees in the Gombe Stream area provided many opportunities to observe a variety of instances and contexts in which inanimate objects and materials were used to facilitate other behaviors, including play. In addition to the generalized throwing of vegetation that characterizes the behavior of chimpanzees when agitated or displaying, on several occasions objects were thrown with apparent intent to make impact with targets, such as baboons or people. When the range was but a few feet, the target was on occasion hit, but usually the throwing skills of the chimpanzees were inaccurate. In other instances chimpanzees in young saplings were able to make them sway with such violence that the branches and tip of the sapling made contact with a presumed target, that is, a human observer.

Careful selection and modification by certain chimpanzees of blades of grass for use in obtaining termites from their heaps is now widely known from van Lawick-Goodall's observations. Generally, the chimpanzees that engage in "termiting," inserting grass or other materials into the holes of termite nests and then carefully withdrawing it so as to be able to eat the termites clinging thereto, did not go further than 10 yards from the site being worked to obtain new materials for tools; however, one male was observed twice to carry a tool for distances over ½ mile, inspecting and rejecting a number of termite heaps as "not ready to work." Additional instances of tools used, for example as probes and as levers to pry at the lids of banana-laden boxes, were reported in number.

For the purpose of this chapter, van Lawick-Goodall's comments on learning and imitation as precursors of successful tool-using behaviors are particularly germane. Infants were frequently observed to manipulate and modify elements as adults of the area did long before they themselves engaged in successful tool-using behaviors. Infants were observed to watch adults as they worked at termiting and also to pick up and manipulate their tools in a comparable manner once the tools were abandoned. Leaves were frequently used by chimpanzees to assist in removing unpleasant materials from their bodies, for example, sticky foods, blood, urine, ejaculate, feces, and so on. Twice a 3-year-old infant was seen to observe his mother wipe fecal material from her rump, then to take leaves and wipe his own rump as though it too were messed.

Just how these various tool-using behaviors were first conceived or conditioned (learned) is difficult to understand, though it is surely the case that opportunity, experience, chance, and certain unlearned behavioral propensities, such as investigating and manipulating, must

FIG. 5. Chimpanzee mother using a stick as a weapon to attack a stuffed leopard that has a chimpanzee doll between its paws (Kortlandt, 1965).

novelly combine to provide for some of the more seemingly improbable ones—such as termiting. Their transmission through a course of generations is apparently ensured by imitative tendencies and skills. Such tendencies are not limited to the great apes, for expansion of the practice of washing sand from potatoes (Kawamura, 1959) prior to eating has been reported among macaques (*Macaca fuscata*). Nonetheless, it is likely that imitative mechanisms are employed with greater versatility and plasticity among the great apes than among macaques. Such is at the very least a working hypothesis of merit.

Wilson and Wilson (1968) have reported that captive chimpanzees at the Holloman Consortium also have been observed to use sticks for striking and throwing at one another. In addition, they have thrown rocks, padlocks, and other manageable units at one another in aggressive encounters. On occasion, objects of manipulation one moment became missiles for throwing in the next. Slapping of one another with burlap bags and spitting were also reported.

Cooper and Harlow (1961) reported one of the most dramatic instances of tool using in a *captive* monkey (*Cebus fatuellus*). This monkey used a stick, described as 13 inches long and ½ inch thick for striking and poking another (*Cebus capucinus*) which had attacked him. The blows were described as forceful and downward in direction.

It is surely the case that all apes use tools, given the opportunity, more in captivity than in the field. Very likely this is a function of their repeated encounter with elements of a given class in what is frequently an otherwise barren, monotonous environment. As an example, the use of tools by the gibbon in the field is at best limited to the apparent intentional dropping of branches toward observers underneath (Carpenter, 1940). Yet in a captive situation, a gibbon kept by the author has spontaneously developed significant tool-using behaviors. The animal was born and raised during its early life in the nursery of the San Diego Zoo. From the age of 1 year to its present age, 4 years, it has lived in the chain-link cage shown in Fig. 6. The cage was equipped with a heated sleeping box, an automatic watering valve, three rods to accommodate the animal's brachiation, and a food basket. A restricted number of plastic toys and balls have been available to the animal over the years.

The first instance of tool using was suggested by the frequent observation that wet segments of cloth were draped over the rods provided for brachiation. It was subsequently observed that these pieces of cloth were soaked by holding them in the same hand used to displace the stem of the automatic watering valve. The animal sucked water from the water-soaked cloth and then draped the cloth over one of the rods, usually the

Fɪɢ. 6. (A) Relative locations of automatic water valve and rod over which the gibbon draped a water-soaked cloth. (B) Water dripping from cloth onto angle-iron strip. Water was sipped by gibbon from small pooling at lower left corner. (C,D,E,F,G) Sequence of gibbon draping rope over rod, equalizing lengths, and swinging through use of it.

middle one. These performances were first observed when the animal was 3 years old. The cloth so used came from the gibbon's "security blanket" which it had been given early in its first year of life. As the animal had been rejected by its mother and raised in a nursery, its highly emotional responsiveness, characteristic of infant gibbons, was

Fig. 6 (B).

most readily attenuated by making available to it a blanket to which it
could cling. As these blankets became tattered and torn, they were replaced by new ones which were readily accepted. The animal's use of
the cloth for modifying its access to water very likely stemmed from its
sustained holding and contact with and manipulation of cloth, but in
all probability the initial soaking of the cloth was accidental.

Still later, but within the third year of life, the animal was observed
to use the cloth in still another way. After thoroughly soaking it, she
positioned it in a *specific place* on the middle rod from which the water

FIG. 6 (C).

dripped onto a 1-inch-wide piece of angle iron. The water ran down and formed a pool in a corner from which the gibbon drank by apparent use of her tongue as well as by sucking. This particular pattern of drinking is of interest for it is *atypical* of the normal drinking pattern of the gibbon in the field, that pattern being one of dipping the fingers and licking water from the hair and skin.

Early in the fourth year, the animal draped two new elements over the rods, a length of hose approximately 2 ft long and a length of rope also about 2 ft long. Where she obtained these elements is not known— perhaps a child dropped them by her cage. Most impressive is what she did with these draped elements. By grasping both ends with her hand, she swung to and fro. With time her skills became more advanced and she would take flying leaps of 5–6 ft, grasp both ends, and then swing wildly, particularly from the rope but also from the hose on rare occasions. Next, she was observed to thread this segment of rope through the chain link that formed the top of her cage and, as before, use it for swinging.

At the age of 4 years, she was given two new pieces of rope, 2- and

Fig. 6 (D).

5-ft lengths. Her preference for the longer rope was immediate and irreversible. Further, the incidents of swinging through use of the rope became an extremely common activity to the point that she spent at least 10% of her waking hours in this way.

These observations are more suggestive than explicative. It seems very likely, however, that the following were all critical in the development of these behaviors: (1) early sustained contact with and manipulation of cloth used as security or for comfort; (2) an automatic watering valve from which the water probably gushed too fast for facile drinking

FIG. 6 (E).

by an immature animal; (3) accidental wetting of the cloth and sucking water therefrom; (4) accidental (?) placement of the cloth so that when wet it would drip water so as to pool in a corner frame or angle iron; (5) access to the short segments of rope and hose; (6) sustained and relatively undisturbed living in a constant environment for several years. These are conditions that can and should be subject to systematic experimental study. That gibbons do not perform accurately in formal test situations underscores our need to study further the interplay between learning skills and tool-using behaviors. Also, it is apparent that we need

FIG. 6 (F).

to seek better understanding of why it is that in *some* captive situations the apes, particularly the gibbon, acquire more skill in the use of tools than their natural, supposedly spontaneous behaviors include in the field.

Schiller's (1952) study of innate constituents of complex responses has contributed greatly to our understanding of implementation behaviors. His work led him to conclude that the *innate* constituents that provide for successful performance in problems such as the box-stacking and hoe-type used by Köhler and Yerkes are *motor patterns,* not per-

Fig. 6 (G).

ceptual organizations. Through a series of experiments he observed that even in the absence of any incentive such as food the chimpanzees manipulated sticks, thrusting them into available openings, even joining them together when it was possible, and otherwise using them in ways basic to their utilization as tools. Similarly, the animals stacked boxes one on top of the other even in the absence of some overhead incentive, such as suspended bananas.

Activities of these kinds were summarized according to what he called a *maturational gradient*. Older chimpanzees were more varied and persistent in their manipulations than were younger chimpanzees. Only to the extent that maturation provides for these basic units of behavior can learning through experience serve to expedite their organization and application in some problem situation in which an incentive might be achieved. Such formulations as provided by Schiller appropriately influence our understanding of the role experience plays in presumed insightful behaviors (Birch, 1945).

Menzel *et al.* (1969) have argued that Schiller's "motor patterns" are not to be accounted for simply as innate, however, for a comparison of the development of tool-using behaviors in wild-born and restriction-reared chimpanzees revealed that the latter were extremely deficient in their performances even though they seemed to have the same repertoire of hand movements as did the wild-born ones. What the restriction-reared chimpanzees appeared to lack was the ability or predisposition to expand upon the basic elements of responsiveness to those refinements of perceptual-motor patterns that are requisites for skillful tool use.

It is tempting to emphasize implementation behaviors of the great apes at the expense of acknowledging its advanced manifestations in certain monkeys. To restore an important balance on this point, be reminded that Klüver (1933) was not at all convinced that instrumentation behaviors necessarily indicate a great gulf between monkey and ape as concluded by Yerkes and Yerkes (1929). To provide empirical support for his position, Klüver presented a detailed account of rather remarkable instances of tool usages by a captive *Cebus* monkey (coded P.-Y. in his program). At least the genus *Cebus* can manifest excellent tool usages, though even Klüver concluded that most of the monkeys, particularly the more primitive ones, do not.

S. DETOUR PROBLEMS

Davis and Leary (1968) determined that on bent-wire detour problems a variety of monkeys were all more inclined to failure and to make errors when the candy Life-Saver incentive had to be *pushed* away rather than *pulled* toward in order to extract it from the bent-wire detour. As children and certain great apes are adept at either response pattern on this type of problem, Davis and Leary suggested that being able to push food away in order to eventually obtain it for ingestion is likely an important component in the evolution of tool using. Davis, McDowell, and Nissen (1957) compared rhesus monkeys with two adult and five adolescent chimpanzees in a series of bent-wire problems. In general, the adult chimpanzees were superior to the adolescent chimpanzees, which in turn were superior to the macaques. As an alternative to Schiller's (1952) maturational gradient which would account for the difference in the two age levels of chimpanzees, Davis and associates suggested that with age *perceptual* organization increases both in complexity and stability, reflecting the cumulation of experience. The authors rejected the possibility that the observed species' difference might have been a reflection of superior motor skills of the apes. This rejection was made on the basis that the slender hands of the monkeys seemed more suited to the tasks than did the relatively thick and blunt fingers of the

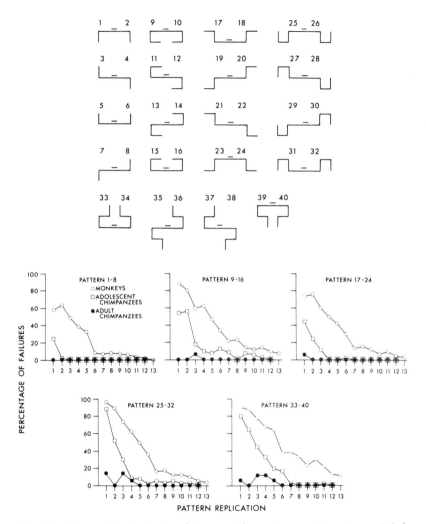

FIG. 7. Upper: The bent-wire detour problems. Lower: Percentage of failures for monkey and chimpanzee groups as coded for pattern groups as a function of practice (from Davis *et al.*, 1957).

chimpanzees and, further, the chimpanzees were additionally handi-capped because they had to work through 2-inch wire mesh which formed their cages in order to work on the problems.

McDowell and Nissen (1959) compared rhesus monkeys 3–9 years of age with juvenile and adult chimpanzees in a task in which the animals had to insert their arms alternately through the spaces defined by ad-jacent rungs spaced 4 inches apart center to center in order to elevate a stylus upward within a vertical slot so as to obtain food. Chimpanzees were much more successful than were the monkeys, though with practice the monkeys did improve to the point where they approached the performance of the chimpanzees. Once again, the chimpanzees had to work through a 2-inch wire mesh rather than just through the 4-inch spaces between the rungs as was the case with the monkeys. Three of the five chimpanzees had *no* failures whatever. The results were attributed to presumed differences in the general perceptual organizations that characterize the species studied.

T. Use of Numbers

One of the most convincing ways of demonstrating whether or not nonhuman primates have the capacity to utilize abstractions that provides at least in part for man's language system would be to demonstrate unequivocally that they are able to utilize the meaning of numbers in the pure sense.

Douglas and Whitty (1941) attempted to teach a mature female chimpanzee to associate a number of clue-shapes to a corresponding number of shapes on a box which contained food reward. In a sense it was a modified match-to-sample task in which the stimulus to be matched consisted of a number of shapes which served to clue which one of two food boxes contained the reward. Training was terminated after more than 5 months of daily testing without any convincing evidence of solution having been obtained.

In a well-controlled experiment, Hicks (1955) demonstrated that rhesus monkeys can acquire the concept of "threeness" to a moderate yet significant degree. Drawing upon earlier studies by both Yerkes and Spence, Rohles and Devine (1966, 1967) have convincingly demonstrated the ability of an adolescent female chimpanzee to acquire the concept of numerical middleness. With as many as 17 stimulus objects, the animal was able to select the middle one at levels better than 75% correct. Even in tests that utilized what the researchers called "asymetrical middleness," in which uncovered or blank wells were interspersed among those covered with the stimulus objects from which the middle one had to be selected for reward to be obtained, the animal was able to reach a 95% correct response level. Though admitting that it was difficult to determine the exact process that provided for the animal's discrimination of the middle object, it was suggested that the relevant process might be in fact a form of primitive counting. Similar studies indicated that the rat could not develop the middleness concept, but that macaque monkeys (*Macaca arctoides*) were able to do so with three or five, but not seven,

DECIMAL	BINARY	LIGHTS
0	000	● ● ●
1	001	● ● ○
2	010	● ○ ●
3	011	● ○ ○
4	100	○ ● ●
5	101	○ ● ○
6	110	○ ○ ●
7	111	○ ○ ○

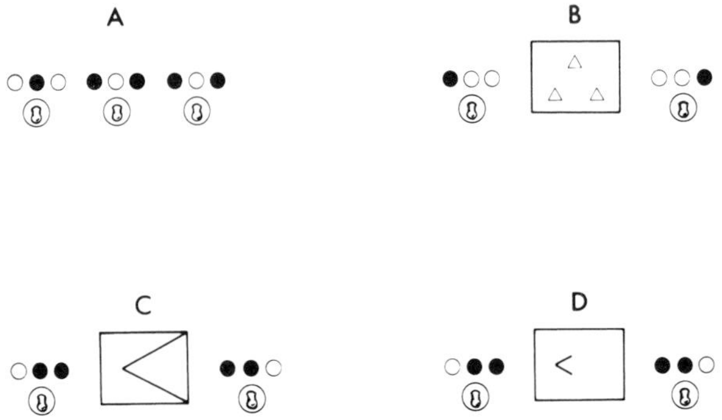

FIG. 8. Upper: Binary equivalents of numbers 0 through 7, with translation into lights turned on (right column). Lower: Matching-to-sample tasks. (A) Subject must match center binary expression. (B) Subject must choose which binary expression is equivalent to number of elements projected on center screen. (C and D) Subject must select larger or smaller binary expression depending upon whether symbol in the center window is large or small (from Ferster, 1964 and Ferster and Hammer, 1966).

objects symmetrically arranged. That the chimpanzee was able to perform accurately with as many as 17 objects suggested sensitivity of this capacity to the variable of phylogeny.

Without question, Ferster and Hammer (1966) have made the most convincing demonstration of arithmetic skills with chimpanzees. In a totally automated situation which provided for literally hundreds of thousands of training trials, the chimpanzees came to be able to "write" by selectively pressing or not pressing each of three keys so as to provide a duplicate of the binary number provided in the sample row. Subsequently, they were trained to write, again with the use of three keys, the binary number equivalent to the number of triangles, squares, or whatever projected onto the sample window (Fig. 8). At the end of 170,000 trials the task had been mastered, and both animals were able to "write" the binary number equivalent to the number of objects. At the very least the animals were demonstrated to have acquired the rudiments of numeric skills. Whether the animals can in fact manage numeric information in the sense of adding or subtracting numeric expressions by means that preclude use of their marked capacity for memory of pictures, as discussed earlier, remains the subject of future research.

U. Language Use and Signing

The basic reason for the study of language-type skills in nonhuman primates should be to determine the relationship between that skill and other behaviors, such as the ability to form concepts, or to the evolution of the central nervous system. Such justification is not incompatible with the less scientific intrigue with the general question whether or not man might be able to communicate through use of his own language with other animal forms.

The ultimately anthropocentric approach to this study is to attempt to teach animals a language system familiar to man, for example, English. Such approaches (see Kellogg, 1968, for a review) have been noted for their frustrations and limited success. With great effort and perserverance, the Hayeses taught their chimpanzee, Viki, to say three words, "papa," "momma," and "cup." To date theirs has been the greatest success, and it is not likely that it will be embarrassed or superceded by the efforts of others. It was extremely difficult for the animal to master the sounds required by English words. Lieberman, Klatt, and Wilson (1969) have evaluated the vowel repertoire of a variety of nonhuman primates, including apes, and have found their capacity for producing the vowels to be very restricted. They concluded that difficulties in teaching nonhuman primates human speech reflect inherent limitations in their vocal mechanisms.

R. Allen Gardner and his wife, Beatrice, of the University of Nevada, are currently engaged in a long-term study to develop sign language with a chimpanzee subject (Gardner and Gardner, 1969). All communication with the chimpanzee, as well as among personnel of the project when with the chimpanzee, is through use of what is essentially the American sign language of the deaf. Human voice usage is limited to imitation of sounds the chimpanzee itself produces. At the age of 3 years, the chimpanzee, Washoe, was well advanced in developing strings of two or more signs, there being 170 of them. This infant chimpanzee by far has developed the most promising channel of communication man has ever had with any animal form; if Lenneberg (1969) is correct that language in children develops *not* by simple imitation but rather by the abstractions of regularities or relations from the language which are then used for building up language for himself, then there is reason for hope that Washoe will come to acquire a language system which will enable it to engage in conversations with its colleague, man. (Washoe's diary, maintained by the Gardners, for the week June 16–22, 1968, augurs well for the future, for the notation reveals that as Washoe was brought over to see a 22-month-old deaf girl carried by her father, Washoe signed "baby" while the little girl signed back "monkey." Most of Washoe's "babies" had been rubber dolls prior to that encounter.) Provided with a language system which its motor capacity can clearly accommodate, only Washoe's natively endowed intellect should stand between it and meaningful, highly informative two-way communication with man.

V. YERKES' VIEWS OF ANTHROPOID LEARNING AND INTELLIGENCE

There are two publications in particular that present in detail the work, ideas, and conclusions of the one man who to date was in his life time more concerned about valid assessment of the learning processes and intelligence of the anthropoid apes than any other. With his wife, Ada, Robert M. Yerkes authored and published *The Great Apes* in 1929. Later (Yerkes, 1943), he published a book entitled *Chimpanzees*. In these two classic works there has been contributed a wealth of history, research, and ideas which relate to the basic nature of the anthropoid apes and their behaviors. From these and other publications of Robert M. Yerkes, it is possible to draw a summary germane to the topic of this chapter.

Yerkes and Köhler had much in common. At least in terms of theoretical bent and method of report, they surely would have made a fine team of researchers. Both placed great emphasis on native endowment, native capacity, and genetically determined tendencies for behaviors to have certain characteristics according to the species studied. Both allowed for the all-important contributions of experience in the manifestation of the highest, most complex forms of intellectual expression. Though both were convinced that trial-and-error behavior accounted for portions of great ape behavior and in fact *could* play an important role in the manifestation of insightful behavior, they were agreed that trial-and-error learning and stimulus-response associationistic processes could not validly describe or account for certain behaviors that they termed insightful and ideational in character. Both were convinced that in order to test for and determine the parameters of insightful learning the subjects must be afforded a perceptual overview of the entire problem situation. Mode of orientation to the problem, the manifestation of pauses and hesitations in behavior, dramatic reorganization of behavioral

patterns with its associated facility and efficiency, and certainly the suddenness of problem solution when it finally occurred, all of these were
taken by both men as probable indices of insightful processes.

Only one of these men, Yerkes, was primarily interested in problems
of assessing the differential behaviors and capacities of the anthropoid
apes. Though there is no question in this author's opinion that Yerkes
had a strong preference for the gorilla, this preference at times appearing to interfere with the drawing of the most straight-forward interpretations and conclusions possible and warranted by available data, he
had great admiration for all apes. The chimpanzee he viewed as more
curious, more imitative than its cousin the gorilla, and to these characteristics he attributed its apparent anthropoid supremacy. The chimpanzee he viewed as sanguine, the orangutan melancholy, and the gorilla
reserved if not out-and-out negativistic and frankly obstinate. The playfulness and inventiveness of the chimpanzee was contrasted with the
more cautious behavior of the orangutan and the cool, calculating, as
well as reserved behavior of the gorilla. The chimpanzee with its love for
applause and approval from the humans with whom it associates in
captivity makes it the most trainable and dependable of the apes. In
terms of profiting from experience that depends upon the higher-order
processes of perceiving relations, however, Yerkes was not confident
that the chimpanzee was the best. Admittedly on the basis of limited
data, he was inclined to rank order them high to low on the basis of
capacity to benefit from *trial-and-error* learning—the chimpanzee, the
orangutan, and last the gorilla; in a similar ranking from high to low on
the basis of ability for *ideational* learning, the order was reversed—first,
the gorilla; second, the orangutan; and third, the chimpanzee. That these
animals do not compete effectively for survival in the field, as might be
expected in view of their advanced capacities for learning, Yerkes attributed to factors other than intelligence. (For that matter he was not at all
convinced that superior intelligence assured human survival.)

Yerkes believed that the apes differed in aptitudes, with the gorilla,
for example, being less curious about novel objects than were the other
apes and, in addition, not particularly skillful in the manipulation of
any kind of mechanical devices, such as hook, hasp, or lock. By contrast, the chimpanzee is far more responsive and inclined toward manipulation of its environment and is also more imitatively inclined. At times,
Yerkes reached for the ultimate, being convinced that it would be possible to relate the remarkable skills and behaviors as viewed at the
empirical level to one of the most unique phenomena, consciousness.

Yerkes was also extremely sensitive to the problems of studying primate behavior and he was convinced that the researcher must know
intimately the characteristics of his animals. Only with such knowledge

could fair test situations be constructed. Too, he fully appreciated the extreme variations in ability presented by a variety of subjects, even of the same species. [For a selected review of his work and conclusions, the reader is encouraged to read Chapters 7 and 10 of Robert M. Yerkes' book, *Chimpanzees* (1943).]

Though Yerkes recognized that the chimpanzee could be more readily exploited behaviorally than any other of the great apes, he was less sure not only about the rank ordering of the great apes so far as intelligence is concerned but also the magnitude of the difference between adjacent ranks. He viewed them all as being remarkable and advanced, clearly superior to the monkeys but clearly inferior to man. And it is important to note that regardless of the close kinship that he properly recognized between the great apes and man he stopped well short of the point of concluding that their ideational processes were *in fact* similar to those of man. He did not support the inference that their experiences were similar to our own. He did insist, however, that man's insight has developed from neural processes not unlike those presumed to exist in the great apes.

II. DISCUSSION

It is apparent that a substantial literature has accumulated over the course of the past 50 years on the subject of *anthropoid learning*. In the interests of future research on this subject, it is appropriate that attention now be given to the summarization of what with reasonable confidence we can hold as true. A survey of the literature has revealed that whereas much work has been done with the chimpanzee, relatively little has been done with the other apes; accordingly, definitive comparisons among the various apes are not yet possible. It would be an egregious error to conclude at this point in time that data are sufficient to allow firm conclusions regarding similarities and differences among the learning skills of the various apes; however, within limits believed justified by our current state of knowledge, tentative conclusions and leads can be formulated.

A. LEARNING SKILLS OF GREAT APES

Experiments to date support the conclusion that differences in the apes' capacities for learning are probably small, if extant. Sufficient work has been done to permit the conclusion that there is great overlap in the distributions of learning capacities for the great apes (Rumbaugh and McCormack, 1967). Without exception, in those studies in which two or more of the great ape genera have been represented and in which appropriate matching or allowance has been made for differences in age and experience, it has been concluded, albeit with qualifications, that great apes regardless of genera do equally well. This is not to deny the existence of great *individual differences* within each genus, for they exist to extreme degrees in all of the great ape genera (for that matter, within all primate species and nonprimate mammalian species as well). As the chimpanzee has been used experimentally far more frequently than either the gorilla or orangutan, there has been much more opportunity to observe its extraordinary skills and performances. Also, as it is probably true that it is easier to induce chimpanzees to cooperate and to otherwise accommodate the conduct of behavioral tests and experiments than it is to do so with either gorillas or orangutans, the likelihood is increased that its high achievements will be observed. It is frequently asserted that this is true of the chimpanzee because it is particularly sensitive to indications of approval and disapproval from humans. If such is the case, in a controlled situation conditioning should be achieved more readily with chimpanzees than either gorillas or orangutans when rewards are provided only through the reactions of humans, such as their vocalizations or applause. It would be well to make such a test.

B. Great Apes Compared with Gibbons

Very little work has been done with gibbons as subjects in formal test situations. None has been reported for its close relative, the siamang (*Symphalangus*). Without exception, however, no research to date supports the conclusion that as a genus *Hylobates* has learning skills commensurate with those of the great apes. For that matter, they do not appear to have abilities consistent with their advanced taxonomic standing, for not infrequently their performances have been inferior to those of a wide variety of both Old and New World monkeys. Intensive study of the learning skills of gibbons is much needed, for they are an enigma.

C. Great Apes, Gibbons, and Monkeys: A Comparison

Evidence to date is overwhelmingly in support of the conclusion that chimpanzees, and probably gorillas and orangutans as well, are superior to both Old and New World monkeys in learning skills. This is not to deny that there are instances in which great apes have not performed better than certain monkeys, such as macaques (e.g., Rumbaugh and

McCormack, 1967). In these instances it is our opinion that the tasks used have either failed to challenge the apes, perhaps even boring them, or have not been complex enough to allow for the apes to excel, that is, the ceilings for performances on the tasks were too low to allow differentiation between apes and monkeys. It is likely that certain *individual* monkey specimens perform better than certain individual great ape specimens, but such observations merely reflect overlap of distributions of skills for the ape and monkey genera compared. Too, a distinction must be made between capacity and performance. It is frequently difficult to motivate a great ape to work well in an experimental situation. Compared to monkeys, apes do not work as readily for food. It is likely that many instances of low performance in ape subjects should be attributed to motivational problems. Motivation of great apes of all ages should be a study area in its own right. Whereas hunger is quite effective in inducing most simian genera to work in formal test situations, at times it can be remarkably ineffective with great apes. It is *as though* their hunger is abated as their attention is diverted from the task to any of a number of apparently trivial stimuli and events, such as the sight of an insect on the floor, hearing of a strange noise, noting a new spot to pick at with a fingernail, discovery of a characteristic of a test apparatus not previously examined, sensing a new odor, and so on. Diversions of these kinds can distract the animal from its testing and make for remarkably poor performance. Nonetheless, it is more *probable* that a well-motivated great ape will succeed in mastery of a complex task than it is that any monkeys will succeed.

As discussed above, all evidence to date supports the unequivocal conclusion that great apes are vastly superior to gibbons so far as performance in formal test situations is concerned. As gibbons are both taxonomically and neurologically more advanced than Old and New World monkeys, it would be reasonable to expect them to be superior to those monkeys in test performances. This is not the case, however. It is likely that if a comparison is made between gibbons and any of a variety of monkey genera that the gibbons will be *inferior* to them. Yet, it is quite possible that such evidence does not validly reflect the gibbon's learning skills. Gibbons, particularly young ones, can be extremely responsive and alert. Perhaps they do poorly by reason of their being prone to emotional display when significantly challenged in any of a number of ways, perhaps even by the demands of a formal test situation. At times they can be remarkably deliberate in their response to a challenge, however. The reader's attention is redirected to the account presented earlier in this chapter which described one gibbon's use of a cloth for water transport and a rope for facilitating its swinging within a cage. In still another instance we have observed a docile 4-year-old male gibbon (*H. lar*) skillfully retrace its path so as to avoid snarling the chain that tethered it. This animal accomplished such retracing by deftly lifting the chain with one hand, apparently noting whether it passed over or under a branch or passed to the left or right of a tree trunk, then making its way accordingly. The animal did not do well when the chain crossed back upon itself, for in such instances it frequently made an incorrect choice of path. Observations as these are in striking contrast to the poor performance of *Hylobates* in formal tests of discrimination learning. As asserted above, continued study of their learning is mandatory.

In general, research to date is consistent with the conclusion that great apes are behaviorally as well as neurologically more similar to man than are either gibbons or monkeys. Compared to monkeys, the great apes excel in situations in which multiple cues must be used, when correct choices must be made on the basis of pictured information. In those instances in which *relational* processes have been suggested (e.g., Gonzales *et al.*, 1954; Robinson, 1960), great apes served as subjects. Such does not deny the existence of relational processes in monkeys, but it seems less likely that such processes would be determined with most monkey forms. Great apes, represented mainly by the chimpanzee, do remarkably well and are superior to monkeys on tasks of delayed response, patterned strings, tool usage, bent-wire detour problems, and numeric skills. It is also surely true that they are superior to monkeys on tasks that entail the use of signs (Gardner and Gardner, 1969). The remarkable performances observed by Crawford (1937) in a situation that required cooperation between pairs of animals are not likely to be replicated with monkey subjects, though an attempt to do so would be totally warranted.

D. INNOVATIVE BEHAVIORS

Another point of prime distinction that separates the great apes from other primate forms is their readiness to devise "games" and to innovate upon other behavior patterns. Chimpanzees frequently have been reported to innovate in ways that would lead to the conclusion that they had learned other than the bare minimum of single instrumental acts and that they had in fact derived intrinsic reward from activities initially required of them in some experimental situation. For example, Menzel *et al.* (1969) reported that one of their wild-born chimpanzees often ignored food it had raked to within reach through use of a stick—*obtaining*, not eating, the food object appeared to suffice. This same animal was observed to throw pieces of food or other objects back onto the stimulus tray only to rake them in again with a tool. A novel, well-organized behavior pattern had been devised.

In an interview reported by Hall (1969) and confirmed by personal communication, Hebb recounted how a chimpanzee he tested solved problems for banana slice incentives. On one particular day she arranged the banana slice rewards in a row instead of eating them. "Apparently she had solved each problem for its own sake. I was out of bananas, but I offered her another problem [the problem being the first of eight problems learned that session—a point clarified by personal communication with D. O. Hebb]. She solved the problem, opened the box and put a slice of banana into it. I took it out and then set the box again. She solved the second problem and put the second slice into the box, I ended up with thirty slices of banana" (Hall, 1969, p. 23). We have observed that whereas monkeys either refuse to take or eat nonpreferred incentives for correct choices in discrimination experiments, great apes on occasion accumulate them and return them to empty foodwells on subsequent trials when correct choices are rewarded with more palatable incentives, such as fruit, that are taken and apparently eaten.

Gardner and Gardner (1969) have reported the following incident which is taken to support the point argued here, that chimpanzees innovate creatively from their experiences. The experience of their chimpanzee, Washoe, came not from a formal testing situation but from procedures of its physical care. Washoe had received regular baths and from her second month with the Gardners she had dolls with which to play. One day, after having been with the Gardners 10 months, she was observed to bathe a doll in the same way she was bathed. She filled a small bathtub with water, immersed, and soaked the doll, and finally dried it. Washoe became the "giver," not the recipient of the bath. To determine the *class* of objects Washoe might "bathe" would be of prime interest.

Hayes and Hayes (1953) reported that their chimpanzee, Viki, innovated upon certain of the test procedures that required of her that she imitate actions portrayed in pictures, both still and moving. Viki was observed to "run" herself in this kind of testing by turning on the projector or going through the still photos and giving the appropriate response in turn.

Yerkes asserted

> that the chimpanzee is naturally the most imitative of all in its association with man, the orangutan although distinctly less given to imitation is by no means negativistic, but the gorilla as exemplified by Congo, seems to show positive resistance to imitative tendency and, as in the case of curiosity, to

be under compulsion or impulsion to act independently and superiorly. These statements apply only to imitation of human acts. My opportunities to study inter-species imitative tendency have been meager except in the case of the chimpanzee, and it may well be that among themselves gorillas are highly imitative by comparison with other apes.

It is in their relative imitativeness of man that we seem to discover the secret of the great apes' peculiar values as performing animals. In this class the chimpanzee easily leads, for it can readily be trained to a considerable variety of skilled acrobatic performances, imitations of human habits or customs and interesting tricks. It loves applause, becomes devoted and strongly attached to its keeper or trainer, and, baring accidents, is loyal and dependable as long as kindly and fairly treated. (Yerkes, 1927, p. 181.)

In summary, the point is suggested that great apes, particularly the chimpanzee, are able to innovate upon their experiences in captivity so as to produce novel, well-organized, repetitive behavioral patterns, which in the vernacular might be considered "games." The resultant behaviors appear to be self-sustaining, not contingent upon receipt of some extrinsic incentive to be ingested. Such behaviors are to be distinguished from other responses that reflect only curiosity and manipulation per se. Though curiosity and manipulation are surely involved in the development of the innovative behaviors here discussed, a capacity for devising behaviors in a manner that suggests imitation is also surely involved. To our knowledge, such highly innovative behaviors, self-indulged and self-sustained, have not been reported for any form of monkey. It seems that contrary to the basic tenet of Thorndikian associationism which posits the development of associations between stimuli and responses (see Bitterman, 1969, for a recent discussion), great apes learn not only stimulus-response associations, but *learn about* those associations and *about* experiences they have had. The experiences might be from a formal test situation or from some other form of treatment they have had, but from those experiences they appear capable of deriving various information vital to the formulation of other highly innovative behaviors.

E. Functional Equivalence and Innovative Behaviors

For such games to be innovated by an ape, it seems that a form of relational or ideational behavior would be required, in accord with conclusions asserted both by Köhler and Yerkes. The formulation of such games seemingly requires of the subject that he be able to perceive and learn about the *functioning roles* of select elements involved in certain experiences. Further, it seems that he must have the capacity to reassign elements with regard to functioning roles. To clarify, a piece of food obtained through use of a tool can become an object other than one to eat—it can become an object to toss *just far enough* so that it can be retrieved once again through use of a tool. Again, uneaten incentives can become something other than elements to leave uneaten—they can become elements to be returned to the experimenter under conditions not now fully understood. A chimpanzee's toy doll can become something other than just an object to manipulate—it can come to be defined as a *functional equivalent* of the chimpanzee itself, and thus receives treatment (e.g., a bath) analogous to treatments experienced formerly by the chimpanzee. And a chimpanzee's bathtub becomes something other than a vessel within which it is bathed, it becomes a vessel within which to bathe other objects which by some dimension *might* be perceived and classified as functional equivalents to the chimpanzee's self.

Without exception, tool usage entails a redefinition of materials—for example, stems of grass and bits of twigs become implements with which to obtain termites or to explore some unfamiliar object or substance; rocks and branches become missiles to throw at presumed predators. Undoubtedly, it is of significance that although baboons associated with the chimpanzees of the Gombe Stream Reserve have frequent opportunity to observe the chimpanzees termiting with use of primitive implements (van Lawick-Goodall, 1968) they do not in fact similarly extract them from their nests. Not only do the baboons appear to lack the ingenuity for termiting—they also appear to lack the capacity to emulate or imitate the model for termiting provided by chimpanzees. Perhaps it is the capacity to flexibly classify and reclassify a wide variety of stimuli and objects with regard to equivalence on a number of dimensions, including utility or function, that provides the basis germane to a broad array of innovative behaviors, including tool usage, for which apes, in contrast to other nonhuman primates, are particularly distinguished.

Admittedly, these considerations are highly speculative, but they are presented with the conviction that they are possibly critical examples of behaviors that reflect some higher-order ideational process quite unlike that involved in the formation of simple stimulus-response associations. Heretofore many examples of the kind just considered have been recounted, primarily for the purpose of amusement it seems, then discarded as having no real significance. Such practice has been probably unfortunate so far as stimulating us to new methods of research which in the final analysis will be required for us to describe the complex learning processes and capacities of the great apes in phylogenetic perspective.

III. TRANSFER INDEX ASSESSMENTS OF
ANTHROPOID LEARNING

It is widely recognized that there are major problems and risks in attempts to obtain equitable measurements of learning for an array of diverse animal forms. "Diversity in receptor-effector mechanisms frequently renders exact comparisons of learning between species and genera questionable and poses major problems when we attempt comparison among orders, classes, and phyla" (Harlow, 1958, p. 269).

Attempts to develop a measurement of capacity for complex learning, such as LS (Harlow, 1949), which would hopefully circumvent or at least minimize risks inherent in testing various primate species in different test apparatuses and comparing their absolute performance levels (such as percentages of responses correct) led Rumbaugh to the formulation of the transfer index (TI). Recent studies thereof will now be discussed.

The phenomenon that provided the foundation for the development of TI was that primates of various genera do not necessarily perform equivalently on the reversal trials in a discrimination reversal training situation *even* though matched for proficiency on the basis of the initial, prereversal trials (see Rumbaugh, 1968, pp. 273–277). As initially defined (Rumbaugh, 1969), TI measurements were obtained by criterionally training subjects to approximately 67% responses correct on each of a series of visual discrimination problems of the kind commonly used in LS studies. Upon achievement of criterion, cue values were reversed so that "+" became "−" and the "−" became "+" for the next 10 trials. This procedure provided subjects of various genera with latitude in reaching criterion, but both a limited and equal opportunity for demonstration of transfer. In general, the correlations between number of trials to reach criterion and performance on the reversal trials *per problem* were not significant. In those instances in which significant correlations were obtained, they were low and, accordingly, of little consequence. It is desirable that the correlations be either nonsignificant or of a low order, for they permit us to discount the possible importance of differences between groups in terms of average number of trials to learn TI problems. In other words, if there is little or no relationship between number of trials to reach criterion and reversal performance, differences in number of trials to reach criterion for a variety of species may be adjudged as being of little consequence. The TI procedure equates subjects of diverse species according to a dimension of fundamental importance—*performance*. Only when they are so equated are the critical performance measures taken via cue reversal. A TI is extracted by dividing the reversal percentage correct (R%) for 10 problems (trial 1 excluded) by the pre-reversal or acquisition criterional percentage (A%), for example, 67%.

There are four main reasons for concluding that TI surely measures that same capacity that provides for Harlow's LS phenomenon: (1) The TI testing procedure is a modified discrimination-reversal training task, which is basically an extension of object-quality LS test procedures, that entails the reversal of cue values at some point; (2) The TI was conceived in part as a consequence of the observation that R%/A% ratios correlated highly with trial-2 LS performances for a variety of apes and macaques (Rumbaugh and McCormack, 1967, p. 300), and TI is, in fact, a specific R%/A% value, for example, where A = 67%; (3) Given continued TI testing, subjects do improve within limits; hence TI, along with classic LS, is an interproblem phenomena of some learning-to-learn process; and (4) TI values lead to conclusions compatible with those arising from LS values so far as a general positive relationship between phylogeny and complex learning capacity is concerned.

Subjects of diverse species are equated with regard to a specific A% value in view of the empirical fact that R% values correlate highly with A% values (between 50 and 95%) for the lowland gorilla (*Gorilla gorilla gorilla*, $r = 0.92$, $p < 0.001$, $df = 23$). Such a fact requires that TI be determined for a restricted A% level, or else inappropriate R% performances for a wide variety of A% levels can render the TI uninterpretable.

As stated above, TI was first defined in terms of the 67% level for A. It is possible of course to obtain TI measurements for any of a variety of A% levels.

The following study was conducted to collect TI measurements at two A levels, the 67 and the 84%. The 67% level for A had been selected initially because it was concluded that the greatest dispersion of R%/A% values was at about that level (Rumbaugh and McCormack, 1967). The 84% level for A was selected because it is approximately half-way between the 67% level and the 100% level. Performance of great apes in comparison to a group of gibbons and two groups of monkeys was desired to determine empirically whether or not there is superiority of the great apes at both levels when compared with other primate groups, and whether or not there is an interaction between genera and A% as reflected in TI measurements. Definition of the two criterion levels is provided in Table I.

All the animals of this experiment were late juveniles or young adults. All of them had received *extensive* object-quality discrimination train-

TABLE I
RESPONSES CORRECT REQUIRED FOR TI 67% AND 84% CRITERIONAL LEVELS

Criterion	Trial no.	Required no. responses correct
67%	11[a]	7 or 8
	14	9
	16	10
	19	12
	22–60	14 within last 21 trials
84%	11	9
	17	14
	21	17 or 18 (on alternate problems)
	22–60	17 or 18 (on alternate problems) within last 21 trials

[a] First trial of training, in which chance prevails, is not included in the count. Hence, trial 11 is the *twelfth* trial of training.

ing in previous experiments. The animals were tested in modified Wisconsin General Test Apparatuses (Harlow, 1949) to which they were thoroughly adapted by reason of their previous training. Two TI measurements were obtained for each specimen at both the 67 and 84% A levels. Measurements at the 67% level were obtained prior to those at the 84% level.

Correlations between number of trials to each criterion and the reversal performances *per problem* are summarized in Table II. It should be noted that the correlations are either nonsignificant or, at best, low. Accordingly, differences between genera in terms of number of trials to reach criterion can be negated as a likely determinant of observed TI measurement. (In earlier studies, when genera were matched on the

TABLE II
PEARSON PRODUCT-MOMENT CORRELATIONS (r) BETWEEN TRIALS TO
CRITERION AND REVERSAL PERFORMANCES (PER PROBLEM)
OBTAINED WITH TI TESTING PROCEDURES

Species	67%	$df\ (n - 2)$	84%	$df\ (n - 2)$
Orangutan (*Pongo*)	−0.22	38	−0.30	18
Gorilla (*Gorilla*)	−0.34[a]	78	−0.39[a]	78
Chimpanzee (*Pan*)	−0.11	138	−0.17	38
Gibbon (*Hylobates*)	−0.30[b]	58	−0.53[a]	58
Vervet (*Cercopithecus*)	−0.08	78	0.09	78
Talapoin (*Miopithecus*)	−0.10	138	−0.10	98

[a] $p < 0.01$.
[b] $p < 0.05$.

basis of absolute numbers of trials to reach criterion, the resultant TI measurements were in complete accord with those in which such matching was not made, that is, it is prereversal performance in terms of *percentage correct* that is the main determinant of reversal performance.)

As stated above, prior to the reversal of cues all subjects of the genera represented were trained criterionally to the 67 and 84% levels. A final indication of success in training them to equivalent levels of mastery is reflected in the fact that when their performances in terms of percentage responses correct was taken for the *last five trials* prior to achieving the criteria as stated the groups did not differ significantly—only 6% where A = 67% (R = 82–76%) and only 3% where A = 84% (R = 95–92%).

The TI values obtained are summarized in Fig. 9. The supremacy of great apes to the other genera is apparent. One specimen of each of the Pongidae genera achieved a TI measurement at the A = 67% level which reflected significant positive transfer from the A to R trials; their R% were significantly higher than the A value of 67% yielding TI≥ 1.15. *Gorilla* and *Pan* performed commensurately at both TI levels. *Gorilla, Pan,* and *Cercopithecus* all held constant in *average* TI as A increased from 67 to 84%; their reversal performance improved as A% was increased (see R% axes at the right of Fig. 9). In contrast, both *Hylobates* and *Miopithecus* had significant decrements in TI as A increased from 67 to 84%; none of their upper range limits challenged those of the great apes.

The TI decrement for the talapoins (*Miopithecus*) is particularly striking. Their extraordinarily low TI measurements at the A = 84% reflect failure of R% to increase with increment of A from 67 to 84%. They had a very persistent tendency to perseverate in the selection of whichever object was initially correct on each problem.

Though *Cercopithecus* tended to perform at lower TI levels than Pongidae, they were not significantly lower than either *Gorilla* or *Pan.* At the A = 84% level, however, while *Cercopithecus* was significantly better than *Miopithecus, Pan* was significantly better than both *Hylobates* and *Miopithecus.*

Though additional numbers of subjects are to be desired for each group represented, the results are consistent with the conclusion that though matched for prereversal performance levels diverse primate forms do not reverse equivalently. Also, results are in keeping with the conclusion that the great apes are able to transfer what they have learned in the prereversal trials to the reversal trials with greater proficiency than gibbons or the two groups of monkeys similarly tested. Perhaps most important of all, TI measurements indicate that the LS capacities of the great apes range to higher levels than any of those yet

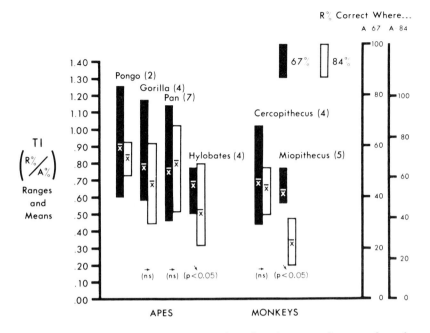

FIG. 9. TI ranges and means (left axis) for selected groups of apes and monkeys where criterional training prior to cue reversal (A%) was at the 67 and 84% levels. TI values at these levels were calculated on the basis of 10-problem blocks, with percentage responses correct on the reversal trials (R%, trial 1 excepted) divided by the appropriate A% value. The two right axes indicate absolute R% values as observed when criterional training had been at the A = 67% and A = 84% levels. TI values dropped reliably for *Hylobates* and *Miopithecus* as A was advanced from the 67 to 84% level, whereas *Gorilla, Pan,* and *Cercopithecus* held constant. No test of significance was calculated for *Pongo* where $N = 2$.

obtained for other nonhuman primate forms. Conclusions regarding differences in TI skills among the great ape genera must await the collection of additional data.

IV. TRENDS: PRESENT AND FUTURE

It is likely that present and future trends of research on anthropoid learning will fall into two major categories—capacity testing and process assessment. With reference to *capacity* testing, much important work remains to be done just to define the apes' presumed *limits* of learning

and performance on various tasks—particularly those that require utilization of several cues concurrently, complex concepts such as numbers,
and linguistic elements such as signs. Though considerable is known
regarding the chimpanzee's capacities, its limits are surely far beyond
the ones presently defined. Little is known regarding the capacities of
gorillas and orangutans, and essentially nothing regarding those of the
enigmatic gibbon. In the final analysis, other than for the rhesus monkey,
relatively little is known regarding the capacities of the very large number of monkey and prosimian species.

The second major trend of research will be directed toward the
assessment of the *processes* whereby various skills and problems are
learned. The phenomena of learning and the processes thereof remain
elusive; to consider that they might differ *qualitatively* among primate
taxa adds a new dimension of complexity to the study of learning. Just
as the species variable has proved to be a very powerful one for a
variety of behaviors ranging from those of locomotion, mother-infant
care patterns, to social behaviors, so it might prove to be equally powerful for the most basic processes of learning with all of its related
phenomena, including transfer, generalization, memory, and so on. In
a broader context, Nissen acknowledged the power of the species variable when he wrote, "Unless there is a continuity or homology of behavioral mechanisms from the lower to the higher animals (including
man), there would be no rationale for the comparative method. This
does not mean, necessarily, that the higher (later) is *merely* an extension or elaboration of the lower (earlier), although this is and has
been a most provocative and fruitful working hypothesis. Quantitative
complication may become so great that it produces, in effect, qualitative
differences with new, 'emergent' properties" (Nissen, 1951, p. 351).

A. BEHAVIORAL PLASTICITY AND ADAPTABILITY

A frequent assertion in the anthropoid literature has been that great
apes possess a plasticity, a problem-oriented adaptibility far beyond that
of monkeys. Such adaptability is presumed to reflect increased reliance
upon the processes of learning and experience as opposed to instinctive
or unlearned behavioral patterning. This proposition is a very attractive
one, for it coincides with those refinements and advancements in adaptability and intelligence presumed to culminate in man. In point of fact,
however, the proposition needs critical inspection and extensive test;
it is too important for any doubt to be left regarding its accuracy and
validity. Instead of our continuing to accept it as either true or probably
true, its implications should be so thoroughly subjected to test that no

doubt remains. (The presumed superiority of the great apes' behavioral plasticity and adaptability is in no apparent way assisting them in avoiding extermination in the wilds; indeed, they fare more poorly than many monkey forms in this regard.) The *presumed* superiority of the great apes to monkeys with regard to adaptability and utilization of learning skills to either cope with or to compensate for *environmental* stresses and limitations should be carefully reexamined, for if such superior capacity and adaptability uniquely characterizes the apes it is one of the most important hallmarks of behavioral primatology and evolution.

B. EARLY EXPERIENCE

Among common report which recurs frequently in the literature and in discussions among researchers is that young apes, so young that they might still be properly termed infants (less than 4 years old), frequently do better than even young adults in formal test situations. As a case in point, Rumbaugh and McCormack's (1967) report indicates that with all tests considered it was a young orangutan (0-2; *Pongo pygmaeus pygmaeus*) that performed the best of all 19 great apes in their study. The orangutan was only 2.5 years old at the beginning of the study program and less than 3.0 years old at its end. Further, in the conventional LS testing portion of the program, among the five orangutans (from 2.3 to 9.0 years of age) there was marked *negative* relation between age and performance.

It is frequently observed, however, that an animal that excels in learning when young *remains* excellent *if* frequently worked with as it grows to adulthood (*at least* 8 years of age) and beyond. Might it be the case that early experience in some manner determines the *avenues* along which intelligent behavior will be manifest? If early experiences are with formal test and learning situations, will the animal's adaptability be maximally manifest as an adult in contexts of that order? Or, if early experience is in the wilds, with the exigencies thereof, will the apes' skills remain essentially untapped in the kinds of tasks typically devised by laboratory researchers? The extent to which emotional conditioning or adjustment to contexts of varied kinds might be the prime determining factor, rather than specialization (aptitude) in learning, is certainly relevant and must be separated for special assessment. The basic point remains, however, that for reasons not now clear immature apes frequently if not typically do better than adults. It is extremely important to understand why this is true, for the answer might have far-reaching implications, especially for attending to the early experiences of the human child.

Without question the most exciting understandings of anthropoid learning and behavior await us. To date we have nothing better than a toehold so far as understanding their complexity is concerned.

As our understanding of anthropoid learning and behavior increases, it will surely be the case that we shall come to view man in a new perspective (Reynolds, 1967). As we do so, the presumed gulf between him and his closest living relative, the apes, will surely diminish.

ACKNOWLEDGMENTS

Research of the author herein reported was supported by grants from the National Science Foundation and facilitated by generous cooperation extended by the Zoological Society of San Diego, the Institute for Comparative Biology, and the San Diego State College Foundation. Critical comments and suggestions offered by Austin H. Riesen and James E. King in response to an earlier draft of this chapter were invaluable and are gratefully acknowledged. Finally, the able assistance of Miss Rosemary Arnold and Miss Judith Burns facilitated completion of this work. Preparation of this chapter was supported by NIH grant RR-00165 to the Yerkes Regional Primate Research Center of Emory University, Atlanta, Georgia.

REFERENCES

Beck, B. B. (1967). A study of problem solving by gibbons. *Behaviour* **28**, 95–109.

Berkson, G. (1962). Food motivation and delayed response in gibbons. *J. Comp. Physiol. Psychol.* **55**, 1040.

Bernstein, I. S. (1961a). Response variability and rigidity in the adult chimpanzee. *J. Gerontol.* **16**, 381–386.

Bernstein, I. S. (1961b). The utilization of visual cues in dimension-abstracted oddity by primates. *J. Comp. Physiol. Psychol.* **54**, 243–247.

Birch, H. (1945). The relation of previous experience to insightful problem-solving. *J. Comp. Physiol. Psychol.* **38**, 367–383.

Bitterman, M. E. (1969). Thorndike and the problem of animal intelligence. *Amer. Psychologist* **24**, 444–453.

Boyer, W. N., Polidora, V. J., Fletcher, H. J., and Woodruff, B. (1966). Monkeys' performance on ambiguous-cue problems. *Perceptual Motor Skills* **22**, 883–888.

Brainard, P. P. (1930). The mentality of a child compared with that of apes. *J. Genet. Psychol.* **37**, 268–293.

Carpenter, C. R. (1940). A field study in Siam of the behavior and social relations of the gibbon, *Hylobates lar. Comp. Psychol. Monogr.* **16**, 1–212.

Connolly, C. J. (1950). "External Morphology of the Primate Brain." Thomas, Springfield, Illinois.

Cooper, L. R., and Harlow, H. F. (1961). Note on a *Cebus* monkey's use of a stick as a weapon. *Psychol. Rep.* **8**, 418.

Cowles, J. T. (1937). Food-tokens as incentives for learning by chimpanzees. *Comp. Psychol. Monogr.* **14**, 1–96.

Cowles, J. T., and Nissen, H. W. (1937). Reward-expectancy in delayed responses of chimpanzees. *J. Comp. Psychol.* **24**, 345–358.

Crawford, M. P. (1937). The cooperative solving of problems by young chimpanzees. *Comp. Psychol. Monogr.* **14**, 88 pp.

Crawford, M. P., and Spence, K. W. (1939). Observational learning of discrimination problems by chimpanzees. *J. Comp. Psychol.* **27**, 133–147.

Crespi, L. P. (1942). Quantitative variation of incentive and performance in the white rat. *Amer. J. Psychol.* **55**, 467–517.

Davenport, R. K., and Rogers, C. M. (1968). Intellectual performance of differentially reared chimpanzees: I. Delayed response. *Amer. J. Ment. Defic.* **72**, 674–680.

Davis, R. T., and Leary, R. W. (1968). Learning of detour problems by lemurs and seven species of monkeys. *Perceptual Motor Skills* **27**, 1031–1034.

Davis, R. T., McDowell, A. A., and Nissen, H. W. (1957). Solution of bent-wire problems by monkeys and chimpanzees. *J. Comp. Physiol. Psychol.* **50**, 441–444.

Diamond, I. T., and Hall, W. C. (1969). Evolution of neocortex. *Science* **164**, 251–262.

Douglas, J. W. B., and Whitty, C. W. M. (1941). An investigation of number appreciation in some subhuman primates. *J. Comp. Psychol.* **31**, 129–143.

Farrer, D. N. (1967). Picture memory in the chimpanzee. *Percept. Mot. Skills* **25**, 305–315.

Ferster, C. B. (1964). Arithmetic behavior in chimpanzees. *Sci. Amer.* **210**, 98–106.

Ferster, C. B., and Hammer, C. E., Jr. (1966). Synthesizing the components of arithmetic behavior. *In* "Operant Behavior: Areas of Research and Application" (W. K. Honig, ed.), pp. 634–676. Appleton-Century-Crofts, New York.

Fiedler, W. (1956). Übersicht über das System der Primates. *In* "Primatologia" (H. Hofer, A. H. Schultz, and D. Starck, eds.), Vol. I, pp. 1–266. Karger, Basel.

Finch, G. (1941). The solution of patterned string problems by chimpanzees. *J. Comp. Psychol.* **32**, 83–90.

Fischer, G. J. (1962). The formation of learning sets in young gorillas. *J. Comp. Physiol. Psychol.* **55**, 924–925.

Fischer, G. J., and Kitchener, S. L. (1965). Comparative learning in young gorillas and orangutans. *J. Genet. Psychol.* **107**, 337–348.

Fletcher, H. J. (1965). Delayed-response problem. *In* "Behavior of Nonhuman Primates" (A. M. Schrier, H. F. Harlow, and F. Stollnitz, eds.), Vol. I, pp. 129–165. Academic Press, New York.

Fletcher, H. J., and Bordow, A. M. (1965). Monkeys' solution of an ambiguous-cue problem. *Perceptual Motor Skills* **21**, 115–119.

Fletcher, H. J., Grogg, T. M., and Garske, J. P. (1968). Ambiguous-cue problem performance of children, retardates, and monkeys. *J. Comp. Physiol. Psychol.* **66**, 477–482.

Gardner, R. A., and Gardner, B. T. (1969). Teaching sign language to a chimpanzee. *Science* **165**, 664–672.

Gonzales, R. C., Gentry, G. V., and Bitterman, M. E. (1954). Relational discrimination of intermediate size in the chimpanzee. *J. Comp. Physiol. Psychol.* **47**, 385–388.

Hall, E. (1969). Hebb on hocus-pocus; a conversation. *Psychol. Today* **3**, 20–28.

Hall, K. R. L. (1963). Observational learning in monkeys and apes. *Brit. J. Psychol.* **54**, 201–226.

Harlow, H. F. (1949). The formation of learning sets. *Psychol. Rev.* **56**, 51–65.

Harlow, H. F. (1958). The evolution of learning. *In* "Behavior and Evolution" (A. Roe and G. G. Simpson, eds.), pp. 269–290. Yale Univ. Press, New Haven, Connecticut.

Harlow, H. F., and Settlage, P. H. (1934). Comparative behavior of primates. VII. Capacity of monkeys to solve patterned string tests. *J. Comp. Psychol.* **18**, 423–435.

Harlow, H. F., Harlow, M. K., Reuping, R. R., and Mason, W. A. (1960). Performance of infant rhesus monkeys on discrimination learning, delayed response, and discrimination learning set. *J. Comp. Physiol. Psychol.* **53**, 113–121.

Harman, P. J. (1957). Paleoneurologic, neoneurologic and ontogenetic aspects of brain phylogeny. James Arthur Lecture on the Evolution of the Human Brain. American Museum of Natural History, New York.

Hayes, K. J., and Hayes, C. (1952). Imitation in a home-raised chimpanzee. *J. Comp. Physiol. Psychol.* **45**, 450–459.

Hayes, K. J., and Hayes, C. (1953). Picture perception in a home-raised chimpanzee. *J. Comp. Physiol. Psychol.* **46**, 470–474.

Hayes, K. J., and Thompson, R. (1953). Nonspatial delayed response to trial-unique stimuli in sophisticated chimpanzees. *J. Comp. Physiol. Psychol.* **46**, 498–500.

Hayes, K. J., Thompson, R., and Hayes, C. (1953a). Discrimination learning set in chimpanzees. *J. Comp. Physiol. Psychol.* **46**, 99–104.

Hayes, K. J., Thompson, R., and Hayes, C. (1953b). Concurrent discrimination learning in chimpanzees. *J. Comp. Physiol. Psychol.* **46**, 105–107.

Hicks, L. H. (1955). An analysis of number-concept formation in the rhesus monkey. *J. Comp. Physiol. Psychol.* **48**, 212–218.

Jarvik, M. E. (1953). Discrimination of colored food and food signs by primates. *J. Comp. Physiol. Psychol.* **46**, 390–392.

Kawamura, S. (1959). The process of subculture propagation among Japanese macaques. *Primates* **2**, 43–60.

Kelleher, R. T. (1957). Conditioned reinforcement in chimpanzees. *J. Comp. Physiol. Psychol.* **50**, 571–575.

Kellogg, W. N. (1968). Communication and language in the home-raised chimpanzee. *Science* **162**, 423–427.

Klüver, H. (1933). "Behavior Mechanisms in Monkeys." Univ. of Chicago Press, Chicago, Illinois.

Köhler, W. (1925). "The Mentality of Apes." Routledge & Kegan Paul, London.

Kortlandt, A. (1965). How do chimpanzees use weapons when fighting leopards? *Yearb. Amer. Phil. Soc.* pp. 327–332.

Leary, R. W. (1958). The learning of ambiguous cue-problems by monkeys. *Amer. J. Psychol.* **71**, 718–724.

Le Gros Clark, W. E. (1959). "The Antecedents of Man." Edinburgh Univ. Press, Edinburgh and London.

Lenneberg, E. H. (1969). On explaining language. *Science* **164**, 635–643.

Levine, M. (1959). A model of hypothesis behavior in discrimination learning set. *Psychol. Rev.* **66**, 353–366.

Levine, M., Levinson, B., and Harlow, H. F. (1959). Trials per problem as a variable in the acquisition of discrimination learning set. *J. Comp. Physiol. Psychol.* **52**, 396–398.

Lieberman, P. H., Klatt, D. H., and Wilson, W. H. (1969). Vocal tract limitations on the vowel repertoires of the rhesus monkey and other nonhuman primates. *Science* **164**, 1185–1187.

McDowell, A. A., and Nissen, H. W. (1959). Solution of a bi-manual coordination problem by monkeys and chimpanzees. *J. Genet. Psychol.* **94**, 35–42.

Mason, W. A. (1968). Scope and potential of primate research. *Sci. Psychoanal.* **12**, 101–118.

Menzel, E. W., Jr., Davenport, R. K., and Rogers, C. M. (1970). The development of tool using in wild-born and restriction-reared chimpanzees. *Folia Primatol.* **12**, 273–283.

Napier, J. R., and Napier, P. H. (1967). "A Handbook of Living Primates." Academic Press, New York.

Nissen, H. W. (1942). Ambivalent cues in discrimination behavior of chimpanzees. *J. Psychol.* **14**, 3–33.

Nissen, H. W. (1951). Phylogenetic comparison. *In* "Handbook of Experimental Psychology" (S. S. Stevens, ed.), pp. 347–386. Wiley, New York.

Nissen, H. W., Carpenter, C. R., and Cowles, J. T. (1936). Stimulus-versus-response-differentiation in delayed reactions in chimpanzees. *J. Genet. Psychol.* **48**, 112–136.

Nissen, H. W., Riesen, A. H., and Nowlis, V. (1938). Delayed response and discrimination learning by chimpanzees. *J. Comp. Psychol.* **26**, 361–386.

Nissen, H. W., Blum, J. S., and Blum, R. A. (1948). Analysis of matching behavior in chimpanzee. *J. Comp. Physiol. Psychol.* **41**, 62–74.

Nissen, H. W., Blum, J. S., and Blum, R. A. (1949). Conditional matching behavior in chimpanzee; implications for the comparative study of intelligence. *J. Comp. Physiol. Psychol.* **42**, 339–356.

Noback, C. R., and Moskowitz, N. (1963). The primate nervous system: Functional and structural aspects in phylogeny. *In* "Evolutionary and Genetic Biology of Primates" (J. Buettner-Janusch, ed.), Vol. I, pp. 131–175. Academic Press, New York.

Parker, C. E. (1969). Responsiveness, manipulation, and implementation in chimpanzees, gorillas, and orangutans. *Proc. 2nd Int. Primatol. Soc. Congr.*, pp. 160–166. Karger, Basel.

Reynolds, V. (1967). "The Apes." Dutton, New York.

Riesen, A. H. (1940). Delayed reward in discrimination learning by chimpanzees. *Comp. Psychol. Monogr.* **15**, 1–54.

Riesen, A. H., and Nissen, H. W. (1942). Nonspatial delayed response by the matching technique. *J. Comp. Psychol.* **34**, 307–313.

Riesen, A. H., Greenberg, B., Granston, A. S., and Fantz, R. L. (1953). Solutions of patterned string problems by young gorillas. *J. Comp. Physiol. Psychol.* **46**, 19–22.

Riopelle, A. J., and Rogers, C. M. (1965). Age changes in chimpanzees. *In* "Behavior of Nonhuman Primates" (A. M. Schrier, H. F. Harlow, and F. Stollnitz, ed.), Vol. II, pp. 449–462. Academic Press, New York.

Robinson, J. S. (1955). The sameness-difference discrimination problem in chimpanzee. *J. Comp. Physiol. Psychol.* **48**, 195–197.

Robinson, J. S. (1960). The conceptual basis of the chimpanzee's performance on the sameness-difference discrimination problem. *J. Comp. Physiol. Psychol.* **53**, 368–370.

Rohles, F. H., and Devine, J. V. (1966). Chimpanzee performance on a problem involving the concept of middleness. *Anim. Behav.* **14**, 159–162.

Rohles, F. H., and Devine, J. V. (1967). Further studies of the middleness concept with the chimpanzee. *Anim. Behav.* **15**, 107–112.

Rohles, F. H., Grunzke, M. E., and Reynolds, H. H. (1963). Chimpanzee performance during the ballistic orbital Project Mercury flights. *J. Comp. Physiol. Psychol.* **56**, 2–10.

Rumbaugh, D. M. (1968). The learning and sensory skills of the squirrel monkey in phylogenetic perspective. *In* "The Squirrel Monkey" (L. A. Rosenblum and R. C. Cooper, eds.), pp. 255–317. Academic Press, New York.

Rumbaugh, D. M. (1969). The transfer index: an alternative measure of learning set. *Proc. 2nd Int. Primatol. Soc. Congr.*, pp. 267–272. Karger, Basel.

Rumbaugh, D. M., and McCormack, C. (1967). The learning skills of primates: a comparative study of apes and monkeys. *In* "Progress in Primatology" (D. Stark, R. Schneider. and H. J. Kuhn, eds.), pp. 289–306. Fischer, Stuttgart.

Schiller, P. H. (1952). Innate constituents of complex responses in primates. *Psychol. Rev.* **59**, 177–191.

Schrier, A. M. (1966). Transfer by macaque monkeys between learning-set and repeated-reversal tasks. *Perceptual Motor Skills* **23**, 787–792.

Schultz, A. H. (1941). The relative size of the cranial capacity in primates. *Amer. J. Phys. Anthropol.* **28**, 273–287.

Schusterman, R. J. (1962). Transfer effects of successive discrimination-reversal training in chimpanzees. *Science* **137**, 422–423.

Schusterman, R. J. (1963). The use of strategies in two-choice behavior of children and chimpanzees. *J. Comp. Physiol. Psychol.* **56**, 96–100.

Schusterman, R. J. (1964). Successive discrimination-reversal training and multiple discrimination training in one-trial learning by chimpanzees. *J. Comp. Physiol. Psychol.* **58**, 153–156.

Schusterman, R. J., and Bernstein, I. S. (1962). Response tendencies of gibbons in single and double alternation tasks. *Psychol. Rep.* **11**, 521–522.

Spence, K. W. (1938). Gradual versus sudden solution of discrimination problems by chimpanzees. *J. Comp. Psychol.* **25**, 213–224.

Spence, K. W. (1941). Failure of transposition in size-discrimination of chimpanzees. *Amer. J. Psychol.* **54**, 223–229.

Spence, K. W. (1942). The basis of solution by chimpanzees of the intermediate size problem. *J. Exp. Psychol.* **31**, 257–271.

Strong, P. N. (1967). Comparative studies in oddity learning: III. Apparatus transfer in chimpanzees and children. *Psychon. Sci.* **7**, 43.

Strong, P. N., and Hedges, M. (1966). Comparative studies in simple oddity learning: I. Cats, raccoons, monkeys, and chimpanzees. *Psychon. Sci.* **5**, 13–14.

Thompson, R. (1954). Approach versus avoidance in an ambiguous-cue discrimination problem in chimpanzees. *J. Comp. Physiol. Psychol.* **47**, 133–135.

Thorndike, E. L. (1898). Animal intelligence; an experimental study of the associative processes in animals. *Psychol. Monogr.* **2**, No. 8.

Thorndike, E. L. (1911). "Animal Intelligence: Experimental Studies." Macmillan, New York.

Tobias, P. V. (1968). Cranial capacity in anthropoid apes, *Australopithecus* and *Homo habilis*, with comments on skewed samples. *S. Afr. J. Sci.* **64**, 81–91.

van Lawick-Goodall, J. (1968). The behavior of free-living chimpanzees in the Gombe Stream Reserve. *Anim. Behav. Monogr.* **1**, 161–301.

Voronin, L. G. (1962). Some results of comparative physiological investigations of higher nervous activity. *Psychol. Bull.* **59**, 161–195.

Warren, J. M. (1966). Reversal learning and the formation of learning sets by cats and rhesus monkeys. *J. Comp. Physiol. Psychol.* **61**, 421–428.

Wilson, W. L., and Wilson, A. C. (1968). Aggressive interactions of captive chimpanzees living in a semi-free ranging environment. Report of the 6571st Aeromedical Research Laboratory, Aerospace Medical Division, Air Force Systems Command, Holloman Air Force Base, New Mexico.

Wolfe, J. B. (1936). Effectiveness of token-rewards for chimpanzees. *Comp. Psychol. Monogr.* **12**, 1–72.

Yerkes, R. M. (1916). The mental life of monkeys and apes; a study in ideational behavior. *Behavior Monogr.* **3**, 1–145.

Yerkes, R. M. (1927). The mind of a gorilla. *Genet. Psychol. Monogr.* **2**, 1–193.

Yerkes, R. M. (1943). "Chimpanzee—a Laboratory Colony." Yale Univ. Press, New Haven, Connecticut.

Yerkes, R. M., and Yerkes, D. N. (1928). Concerning memory in the chimpanzee. *J. Comp. Psychol.* **8**, 237–271.

Yerkes, R. M., and Yerkes, A. W. (1929). "The Great Apes; A Study of Anthropoid Life." Yale Univ. Press, New Haven, Connecticut.

Primate Status Hierarchies

Irwin S. Bernstein

Yerkes Regional Primate Research Center
Emory University, Atlanta, Georgia

I. INTRODUCTION: THE SIGNIFICANCE OF THE CONCEPT AND ITS MEASUREMENTS

Monkeys and apes are basically social animals; almost all live as members of a social group, but the specific organization of the group varies from one primate taxon to another. Some live in a "family" unit consisting of a single adult pair and their associated offspring; others live as members of more extensive social units consisting of multiple adult males

71

and females and associated immature animals. Another type of organization is based on one-male units consisting of an adult male with multiple adult females and young. In some taxa, one-male units are often united by social bonds into higher levels of social organization. Many troops have stable membership for extended periods of time, whereas others may divide into consistent subgroups during the day or in response to changing ecological conditions. In still other primate taxa, the primary social unit consists of a loosely organized community which is constantly reorganizing into small bands of varying composition.

Considering the wide variety of social organizations to be found, it is not surprising that no single unifying concept has been found to explain all primate sociality. Certainly primate social groups serve many important functions, such as predator defense, infant protection, and the facilitation of information transmission. Living in permanent social units also facilitates reproduction, but the units are not simply breeding assemblages. Many primates breed seasonally, and mating accounts for only a small segment of social interactions. Some investigators have suggested that rather than sex it is aggression—or at least dominance relations— that units primate societies. Strict status hierarchies are strikingly prevalent among macaques and baboons. Angermeier *et al.* (1968), Carpenter (1950, 1954), Chance (1961, 1967), Hamburg (1968), and West (1967) have all concluded that the status hierarchy is fundamental to primate societies. Considering the available literature on primate social organization, it is easy to see how West came to his conclusions. Hamburg, however, was apparently less confident that the status hierarchy model would prove fundamental to all primate societies and theorized that ground living enhanced the expression of dominance relations. Such a view finds support in Carpenter's (1940, 1965) statements that the dominance gradient in gibbon and howler monkey troops may be very low and the expression of relative status very subtle. Carpenter (1954), nevertheless, concluded that dominance relationships influence the probability of all types of social behavior and therefore have a pervasive effect on group organization. Other investigators (Mason, 1964), however, have concluded that neither sex nor dominance are the source of primate social organization.

A. DEFINITIONS AND PRIORITY TO INCENTIVES

In various discussions of dominance, however, the word has undergone several changes in meaning. Accepted expressions of dominance and their measures started with the original "peck order" concept published by Schjelderup-Ebbe (1913, 1935) and gradually changed to stress "priorities to incentives" as the basic expression and measure. Boelkins (1967) used water deprivation and drinking scores with some success, but admits less than satisfactory reliability in test-retest, which he suggests reflects the fluidity of the relationships. Carpenter (1950, 1954), Crawford (1940), Jay (1965b), and Zuckerman (1932) have all stressed priority to incentives as the key criterion, be the incentive food, access to sexual partners, entry into another's cage, or any other activity deemed desirable. In Hall's (1964) view, animals may use aggression to obtain priority to incentives; spacing and other mechanisms serve to limit the form and scope of aggression instigated by competition within a social unit. A social code is apparently established and Hall believed that most aggression within the social group was directed at punishing nonconformist behavior. Chance (1961, 1967) focuses on the spacing mechanism and the ability of dominant animals to supplant subordinates. He concludes that it is these displacements and the equilibration of spatial relations that are the crucial attributes of dominance hierarchies. All other related social activities stem from the ability of animals to control space. Ploog et al. (1963), however, found that none of these measures was a reliable indicator of dominance relationships among squirrel monkeys. Believing that stable status relationships were basic to primate societies, they explored alternate measures and for awhile thought that genital presentations were consistently unidirectional and could be used as a dominance measure. Later, Winter and Ploog (1967) found that they could not demonstrate clear status hierarchies in other squirrel monkey groups even with the use of this measure.

Feeding order and the acquisition of presumed incentives has been a popular measure in laboratory studies of dominance and social relationships between pairs of animals. This measure is a simple objective one, but it may not be clear exactly what is being measured inasmuch as this measure may correlate poorly with other presumed dominance measures, such as the direction of agonistic episodes. Even where significant positive correlations are demonstrated, these correlations are not high enough to suggest a unitary phenomenon being measured by two independent scoring techniques. Warren and Maroney (1958) demonstrated a correlation of 0.77 between priority of access to food and agonistic dominance, and our own work (Bernstein, 1969) confirms their findings, within the context or larger social groups. A correlation of this magnitude certainly indicates that the measures are related to one another, but inasmuch as test-retest reliability using the direction of agonistic encounters is near 1.0, there seems to be no reason to use the less precise and indirect measure based on priority to food incentives. In fact, in less controlled situations the correlation between success in obtaining food incentives and status seems much lower, and Hall (1968) concluded that food tests correlated so poorly with social rank that they were not a useful measure even in approximating the dominance relationships in a group.

Nowlis (1941) discovered correlations between status and age of 0.61 and between status and weight of 0.85 using chimpanzee subjects. Schaller (1963) indicates that dominance in gorillas is also closely related to size. He used displacements as his measure of dominance but the rate of such activity was only 0.23 episodes per hour of troop observation. Emlen (1962) suggested that one of the vocalizations used by gorillas may relate to dominance assertion when intruders are encountered. Correlations between status and weight have also been reported by Maroney et al. (1959) using rhesus monkey subjects. Inasmuch as these correlations often surpass those obtained between food tests and agonistic dominance using scores obtained from macaque subjects, it seems clear that these correlations are a result of mutual reliance on other unspecified variables rather than a result of measurement of different attributes of the same phenomenon.

B. CONTROL ROLE

With all the confusion engendered in the multiplicity of definitions and presumed dominance measures, and with the required subjective determination of which objects and acts an animal considers desirable and therefore incentives, some investigators have directed their attention to other biologically important attributes associated with dominance which may be relevant to the evolution of dominance relationships. The "control role" as described by Bernstein (1964, 1966) has essentially also been described by DeVore and Hall (1965), Hall and DeVore (1965), Kaufman (1967), Kummer (1968b), Rowell (1967), Schaller (1965), Struhsaker (1967b, relating it to territorial behavior), and Varley and Symmes (1966). This social role, defending group or troop members against internal and external sources of disturbance, has been recognized in almost all species showing strong dominance relationships within the society and has also been recognized in groups in which status relationships are either difficult to detect or so poorly developed that they are socially not meaningful. Carpenter (1935), for example, reports that although the spider monkey males he observed defended the troop they were not "true leaders," perhaps a reference to the lack of observable dominance relationships. The extent to which the control role is characteristic of monkeys and apes is, however, still unknown. Hall (1967), for example, found that the male patas monkey seemed more concerned with conspecific males than troop defense. The social roles in a patas troop are still imperfectly understood, but it seems that many of the control role functions are performed by adult females.

C. Status Measurement

Status relationships are, nevertheless, well structured in many primate taxa and are of great social significance. In many cases, a matrix showing the direction of agonistic encounters reveals virtually unidirectional relations between every possible dyad in the troop or group, for example, *Macaca fuscata* (Alexander and Bowers, 1969; Imanishi, 1957) and *M. mulatta* (Carpenter, 1942; Marsden, 1968; and Varley and Symmes, 1966, who incidentally found priority to food incentives to be a poor measure of status). The direction of agonistic episodes has been used as the criterion to determine social rank by a number of investigators, some of whom could only detect stable hierarchial relations among the males in a troop. DeVore (1963, 1965) determined the male hierarchy in the Nairobi Park baboons by observing agonistic episodes. He also describes the results obtained when food offers were used and rejects success in obtaining food as an indicator of relative status. His theories and evidence concerning the central hierarchy, however, make it difficult to understand how he finally determined relative status among the adult males; the agonistic evidence alone is confusing. Hall and DeVore (1965) and Hall (1965) report that dominance hierarchies are not well developed in chacma baboons, and perhaps this may also account for some of the observations described for Nairobi Park baboons. Hall (1965) then departs from a status orientation and goes on to stress role functions.

Rowell (1966a, 1967) states that the wild baboons she studied in a Uganda gallery forest showed little sign of dominance interactions, whereas caged animals showed clear dominance orders and even the females (Rowell, 1968) showed very stable dominance relationships. In a detailed study of status measures, Rowell (1966b) also concluded that approach and withdrawal episodes were the best indicators of relative status. Other agonistic measures correlated well but not perfectly, for example, threat 0.76, chase and attack 0.78, flee 0.84 lipsmack—0.85, bite 0.99, grimace 0.97, avoidance 0.99, and supplanting 0.97. We can see from these results why she concluded that the responses associated with the subordinate animal were better indicators of relative status than were the aggressive responses. Her data may also account for many of the difficulties encountered in attempting to measure status relationships in savannah baboons. Some responses ordinarily interpreted as dominant or aggressive responses may be used with other communication significance by savannah baboons engaged in certain social interactions.

Studies of *M. mulatta* almost invariably demonstrate strong status hierarchies, although Kaufman (1965) could only demonstrate this for males. Biernoff *et al.* (1964) relate aggressive interactions with food and lever use measures and come up with the interesting observation that more direct aggression is expressed between dyads of similar rank than between dyads of more disparate ranks, a finding also noted by Bernstein and Sharpe (1965). Kaufman (1967) noted that highly aggressive animals were not necessarily the highest ranking animals. Thus despite the decreasing number of available subordinate targets for successively lower animals in the hierarchy, a lower-ranking individual might account for the largest number of initiated aggressive episodes. Such an outcome is unlikely in very small groups and could easily be obscured. Warden and Galt (1943), in a study of interspecific dominance relations, used aggressive scores as their measure and noted interspecific differences in the quality of aggression but found that differences in aggressive quality did not mean that the more aggressive animal would be the more dominant. Koford (1963b) has used both incentives and threats to measure social rank in the rhesus monkeys of Cayo Santiago, but Sade (1967) reported that the direction of agonistic episodes was the best indicator.

Kummer (1968b) has used agonistic episodes to study dominance in hamadryas baboons, but Kummer and Kurt (1963) indicate that such hierarchies are poorly developed, if at all, between members of different harems. Crook (1967) reports the same situation in gelada baboons. Poor evidence of dominance relations are also reported for gibbons (Carpenter, 1940), spider monkey (Eisenberg and Kuehn, 1966), Callicebus monkeys (Mason, 1968), and *Presbytis entellus* (Yoshiba, 1968), although Jay (1963, 1965a,b) states that in the latter species it was only among the females that stable relationships could not be found. Furuya (1961–1962) indicated the presence of dominance hierarchies in *Presbytis cristatus* but this was not confirmed by Bernstein (1968b). Unstable or absent status hierarchies are also reported for bonnet macaques (Simonds, 1965) and *Macaca fascicularis* (Shirek-Ellefson, 1968). Similarly, Schenkel and Schenkel-Hulliger (1967) reported that status relationships exist only between male *Colobus* monkeys. Van Lawick-Goodall (1968) reports the existence of dominance relationships among wild chimpanzees but indicates that orthodox measures of relative status did not reveal these relationships. The indicator responses finally selected are described at length, but the interpretation of social rank relations, in chimpanzee bands in which membership is fluid within the larger community, appears very complex. In such situations a preselected model might dictate the measures on the basis of whether or not the data obtained from any particular measure matched the model. The availability of multiple measures and the complexity of status relationships is illustrated by Struhsaker's (1967a) data on vervets in which the outcome of some encounters is ambiguous and temporary coalitions may neutralize the ordinary status relationships described.

Related to all this are discussions of alliances and their function, dependent versus independent rank, and the manner in which individuals earn or acquire their rank standings. If the direction of agonistic episodes is used as the basic measure of dominance, then to what other measures does this correlate? If dominant animals can claim priority to any incentive, what kinds of things do they use their rank position to claim? Is being groomed considered "desirable," and is it more desirable than being the groomer? Is mounting another animal or copulating with an estrous female a highly motivated action for which animals compete? If these are incentives, then the dominant animal ought to be able to claim mounting or grooming rights, and matrices showing the directionality of these activities should correlate highly with matrices of social rank relations as measured by aggressive scores and/or received submission scores. Considerable information has already been collected and several theoretical accounts have been published concerning the interrelationships between agonistic dominance and grooming and sexual patterns.

D. SEXUAL BEHAVIOR CORRELATES

Sexual behavior may be readily divided into those sequences with and those without reproductive potential. Copulations with estrous females, presumed to be at maximum fertility, have the utmost biological importance, and investigators (Carpenter, 1942; Hall and DeVore, 1965; Jay, 1963; Koford, 1963a; Struhsaker, 1967a) have reported that mating patterns in a troop maximize the probability of reproductive success for dominant males. In fact, Etkin (1964) asserts that sex is the basis for the establishment of dominance competition, a view expressed less strongly by Maslow (1936a,b; Maslow and Flanzbaum, 1936) and in slightly different form by Carpenter in 1954. Yerkes (1943) wrote of "sex rights" when discussing dominance, and Chance (1961) also hypothesized a strong correlation. DeVore (1965) says that it is completed copulations that must be considered rather than all attempted copulations, and that in his baboon studies the dominant males had the highest ratio of completed copulatory attempts. Van Lawick-Goodall (1968) states that in the chimpanzee it is sexual presentation that reveals submission rather than mounting indicating dominance, a view shared also by Hall and DeVore (1965) with reference to baboons.

The cyclical states of female primates can be ascertained through visual cues in those species that show cyclical edema or color changes. In others the sexual cycle stage can only be guessed, without close examination of the females, and it is more difficult to decide what is the reproductive potential of any particular mounting sequence. Of course, male-to-male mountings, female-to-female mountings, female-to-male mountings, mountings involving immature animals and improperly oriented mountings can be assumed to have no reproductive potential. It is probably also true that the same motor patterns can be used to communicate different social messages regardless of the reproductive potential of the act. As such, it is an extremely difficult job to interpret and correctly identify each and every mounting as a sexual, or a status, or a greeting response. At least some rhesus monkey mountings have been identified as status mounts (Altmann, 1962b; Carpenter, 1942; Koford, 1963a; Neville, 1968), but both Altmann and Carpenter have acknowledged that some mountings, which were presumed to have status significance, occurred in reciprocal sequences. In rhesus monkeys for which a distinct breeding season exists, it is possible to eliminate reproductive mountings from a consideration of the directionality of mounting responses by restricting observations to the nonbreeding reason. Kaufman (1967) has correlated mountings without reproductive potential with the status hierarchy but found that only 75% of mountings were in the same direction as the agonistic status relationships observed. Similarly, Simonds (1965) found that only two-thirds of the mountings in wild bonnet macaques were in the direction of dominance relations, but Itani (1961); Kawai (1960), and Miyadi (1964, 1965) indicate that in Japanese macaques dominance relationships determine the direction of virtually every mounting. DeVore and Hall (1965) report that status determines the directionality of baboon mountings and, further, that males compete among themselves for access to estrous females. Schaller and Emlen (1963) report reciprocal mountings in gorillas, but the data on gorillas are very scanty. Crawford (1942a,b), however, reports reciprocal mountings in chimpanzees even though the dominant animal presumably mounts first and more often.

In considering only mountings of estrous females, it might be hypothesized that if males were to compete for access to a limited number of estrous females then the best males would account for the majority of reproduction, a genetically desirable outcome. The success of a male might depend on his physical abilities and/or his social skills in forming alliances. If prestige factors allow a male past his prime to retain high status, there is still no genetic loss, since at one point in the past he proved his superior abilities in the troop.

Still, when we examine the data it is not clear whether or not status is directly related to reproductive success in all the primate taxa studied. Simonds (1965) found no evidence for it in bonnet macaques; DeVore (1965) found evidence for it in one troop of baboons but not in another; and whereas Jay (1963) reports a correlation in a langur troop, Yoshiba (1968) failed to find one. Similarly, Reynolds (1963) failed to find a correlation between chimpanzee social rank and reproductive activities, and Southwick *et al.* (1965) reported that only among juvenile rhesus monkeys was relative status correlated with the direction of observed mountings. In Kaufman's (1965) analysis of rhesus copulation, he noted that whereas high-ranking males tended to participate in copulatory activities more frequently than lower-ranking males, frequency of copulatory activity was not related to female rank. The question thus arises why status should fail to influence female reproductive potential if a strong effect exists among males.

The relationship of female status to estrous activity is further confounded by possible interactions between the two variables. Altmann (1962a,b) Carpenter (1942), Crawford (1940), DeVore and Hall (1965) and Jay (1965a) all noted an apparent rise in a female's status position during estrus. This was in part attributable to temporary consort relationships with high-ranking males. The females closely associated with their consorts and supported and enlisted the male's aid in agonistic episodes. In such circumstances the criteria for determining status becomes critical. If access to incentives is used, and if a powerful male is more tolerant of estrous females, then the females can obtain access to incentives while with this male rather than having to compete with other animals solely on their own abilities. If aggressive activity is the criterion, and if estrous females are simply more active, their aggressive scores will be elevated. If, however, dyadic interactions are used as the major criterion, little apparent change can be seen and the hierarchy can be judged stable (Bernstein, 1963).

E. GROOMING BEHAVIOR AND STATUS

Another major social activity in primate societies which has been hypothesized to be correlated with status relationships is grooming. If being groomed is a pleasurable experience, then might not a dominant animal choose to be groomed and might not a subordinate groom a dominant animal to placate it? However, what if grooming is as pleasurable as being groomed? What other factors influence grooming episodes, for example, age, sex, genealogical relationships?

Crawford (1941) found that dominant chimpanzees entered the partner's living quarters, showed more attack responses, and were groomed more. Sade (1965) states that subordinate rhesus monkeys may placate potential aggressors by grooming them, although the relationship between grooming and genealogy confounds this effect. Falk (1958), however, has demonstrated that grooming can be used as a positive reinforcer for chimpanzees, thereby suggesting that grooming as well as being groomed is a social reward, a situation that Schenkel and Schenkel-Hulliger (1967) say pertains to *Colobus* monkeys as well. Maslow and Flanzbaum (1936) assumed that being groomed was a positive reinforcer and Struhsaker (1967a) describes vervets as supplanting subordinates in grooming situations, although a brief reciprocal grooming session may initiate a longer session.

Hall (1962) indicated that among baboons the males and females have equal grooming frequencies, but again, they may be differentiated on the duration of the grooming, the females doing far more. In contrast, Kummer (1968a) reports that male hamadryas baboons do the most grooming, and his descriptions suggest that this is one of the mechanisms that holds the one-male social units together. Michael and Herbert (1963) indicate that in rhesus monkeys it is the female that does the most grooming and that her activitives are related to her sexual cycle, grooming being more frequent when the female is in estrus and remaining in close association with a male for a prolonged period. Male-female grooming combinations are more common than unisexual grooming pairs in mangabeys (Chalmers, 1968), but in these animals the male of highest status does the most grooming. Similar negative correlations between receiving grooming and high status have been reported by Warden and Galt (1943), by Eisenberg and Kuehn (1966) for spider monkeys, and by Rowell (1966b, 1968) for baboons, for which the correlation was —0.65. Although Jay (1965b) writes of grooming rights among langurs, she too (Jay, 1965a) reports a negative correlation. Kaufman (1967) found a positive correlation for rhesus monkeys only when he examined pairs made up of peers. No correlations could be found by Nolte (1955) or Rosenblum et al. (1966), or Simonds (1965) for bonnets, nor did Rosenblum et al. find any in their pigtail macaque groups, although they indicate that adult males do relatively little grooming—perhaps a function of age and sex rather than of their status. Koford (1963a) could find no significant correlation in rhesus monkey groups, and Varley and Symmes (1966) suggest that grooming relations are determined more by habitual patterns of association and physical proximity than by other social mechanisms such as status. Loizos (1967) looks at grooming in another light and suggests that in groups maintaining a strong status hierarchy adults seldom engage in play but instead seem to interact socially through frequent grooming bouts. Grooming and play could then be viewed as indicators of social bonding rather than as related to status or the acquisition of social rank.

F. Unanswered Questions

Considering the data available, several important questions remain in considering the importance of status relationships in primate societies and their impact on other social relationships. Among these questions are; (1) Is dominance a universal attribute of primate societies? (2) Does dominance determine which partner in any nonagonistic social interaction will be active or passive? (3) Is grooming an expression of submission and service to a superior? (4) Are grooming relationships otherwise directly determined by status relationships? (5) Do sex, age, and other factors override the importance of status in grooming relationships? (6) Do higher-status males copulate more frequently with estrous females than do lower-status males? (7) Is assumption of the male role in mounting patterns an assertion of dominance? (8) Does a female in estrus change her rank in the status hierarchy? (9) Are the various kinds of sexual and sexually related response patterns, grooming relationships, and status relationships all determined by a multitude of overlapping factors such that some may be related to independent common factors (e.g., sex, age, size, geneaology, alliances, proximity, other social skills), or does any one of them determine the directionality of any or all of the other patterns?

Obviously, these are basic questions concerning the organization of primate societies, and the answers may vary from one taxon to another depending in part on the particular expression of social organization typical of the taxon under consideration. These broad questions cannot all be answered in any single study, nor can these problems be adequately treated without examining a host of other key problems in primate societies. The balance of this chapter is used to present the results of one study designed to examine certain aspects of these problems.

II. RESEARCH APPROACH

A. Groups

The Field Station of the Yerkes Regional Primate Research Center has a number of outdoor compounds with provision for shelter and year-round maintenance of monkey groups with minimal disturbance. These facilities have allowed us to study small groups, representing selected primate taxa, over long periods of time. Group composition is initially determined by attempts to approximate that reported in small natural troops, but reproduction and mortality thereafter determine the composition of the group. Seven of these groups were selected for a research program concerned with status hierarchies, grooming relationships, and mounting patterns. At the start of the program, the groups consisted of 26 *Macaca nemestrina* (pigtail macaque), 25 *M. fascicularis* (crabeating macaque), 11 *Theropithecus gelada* (gelada baboon), 9 *Cercopithecus aethiops sabaeus* (west African variety of green monkey), 11 *Cynopithecus niger* (Celebes black ape), which were living with a juvenile *Macaca maura* (Celebes or moor macaque), and two groups, one of 10 and one of 8 *Cercocebus atys* (sooty mangabey). Figures 1 through 6 show geladas, green monkeys, pigtail macaques, and mangabeys in the areas in which they were observed. Births occurred in all except the second mangabey group during the course of observations, and all groups, with the exception of the mangabey and green monkey groups, increased in size during the study year as a result of favorable natality-to-mortality ratios. Data collection on agonistic relations started shortly after group formation in each case and in the pigtail macaque group had

Fig. 1. Gelada (*T. gelada*) compound showing observation windows and vegetation killed by overbrowsing.

Fig. 2. Geladas (*T. gelada*) at Yerkes Field Station. Note stripped trees and bare earth with concrete culverts substituting for cover and lumber trellis in background replacing dead trees.

been accumulating for more than 6 years. Most of the other groups were formed within the year previous to the present study except that the green monkey group was formed 7 years previously and the second mangabey group was formed only 5 months before the completion of data collection. Data on grooming and mounting episodes were collected systematically for 9 months for all except the second mangabey group, but the first month of data collection for the crabeating macaque group was discarded because of technical difficulties. Data on subjects and group size and composition are presented in Table I.

Significant social events in the groups that might have influenced data collection during this period were: (1) The *M. fascicularis* group

Fɪɢ. 3. Confrontation of two gelada males (*T. gelada*). Redwood structures re-place trees destroyed by overbrowsing.

had been transferred intact from the Monkey Jungle in Florida to the Yerkes Field Station in August 1968. This group had detached itself from the main group which, according to Mr. F. DuMond, the owner, was established in 1933. (2) The *C. niger* group and one *C. atys* group were in the process of formation at the start of the study and animals were added to both during the first months. Social relations within these groups seemed to stabilize rapidly, and infants were born into both groups within the first year. (3) The *C. aethiops sabaeus* group was established in 1961 with a single breeding pair. All members were the offspring of this pair, but both parents died during the year of the study, partially as a result of fighting between the parents and their newly adult offspring. With the possible exception of this group, no abnormal morbidity or mortality occurred.

B. Data Collection and Analysis

The basic data collection technique consisted of recording the partici-pants and direction in each agonistic, mounting, or grooming sequence. Data were collected systematically during fixed time periods for which the time of onset and cessation of each episode was recorded. Grooming and mounting data were collected for each group for 1000 seconds dur-ing each working day for 9 months. Agonistic episodes were scored

FIG. 4. Green monkeys (*C. aethiops sabaeus*) entering tree at Yerkes Field Station. The lower center female is supporting a new infant between her thigh and abdomen.

FIG. 5. Pigtail monkeys (*M. nemestrina*) shortly after introduction to new compound at the Yerkes Field Station.

during feeding sessions designed to increase the probability of agonistic episodes, as well as during undisturbed periods. At least two observers always worked together during formal sessions. Observations of grooming and mounting noted during other scoring sessions, or as a result of informal observation, were scored only to indicate directionality and were considered as informal supplements. Formal sessions were analyzed to show frequency rates and the percent of time each individual in the group was involved in each social pattern. Separate scores were maintained for initiator and recipient roles. Solitary or self-directed activities were not scored. Agonistic scores were treated to show aggressive and submissive scores as reciprocals. A matrix was prepared for each of the three categories for each group. Each matrix was ordered monthly to show the "highest-ranking" animal in the top position and the others were placed accordingly (see Bernstein, 1968a). On the agonistic chart, an animal who won every agonistic encounter with every possible part-

FIG. 6. Female sooty mangabeys (*C. atys*) in alert postures. Note the white eye-lid displays of the female with infant and the lower female in the tree.

ner in the group would thus have first rank. His total wins and losses did not necessarily reflect his rank, however, as these totals often reflected intensive interactions with only a few group members. The mounting and grooming matrices were similarly ordered to place the animal functioning in the male mounting role and the animal being groomed in the highest rank positions on each of the respective matrices. The matrices obtained from agonistic, mounting, and grooming responses were cross-correlated monthly to determine the relationships among the three rank orders obtained. In order to measure the reliability and stability of the ranks obtained for each of the three responses, each monthly score was correlated with the corresponding score obtained during the previous month.

TABLE I

SUBJECTS USED IN STUDY OF GROOMING AND MOUNTING RESPONSES

Group	Date group was first assembled	Range of group size during study	Total number of animals on which data were collected	Number of births recorded in group	Adult males	Immature males	Adult females	Immature females
Macaca nemestrina	October 1963	26–31	33	27	3	10	12	8
Macaca fascicularis	1933[a]	25–27	27	27+	1	15	9	2
Cynopithecus niger	July 1968	11–13	16	2	6	3	5	2
Theropithecus gelada	May 1968	11–15	16	5	3	4	8	1
Cercocebus atys, I	September 1968	8–10	11	1	2	0	8	1
Cercocebus atys, II	February 1969	5–8	8	0	2	1	5	0
Cercopithecus aethiops sabaeus	1956[b]	7–9	9	11	2	3	3	1

[a] Indicates group was transferred to Yerkes Field Station September 28, 1968.
[b] Indicates group was transferred to Yerkes September 1, 1964. The *Cynopithecus* and *Cercocebus* group number fluctuations reflect replacements shortly after group formation. No animals were added to or removed from any other group, or following initial formation of these groups, except through births and deaths.

As an adjunct to this analysis, all possible dyads in each group were listed and their dyadic relationship for each month was shown indicating whether, for example, all agonistic episodes had gone in the same direction or not. This was shown for each month, and the consistency of dyadic relations was measured by counting dyads that had unchallenged unidirectional relations, dyads that maintained the same average directional relationships during each successive 2 months despite occasional reciprocal scores, dyads that reversed relationships, and the frequency of reverals in monthly dyadic total scores.

The only modification necessary, once data collection and analysis had begun, concerned newborn animals. Because of their lack of participation in some patterns as either active or passive partners, they were dropped from rankings on the matrices. Their dyadic interactions were also often nonexistent and this was reflected in the number of dyad months for which no interactions were reported.

The results obtained thus pertain to the following questions: (1) Are agonistic, mounting and/or grooming relationships stable in the six taxa studied? (2) Do any of these three social interaction patterns correlate significantly with either of the other two? (3) What type of influence does taxonomic affiliation have on the frequency and duration of participation in these three social response patterns? (4) What influence does sex have on participation as initiator or recipient of grooming, and in sexual interactions?

III. RESULTS AND DISCUSSION

A. AGONISTIC RANKINGS

Agonistic data included more than 22,000 scored interactions in the seven groups. Table II shows the number of month-to-month rank order correlations that were significant at the 0.01 level of confidence. Data available from all seven groups are shown. Inspection of the data for agonistic episodes reveals that "dominance orders" existed and were stable in all the groups except the green monkeys. The remarkable stability of the status hierarchy in the pigtail macaque group during the previous 5 years has been reported by Bernstein (1969). It is to be noted also that whereas multiple male troop organizations existed in these groups in the *T. gelada* group the organization consisted of harem-type units typical of this taxon. Few if any interactions between the females of different harems were detected. This difference in social organization perhaps accounts for the fact that whereas month-to-month rank order

TABLE II
RANK ORDER CORRELATIONS BETWEEN MONTHLY MEASURES AND FROM MONTH TO MONTH

Behavioral categories tested	Pigtail	Crabeater	Celebes	Gelada	Mangabey, I	Mangabey, II	Green
Correlations significant at 0.01 level							
Month to month							
Agonistic	8 of 8	8 of 8	10 of 10	9 of 15	8 of 9	3 of 3	8 of 22
Mounting	8 of 8	7 of 7	7 of 8	4 of 8	6 of 8	1 of 3	1 of 5
Grooming	7 of 8	4 of 7	1 of 8	5 of 8	1 of 8	0 of 3	2 of 8
Within months							
Agonistic to grooming	0 of 9	−1 of 8	0 of 9	0 of 9	0 of 9	0 of 4	1 of 9
Agonistic to mounting	0 of 9	0 of 8	0 of 9	0 of 9	0 of 9	1 of 4	0 of 5
Mounting to grooming	0 of 9	−3 of 8	0 of 9	0 of 9	0 of 9	0 of 4	0 of 6
Positive correlations							
Month to month							
Agonistic	24 of 24	8 of 8	10 of 10	15 of 15	9 of 9	3 of 3	30 of 31
Mounting	8 of 8	7 of 7	8 of 8	8 of 8	8 of 8	3 of 3	4 of 5
Grooming	8 of 8	7 of 7	8 of 8	8 of 8	8 of 8	2 of 3	6 of 8
Within months							
Agonistic to grooming	9 of 9	0 of 8	1 of 9	6 of 9	5 of 9	0 of 4	5 of 9
Agonistic to mounting	6 of 9	8 of 8	7 of 9	9 of 9	9 of 9	4 of 4	4 of 5
Mounting to grooming	4 of 9	0 of 8	7 of 9	4 of 9	0 of 9	3 of 4	2 of 6

correlations were significant at the 0.01 level in every case for the pigtail and crabeating macaques, the Celebes group, and 11 of the 12 mangabey comparisons, in the gelada group only 9 of 15 were significant at the 0.01 level; all 15 were, however, significant at the 0.05 level. Only a little more than one-third of the month-to-month rank order correlations achieved significance in the green monkey data. It should be noted that almost all correlations were positive, even in this last group, which indicates that the monthly rank data obtained were not random, although relative ranks could not be reliably predicted from data on previous rank relationships.

Examination of the patterns of agonistic dyadic interactions (summarized in Table III) confirms these findings. Because many dyads had no scored interactions during a particular month (this was especially true where at least one member of the pair was an infant), the number of changed relationships was related only to the number of relationships for which sufficient data existed to detect a change. In all groups except the green monkeys the number of changes occurring represented less than 6% of the changes possible. The number of changes in the green monkey group represented 17% of the changes possible. Data had been collected on the green monkey group for a full year prior to the initiation of the present study and the comparable figure for that year was, coincidentally also 17%.

The percentage of dyads maintaining stable relationships during the study was also lowest among the green monkeys. and during the first year more than one-half of the dyadic relationships revealed at least one reversal. When the distribution of directional changes among dyads was examined in the other groups, it was found that only one change was typical in dyads showing changes, perhaps revealing new relationships developing as a function of age. Among the green monkeys, however, a dyad was more likely to reverse roles several times in the course of a year.

More detailed analysis of dyadic interaction patterns reveals that agonistic episodes were unidirectional in one-half or more of the stable dyads in all groups except the green monkey group in which less than 27% of the pairs showing stable monthly relationships never showed reversed roles in any agonistic episode. During the previous year the number of unidirectional pairs was higher, but was still only 34%.

B. MOUNTING EPISODES

An analysis of mounting episodes was performed in the same fashion as that described for agonistic relations. Data were available from 3100

TABLE III

AGONISTIC RESPONSES, DYADIC ANALYSIS

Measure	Pigtail	Crabeater	Celebes	Gelada	Mangabey, I	Mangabey, II	Green	Green (previous year)
Episodes analyzed	8729	2482	3357	2122	3081	840	1546	N/A
Dyads in group	458	351	99	104	51	28	36	36
Percent stable dyads of those with sufficient data to have shown a change	77	88	69	85	78	91	67	44
Percent of stable dyads maintaining unidirectional relations at all times	63	81	52	51	56	91	27	34
Percent directional changes in successive monthly dyadic scores	6	5	5	3	6	5	17	17

mounting episodes. The month-to-month correlations that were significant at the 0.01 level are also shown in Table II. The number of mounting episodes observed was far less than the number of agonistic episodes observed and this was at least partially accounted for by the fact that agonistic episodes were frequently stimulated by feeding situations, and informal supplements were more likely when noisy outbreaks attracted the attention of observers who had been otherwise occupied. Because of the relatively small number of mountings recorded, many animals received tied ranks inasmuch as many of the cells were empty in the monthly matrices obtained. In some groups so few mountings occurred that ranks were virtually meaningless for some months.

The month-to-month mounting data correlations achieved significance at the 0.01 level in almost every case for the macaques and Celebes. As seen in Table II, the gelada, mangabey, and green monkey correlations failed to achieve significance in many cases. When the tests were run using the 0.05 level, only the green monkeys continued to show no significant month-to-month correlation. The green monkey group, however, had the lowest mounting frequency rate and no comparisons could be run for three of the data months because no episodes were recorded during these months.

Considering that mountings with reproductive potential were included in the data, it should not be surprising that almost every month-to-month correlation of the rank orders obtained was positive. The failure of many groups to achieve consistent significant monthly correlations may be attributed to the large number of empty cells in the matrices, but an analysis of dyadic interaction patterns reveals greater change in some groups than was found with regard to agonistic relationships despite the bias introduced by reproductive mountings (see Table IV). The percentage of stable dyads with regard to mounting roles was slightly less in the pigtail, gelada, and mangabey groups as compared with the same percentages with regard to agonistic roles. The percentage of directional changes in successive monthly mounting scores was considerably higher in four of the groups as compared to the scores obtained for agonistic relations, and in the pigtail monkey group 20% of dyadic scores reversed from month to month. This contrasts sharply with the small number of changes seen in the crabeating macaque group. The fact that from 53 to 100% of dyads maintaining stable relationships never reversed mounter-mountee roles can probably be largely attributed to sex-associated reproductive roles.

The explanation for the contrast between the two macaques is partially explained by Table IV where it is revealed that in more than one-half of the mounting episodes scored for pigtail macaques one or more par-

TABLE IV

MOUNTING RESPONSES, DYADIC ANALYSIS[a]

Measure	Pigtail	Crabeater	Celebes	Gelada	Mangabey, I	Mangabey, II	Green
Episodes analyzed	1225	366	592	271	497	219	22
Dyads in group	496	351	78	118	51	28	36
Percent stable dyads of those with sufficient data to have shown a change	60	97	81	83	65	83	100
Percent of stable dyads maintaining unidirectional relations at all times	53	83	68	83	65	83	100
Percent directional changes in successive monthly dyadic scores	20	2	11	5	2	9	0
Male-male mounts	271	117	132	37	15	0	1
Female-female mounts	329	0	0	8	90	27	1
Female-male mounts	50	0	0	0	0	0	0
Male-female mounts	575	249	460	226	392	192	20
Percent "normal" mountings	47	69	78	83	79	88	91

[a] The number of dyads in each group is not identical to that shown in Table II because of some minor differences in the dates of initiation of data collection for mounting and agonistic response data. Differences in the number of dyads in the group reflect births and/or deaths during the nonoverlapping months of data collection.

ticipants was engaged in a role pattern inappropriate for its sex with regard to reproduction. This was the only group in which females were scored mounting males, although such role reversals had been seen on rare occasions in other groups, albeit not during formal scoring sessions. It should also be noted that a female mounting a female had not been observed in any of the data-collecting sessions for the crabeating macaque and Celebes groups but that such interactions were seen in all other groups and most commonly in the pigtail macaque group.

The hourly rate of involvement in mounting behavior (Table V) was lowest among the green monkeys where the average individual was observed engaged in such interactions only 0.03 times per observation hour. This group also had the lowest frequency of mountings with no possible reproductive potential. The other groups are all more in agreement, and it might be noted that in all groups the sex of the participant determined mounter or mountee roles in the majority of interactions, even in the pigtail macaque group in which the least effect could be seen.

Neither male nor female mounting data revealed any suggestion of seasonal effects, except in the green monkey group. The past history of births in this group also suggests seasonal breeding. The small number of females in the group precludes definite analysis despite the relative absence of mountings with no reproductive potential. The two homosexual mountings observed also occurred during the months of peak female sexual activity and it may be significant that agonistic episodes and grooming frequencies were also highest during the period of increased mounting activity.

In the other groups male and female participation in either male or female mounting roles seemed consistent from month to month. The one exception appears in the gelada data, but the abrupt appearance of male-male mounting can be attributed entirely to the activity of the bachelor male and three infant males born into the group shortly before the abrupt rise in male-to-male mounting episodes.

C. Grooming Patterns

Over 16,000 grooming interactions were observed and scored. The analysis of grooming interactions revealed that in the green monkey and second mangabey groups, month-to-month correlations were actually negative in some cases, and in the latter a negative correlation was significant at the 0.05 level. Little consistency in grooming roles from month to month could be seen in the mangabey, green monkey, or Celebes groups and even in the other groups month-to-month correlations tended to be low.

TABLE V

FREQUENCY RATES FOR MOUNTING PATTERNS[a]

Average hourly rate measures for	Pigtail	Crabeater	Celebes	Gelada[b]	Mangabey, I[b]	Mangabey, II	Green[b]
Male mounts	0.6	0.3	0.8	0.6	1.0	0.8	+
Female mounts	0.2	0.0	0.0	+	0.1	0.1	+
Average group member mounts	0.4	0.2	0.4	0.2	0.4	0.3	+
Male receives mounting	0.2	0.1	0.1	0.1	+	0.0	+
Female receives mounting	0.5	0.4	0.7	0.3	0.5	0.4	+
Average group member receives	0.4	0.2	0.4	0.2	0.4	0.3	+

[a] Average group member scores reflect the male-to-female ratios in the study groups and are not simply male plus female averages divided by two.

[b] A + indicates a rate less than one in 10 hours.

The analysis of dyadic relations confirms the lack of consistent month-to-month relations (Table VI). When data from each month were compared with data from the previous month, role reversals were seen in 74% of the Celebes dyads and in more than one-half the dyads in all except the pigtail macaque and second mangabey groups. The relative number of changes that occurred compared to the number of possible changes varied from a low of 17% in the pigtail macaque group to a high of 41% in the second mangabey group. Coincidentally, the same two groups showed the greatest percentage of dyadic stability, that is, these groups had the greatest number of dyads that maintained the same relationships from month to month.

It should be noted that grooming is essentially a reciprocal response pattern—the vast majority of dyads exchanged roles either during a single grooming bout or from time to time. The grooming of young infants was one of the few cases in which the animal grooming seldom if ever received grooming in return. Pairs with stable directional grooming patterns usually maintained reciprocal grooming relationships. In the two macaques, 38 and 24% of stable relationships were unidirectional; comparable figures for the other groups varied from 5 to 15%. The two macaque groups also had the greatest percentage of infants.

Examination of individual patterns revealed that whereas some animals did a lot of grooming others received an inordinate share of grooming responses in the group and some pairs were more frequently involved than would be expected by chance. The effect of genealogy was not examined but it may account for many of the consistent grooming associations.

The sex of the animal was found to have a profound influence on grooming patterns even when data for animals of all ages were pooled (Table VII). Females spent more time grooming than did males and this was true in every group and for every month except for one in the green monkey group. The frequency rate for initiating grooming was also highest in females in all groups except one mangabey group. (See Chalmers, 1968, field data reporting the same observation in wild mangabeys.)

When data were analyzed to indicate the sex of the animal being groomed, it was found that even in the two mangabey groups males received more grooming than did females, although there were some month-to-month reversals. The frequency with which mangabey males received grooming was not greater than the frequency with which mangabey females received grooming. The larger percentage of time that males received grooming was therefore attributable to longer durations of grooming episodes in which the males were being groomed. In

TABLE VI
GROOMING RESPONSES, DYADIC ANALYSIS

Measure	Pigtail	Crabeater	Celebes	Gelada	Mangabey, I	Mangabey, II	Green
Episodes analyzed	4263	3257	2519	4112	1664	457	573
Dyads in group	496	351	78	118	51	28	36
Percent stable dyads of those with sufficient data to have shown a change	55	48	26	46	41	73	27
Percent of stable dyads maintaining unidirectional relations at all times	38	24	8	15	5	9	6
Percent directional changes in successive monthly dyadic scores	17	27	30	19	25	41	32

TABLE VII

HOURLY RATE AND DURATION DATA ON GROOMING

Measure	Pigtail	Crabeater	Celebes	Gelada	Mangabey, I	Mangabey, II	Green
Hourly rate, grooming							
Male grooms	1.1	1.7	2.6	2.3	2.9	1.9	0.9
Female grooms	3.0	4.2	4.6	6.6	2.5	2.8	2.2
Mean	2.2	2.8	3.5	4.9	2.6	2.5	1.4
Male is groomed	1.5	1.8	3.3	3.1	1.2	2.3	0.9
Female is groomed	2.7	4.1	3.9	5.9	3.1	2.6	2.1
Mean	2.2	2.8	3.6	4.9	2.6	2.5	1.5
Months female grooms more than male	9 of 9	8 of 8	9 of 9	9 of 9	5 of 9	4 of 4	9 of 9
Months female is groomed more than male	9 of 9	8 of 8	8 of 9	9 of 9	9 of 9	2 of 4	9 of 9
Months male grooms less than receives	8 of 9	3 of 8	8 of 9	6 of 9	1 of 9	3 of 4	7 of 9
Percent time grooming							
Male grooms	2.2	3.6	6.8	4.2	2.1	0.7	1.4
Female grooms	8.5	13.7	10.2	16.1	5.2	10.9	4.6
Mean	5.5	7.8	8.4	11.5	4.4	7.5	3.1
Male is groomed	5.1	5.5	8.4	6.9	5.2	11.0	1.9
Female is groomed	6.7	11.0	8.5	14.4	4.3	5.7	4.6
Mean	6.1	7.8	8.4	11.6	4.4	7.5	3.1
Months female grooms more than male	9 of 9	8 of 8	9 of 9	9 of 9	9 of 9	4 of 4	8 of 9
Months female is groomed more than male	9 of 9	8 of 8	5 of 9	9 of 9	6 of 9	0 of 4	7 of 9
Months male grooms less than receives	9 of 9	8 of 8	7 of 9	9 of 9	6 of 9	4 of 4	8 of 9
Mean duration of episodes, minutes							
Male grooms	1.2	1.3	1.6	1.1	0.4	0.2	1.0
Female grooms	1.7	2.0	1.3	1.5	1.3	2.4	1.3
Male is groomed	2.0	1.9	1.6	1.3	2.6	2.8	1.2
Female is groomed	1.5	1.6	1.3	1.5	0.8	1.3	1.3

all other groups females not only did more grooming than did males but also received more grooming, albeit in the Celebes group the two scores were nearly identical. As can be seen from the mean duration scores in the mangabey groups, males tended to groom briefly whereas bouts in which males were being groomed tended to be prolonged.

Although there were some contradictions in all groups except the first mangabey group and the crabeating macaque group in which scores were nearly identical, males typically received grooming more frequently than they groomed. The time they spent grooming was almost always less than the time they spent being groomed. Mean duration times for grooming episodes tends to follow the mangabey pattern, but in no other group is the difference so clear.

Examination of the data across reveals variable frequency and duration measures from group to group. The green monkeys were engaged in social grooming during only 6% of the observations, whereas the geladas spent more than 23% of their time engaged in social grooming. The frequency rate for initiation of grooming interactions reflects the same general picture but with less dramatic differences. Most of the gelada score was accounted for by female activity; male grooming activity was highest in the Celebes group. The crabeating macaque females also spent a lot of time in social grooming, but neither male nor female pigtail macaques were notable for the time they spent in this interaction.

The data collected in the present study closely parallel data collected in another study using an independent group-scoring technique with the same groups (Bernstein, in press). There are only minor variations in the percentage figures and the general relationships among groups remains the same.

D. Cross Correlations

The comparison of rank orders based on the direction of agonistic encounters, mounting sequences, and grooming relations failed to reveal a close correlation between any two of the three response patterns studied. No one pattern could thus be a reliable indicator of what relationship would prevail in either of the other two response patterns, but some low correlations exist which suggest that additional factors may operate to influence more than one of these social interaction patterns. Such factors might be sex, size, age, or some such physical characteristic, but analyses for these factors was not done in systematic fashion, except in the case of sex influences where it was found that males typically did less grooming than they received and also did more mounting than they received. There was pronounced sexual dimorphism in all of the taxa studied, both with regard to body size and canine tooth size. The highest rank position based on agonistic relationships was almost always held by an adult male, the exception occurring in the green monkey group in which ranks based on agonistic encounters were unstable and a female often had top position in the group. These sex effects, and others like them, may account for whatever low level of correlation is suggested among any of the three social rankings examined.

1. Agonistic and Mounting Correlations

In each month of scoring, rank orders obtained using agonistic episodes were compared with those obtained based on the directionality of observed mountings. In only one of the total of 53 correlations was a relationship significant at the 0.01 level. These were found in one mangabey group.

At the 0.05 level, 9 of the 53 correlations were found to be significant; 4 of 9 correlations for the gelada group, 3 of 8 correlations for the crabeating macaque group, and 1 of 9 correlations for the Celebes group. It should be noted, however, that two correlations in the Celebes group, three in the pigtail macaque group, and one in the green monkey group were negative. Consistent positive, albeit low, correlations were therefore found only in the crabeating macaque, gelada, and mangabey groups.

Dyadic agreements (Table VIII) confirm these suggestions of low positive correlations which could be at least partially accounted for by adult males of high agonistic rank mounting females. The percentage of dyadic agreements ranged from 60 to 92%, with the exception of the green monkey group for which it reached 45% total agreement. The percentage of dyads for which sufficient data were available for comparison was admittedly very low in this group, but it was also low in several other groups.

2. Agonistic and Grooming Correlations

In only 2 of 57 cases was any correlation between these two factors significant at the 0.01 level, and in one of the cases it was a negative correlation. Nine of the 16 correlations considered significant at the 0.05 level were negative correlations, that is, the dominant animal did the most grooming. Only in the pigtail macaque group were all month-to-month correlations positive. By contrast, all month-to-month correlations in the cogeneric crabeating macaque group were negative. Examination of the dyads reveals agreements ranging from 33 to 75%, suggesting a chance distribution.

TABLE VIII

CROSS CORRELATIONS, DYADIC ANALYSIS

Measures	Pigtail	Crabeater	Celebes	Gelada	Mangabey, I	Mangabey, II	Green
Agonistic, mounting							
Dyad months with data sufficient for analysis	311	120	119	52	92	38	11
Percent dyadic agreement	60	88	71	92	71	82	45
Agonistic, grooming							
Dyad months with data sufficient for analysis	662	412	299	166	193	59	85
Percent dyadic agreement	67	33	40	52	75	34	44
Mounting, grooming							
Dyad months with data sufficient for analysis	313	127	126	51	84	33	9
Percent dyadic agreement	48	23	43	37	19	45	44

3. Grooming and Mounting Correlations

There was no significant correlation between the directionality of these two patterns. This is not to say that animals who mounted one another frequently did not also groom one another frequently. That may well be the case without implying that the directionality of their participation in the one type of pattern determined the directionality in the other type of pattern.

Mounting and grooming rank systems achieved significance at the 0.01 level only in the crabeating macaque group and here a negative correlation was demonstrated three times. At the 0.05 level, mounting and receiving grooming ranks were significantly negatively correlated in three other groups, but still only 9 of 57 correlations were significant. In the crabeating macaque group and one mangabey group, these two measures were always negatively correlated, whereas in the other groups they fell on either side of zero equally often.

Dyadic agreements support the data on rank orders. The least dyadic aggreement was seen in the crabeating macaque and second mangabey groups, 23 and 19%, respectively, whereas all other groups varied between 37 and 48% agreement. It is notable that even here disagreement was more common than agreement although differences were not significant.

All this does not prove that the various interaction patterns do not influence each other. In fact, the directionality of one pattern may indeed be influenced by the directionality of another, but from the data available other influences seem to obscure any such relationships that might exist. Age, sex, estrus, and even environmental effects may have such pervasive influences as to obscure these other effects. Control of all these factors, and recognition of the intent of a mount or grooming episode as assertive or placatory, is so difficult that no effort was attempted in the present program. We can thus only say that if these three factors do influence one another in any way then their influence is less important than other variables which can apparently completely obscure any such relationships.

IV. SUMMARY AND CONCLUSION

No single operating principle could be demonstrated to relate the aggressive role in agonistic relationships, the male role in mounting sequences, or the recipient role in grooming interactions in seven study groups representing six distinct primate taxa. Sex membership did, however, influence individual participation in these interacting roles and females in most groups both groomed and received more grooming than did males, most of which received more grooming than they groomed. As to be expected, females typically were mountees rather than mounters although considerable intertaxa variation existed. Reverse roles were especially common in the pigtail macaque group. All six taxa were sexually dimorphic and adult males usually had the highest social rank, interpreted on the basis of the direction of agonistic relationships.

The direction of agonistic relationships was stable in all except the green monkey group. Month-to-month rank order correlations were generally high with few changes in relationships between pairs. In many cases agonistic relationships within dyads were unidirectional for prolonged periods of time.

Mounting roles changed somewhat more often and rank orders based on these roles were less highly correlated from month to month than was the case for agonistic relationships. Grooming relationships, however, seemed more typically to be reciprocal relationships with frequent reversals within dyads, both with reference to a grooming bout and with references to total grooming during a month. The exceptions were largely attributable to the sex differences in grooming activity and to age effects in the case of infants.

Cross correlations failed to reveal significant correlations among the three social response rank systems. Whatever correlation exists must be low and can easily be attributed to sex identification, age, and similar factors.

It was therefore concluded that these three response relationships are not derived from any single social mechanism. None of these response relationships should be considered a necessary component of all primate societies, or as necessarily predictive of other social relationships. They may be independently determined by a variety of factors influencing animals in a social group, and different mechanisms may apply to different primate taxa.

Acknowledgments

This research was supported by Public Health Service Grant No. 1 RO1 MH13864 from the National Institutes of Mental Health and National Institutes of Health Grant No. FR-00165 from the Animal Resources Branch.

References

Alexander, B. K., and Bowers, J. M. (1969). Social organization of a troop of Japanese monkeys in a two-acre enclosure. *Folia Primatol.* **10**, 230–242.
Altmann, S. A. (1962a). The social behavior of anthropoid primates: An analysis of

some recent concepts. In "Roots of Behavior" (E. I. Bliss, ed.), pp. 277–285. Harper & Row (Hoeber), New York.

Altmann, S. A. (1962b). A field study of the sociobiology of rhesus monkeys, *Macaca mulatta*. *Ann. N. Y. Acad. Sci.* **102**, 338–435.

Angermeier, W. F., Phelps, J. B. Murry, P. S., and Hosanstine, J. (1968). Dominance in monkeys: Sex differences. *Psychonomic Sci.* **12**, 344.

Bernstein, I. S. (1963). Social activities related to rhesus monkey consort behavior. *Psychol. Rep.* **13**, 375–379.

Bernstein, I. S. (1964). The role of dominant male rhesus in response to external challenges to the group. *J. Comp. Physiol. Psychol.* **57**, 404–406.

Bernstein, I. S. (1966). Analysis of a key role in a capuchin (*Cebus albifrons*) group. *Tulane Stud. Zool.* **13**, 49–54.

Bernstein, I. S. (1968a). Social status of two hybrids in a wild troop of *Macaca irus*. *Folia Primatol.* **8**, 121–131.

Bernstein, I. S. (1968b). The lutong of Kuala Selangor. *Behaviour* **32**, 1–16.

Bernstein, I. S. (1969). Stability of the status hierarchy in a pigtail monkey (*Macaca nemestrina*) group. *Anim. Behaviour* **17**, 452–458.

Bernstein, I. S. (in press). Group behavior. In "Behavior of Nonhuman Primates" (A. M. Schrier, ed.), Vol. 3. Academic Press, New York.

Bernstein, I. S., and Sharpe, L. G. (1965). Social roles in a rhesus monkey group. *Behaviour* **26**, 1–2.

Biernoff, A., Leary, R. W., and Littman, R. A. (1964). Dominance behavior of paired primates in two settings. *J. Abnorm. Soc. Psychol.* **68**, 109–113.

Boelkins, C. R. (1967). Determination of dominance hierarchies in monkeys. *Psychonomic Sci.* **7**, 317–318.

Carpenter, C. R. (1935). Behavior of red spider monkeys in Panama (*Ateles geoffroyii*). *J. Mammal.* **16**, 171–180.

Carpenter, C. R. (1940). A field study in Siam of the behavior and social relations of the gibbon (*Hylobates lar*). *Comp. Psychol. Monogr.* **16**, 1–212.

Carpenter, C. R. (1942). Sexual behavior of free ranging rhesus monkeys. II. Periodicity of estrus, homo and autoerotic and nonconformist behavior. *J. Comp. Psychol.* **33**, 147–162.

Carpenter, C. R. (1950). Social behavior of non-human primates. Structure et physiologie des societies animals. *Colloq. Int. Centre Nat. Rech. Sci.* **34**, 227–246.

Carpenter, C. R. (1954). Tentative generalizations on the grouping behavior of non-human primates. In "The Non-Human Primates and Human Evolution" (J. A. Gavan, ed.), pp. 91–98. Wayne Univ. Press, Detroit, Michigan.

Carpenter, C. R. (1965). The howlers of Barro Colorado Island. In "Primate Behavior. Field Studies of Monkeys and Apes" (I. DeVore, ed.), pp. 250–291. Holt, Rinehart & Winston, New York.

Chalmers, N. R. (1968). The social behavior of free living mangabeys in Uganda. *Folia Primatol.* **8**, 263–281.

Chance, M. R. A. (1961). The nature and special features of the instinctive social bond of primates. In "Social Life of Early Man" (S. L. Washburn, ed.), pp. 17–33. Aldine, Chicago, Illinois.

Chance, M. R. A. (1967). Attention structure as the basis of primate rank orders. *Man* **2**, 503–518.

Crawford, M. P. (1940). The relation between social dominance and the menstrual cycle in female chimpanzees. *J. Comp. Psychol.* **30**, 483–513.

Crawford, M. P. (1941). Relation between dominance and noncompetitive behavior in female chimpanzees. *Psychol. Bull.* **38**, 697.

Crawford, M. P. (1942a). Dominance and the behavior of pairs of female chimpanzees when they meet after varying intervals of separation. *J. Comp. Psychol.* 33, 259–265.

Crawford, M. P. (1942b). Dominance and social behavior in chimpanzees in a noncompetitive situation. *J. Comp. Psychol.* 33, 267–277.

Crook, J. H. (1967). Evolutionary change in primate societies. *Science J.* 3, 66–70.

DeVore, I. (1963). A comparison of the ecology and behavior of monkeys and apes. *In* "Classification and Human Evolution" (S. L. Washburn, ed.), pp. 301–319. Aldine, Chicago, Illinois.

DeVore, I. (1965). Male dominance and mating behavior in baboons. *In* "Sex and Behavior" (F. A. Beach, ed.), pp. 266–289. Wiley, New York.

DeVore, I., and Hall, K. R. L. (1965). Baboon ecology. *In* "Primate Behavior. Field Studies of Monkeys and Apes" (I. DeVore, ed.), pp. 20–52. Holt, Rinehart & Winston, New York.

Eisenberg, J. F., and Kuehn, R. E. (1966). The behavior of *Ateles geoffroyi* and related species. *Smithson. Misc. Collect.* 151, 1–63.

Emlen, J. T. (1962). The display of the gorilla. *Proc. Amer. Phil. Soc.* 106, 516–519.

Etkin, W. (1964). Types of social organization in birds and mammals. *In* "Social Behavior and Organization among Vertebrates" (W. Etkin, ed.), pp. 256–297. Univ. Chicago Press, Chicago, Illinois.

Falk, J. L. (1958). The grooming behavior of the chimpanzee as a reinforcer. *J. Exp. Anal. Behav.* 1, 83–85.

Furuya, Y. (1961–1962). The social life of silvered leaf monkeys. *Trachypithecus cristatus. Primates* 3, 41–60.

Hall, K. R. L. (1962). The sexual agonistic and derived social behavior patterns of the wild chacma baboon *Papio ursinus. Proc. Zool. Soc. London* 139, 283–327.

Hall, K. R. L. (1964). Aggression in monkey and ape societies. *In* "The Natural History of Aggression" (J. D. Carthy and F. J. Ebling, eds.), pp. 51–64. Academic Press, New York.

Hall, K. R. L. (1965). Social organization of the old-world monkeys and apes. *Symp. Zool. Soc. London* 14, 265–289.

Hall, K. R. L. (1967). Social interactions of the adult male and adult females of a patas monkey group. *In* "Social Communication among Primates" (S. A. Altmann, ed.), pp. 261–280. Univ. of Chicago Press, Chicago, Illinois.

Hall, K. R. L. (1968). Social learning in monkeys. *In* "Primates" (P. C. Jay, ed.), pp. 383–397. Holt, Rinehart & Winston, New York.

Hall, K. R. L., and DeVore, I. (1965). Baboon social behavior. *In* "Primate Behavior. Field Studies of Monkeys and Apes" (I. DeVore, ed.), pp. 53–110. Holt, Rinehart & Winston, New York.

Hamburg, D. A. (1968). Evolution of emotional responses: Evidence from recent research on nonhuman primates. *Sci. Psychoanal.* 12, 39–54.

Imanishi, K. (1957). Social behavior in Japanese monkeys, *Macaca fuscata. Psychologia* 1, 47–54.

Itani, J. (1961). The society of Japanese monkeys. *Jap. Quart.* 8, 10.

Jay, P. (1963). The Indian langur monkey (*Presbytis entellus*). *In* "Primate Social Behavior" (C. H. Southwick, ed.), pp. 114–123. Van Nostrand, Princeton, New Jersey.

Jay, P. (1965a). The common langur of North India. *In* "Primate Behavior. Field Studies of Monkeys and Apes" (I. DeVore, ed.), pp. 197–249. Holt, Rinehart & Winston, New York.

Jay, P. (1965b). Field studies. *In* "Behavior of Nonhuman Primates" (A. M.

Schrier, H. F. Harlow, and F. Stollnitz, eds.), pp. 525–592. Academic Press, New York.

Kaufmann, J. H. (1965). A three-year study of mating behavior in a free ranging band of rhesus monkeys. *Ecology* **46**, 500–512.

Kaufmann, J. H. (1967). Social relations of adult males in a free-ranging band of rhesus monkeys. *In* "Social Communication among Primates" (S. A. Altman, ed.), pp. 73–98. Univ. of Chicago Press, Chicago, Illinois.

Kawai, M. (1960). A field experiment on the process of group formation in the Japanese monkey (*Macaca fuscata*) and the releasing of the group at Ohirayama. *Primates J. Primatol.* **2**, 181–253.

Koford, C. B. (1963a). Group relations in an island colony of rhesus monkeys. *In* "Primate Social Behavior" (C. H. Southwick, ed.), pp. 136–152. Van Nostrand, Princeton, New Jersey.

Koford, C. B. (1963b). Rank of mothers and sons in bands of rhesus monkeys. *Science* **141**, 356–357.

Kummer, H. (1968a). "Social Organization of Hamadryas Baboons," 189 pp. Univ. of Chicago Press, Chicago, Illinois.

Kummer, H. (1968b). Two variations in the social organization of baboons. *In* "Primates" (P. C. Jay, ed.), pp. 293–312. Holt, Rinehart & Winston, N. Y.

Kummer, H., and Kurt, F. (1963). Social units of a free-living population of hamadryas baboons. *Folia Primatol.* **1**, 4–19.

Loizos, C. (1967). Play behavior in higher primates: a review. *In* "Primate Ethology" (D. Morris, ed.), pp. 176–218. Weidenfeld & Nicolson, London.

Maroney, R. J., Warren, J. M., and Sinha, M. M. (1959). Stability of social dominance hierarchies in monkeys (*Macaca mulatta*). *J. Soc. Psychol.* **50**, 285–293.

Marsden, H. M. (1968). Agonistic behavior of young rhesus monkeys after changes induced in social rank of their mothers. *Anim. Behaviour* **16**, 38–44.

Maslow, A. H. (1936a). The role of dominance in the social and sexual behavior of infrahuman primates: Observations at Villas Park Zoo. *J. Genet. Psychol.* **48**, 261–277.

Maslow, A. H. (1936b). A theory of sexual behavior in infrahuman primates. *J. Genet. Psychol.* **48**, 310–336.

Maslow, A. H., and Flanzbaum, S. (1936). An experimental determination of the behavior syndrome of dominance. *J. Genet. Psychol.* **48**, 278–309.

Mason, W. A. (1964). Stability and social organization in monkeys and apes. *In* "Advances in Experimental Social Psychology" (L. Berkowitz, ed.), Vol. 1, pp. 277–305. Academic Press, New York.

Mason, W. A. (1968). Use of space by Callicebus groups. *In* "Primates" (P. C. Jay, ed.), pp. 200–216. Holt, Rinehart & Winston, New York.

Michael, R. P., and Herbert, J. (1963). Menstrual cycle influences grooming behavior and sexual activity in the rhesus monkey. *Science* **140**, 500–501.

Miyadi, D. (1964). Social life of Japanese monkeys. *Science* **143**, 783–786.

Miyadi, D. (1965). Social life of Japanese monkeys. *In* "Science in Japan" (A. H. Livermore, ed.), pp. 315–334. *Amer. Ass. Advan. Sci.*, Washington, D. C.

Neville, M. K. (1968). A free-ranging rhesus monkey troop lacking adult males. *J. Mammal.* **49**, 771–773.

Nolte, A. (1955). Field observations on the daily routine and social behavior of common Indian monkeys, with special reference to the bonnet monkey (*Macaca radiata* Geoffrey). *J. Bombay Natur. Hist. Soc.* **53**, 177–184.

Nowlis, V. (1941). The relation of degree of hunger to competitive interaction in chimpanzees. *J. Comp. Psychol.* **32**, 91–115.

Ploog, D. W., Blitz, J., and Ploog, F. (1963). Studies on social and sexual behavior of the squirrel monkey (*Saimiri sciureus*). *Folia Primatol.* **1**, 29–66.

Reynolds, V. (1963). An outline of the behaviour and social organization of forest living chimpanzees. *Folia Primatol.* **1**, 95–102.

Rosenblum, L. A., Kaufman, I. C., and Stynes, A. J. (1966). Some characteristics of adult social and autogrooming patterns in two species of macaque. *Folia Primatol.* **4**, 438–451.

Rowell, T. E. (1966a). Forest living baboons in Uganda. *J. Zool.* **149**, 344–364.

Rowell, T. E. (1966b). Hierarchy in the organization of a captive baboon group. *Anim. Behaviour* **14**, 430–443.

Rowell, T. E. (1967). A quantitative comparison of the behaviour of a wild and a caged baboon group. *Anim. Behaviour* **15**, 499–509.

Rowell, T. E. (1968). Grooming by adult baboons in relation to reproductive cycles. *Anim. Behaviour* **16**, 585–588.

Sade, D. S. (1965). Some aspects of parent-offspring and sibling relations in a group of rhesus monkeys, with a discussion of grooming. *J. Phys. Anthropol.* **23**, 1–18.

Sade, D. S. (1967). Determinants of dominance in a group of free-ranging rhesus monkeys. *In* "Social Communication among Primates" (S. A. Altmann, ed.), pp. 99–114. Univ. of Chicago Press, Chicago, Illinois.

Schaller, G. B. (1963). "The Mountain Gorilla. Ecology and Behavior," 431 pp. University of Chicago Press, Chicago, Illinois.

Schaller, G. B. (1965). The behavior of the mountain gorilla. *In* "Primate Behavior. Field Studies of Monkeys and Apes" (I. DeVore, ed.), pp. 324–367. Holt, Rinehart & Winston, New York.

Schaller, G. B., and Emlen, J. T., Jr. (1963). Observations on the ecology and social behavior of the mountain gorilla. *In* "African Ecology and Human Evolution" (F. C. Howell and F. Bourliere, eds.), pp. 368–383. Aldine, Chicago, Illinois.

Schenkel, R., and Schenkel-Hulliger, L. (1967). On the sobiology of free-ranging *Colobus* (*Colobus guereza caudatus* Thomas 1885). *In* "Progress in Primatology" (D. Starck, R. Schneider, and H.-J. Kuhn, eds.), pp. 185–194. Fischer, Stuttgart.

Schjelderup-Ebbe, T. (1913). Hönsenes stemme. Bidvag til hönsenes psykologi. *Naturen* **37**, 262–276.

Schjelderup-Ebbe, T. (1935). Social behavior of birds. *In* "Murchison: A Handbook of Social Psychology" pp. 947–972. Clark Univ. Press, Worcester, Massachusetts.

Shirek-Ellefson, J. (1968). Social organization and social communication in the genus *Macaca*. *Wenner-Gren Found. Anthropol. Res., A. Burg Wartenstein Symp.* pp. 1–15.

Simonds, P. E. (1965). The bonnet macaque in South India. *In* "Primate Behavior. Field Studies of Monkeys and Apes" (I. DeVore, ed.), pp. 175–196. Holt, Rinehart & Winston, New York.

Southwick, C. H., Beg, M. A., and Siddiqi, M. R. (1965). Rhesus monkeys in North India. *In* "Primate Behavior. Field Studies of Monkeys and Apes" (I. DeVore, ed.), pp. 111–159. Holt, Rinehart & Winston, New York.

Struhsaker, T. T. (1967a). Social structure among vervet monkeys (*Cercopithecus aethiops*). *Behaviour* **29**, 83–121.

Struhsaker, T. T. (1967b). Behavior of vervet monkeys and other *Cercopithecus*. *Science* **156**, 1197–1203.

van Lawick-Goodall, J. (1968). A preliminary report on expressive movements and communication in the Gombe Stream chimpanzees. *In* "Primates" (P. C. Jay, ed.), pp. 313–374. Holt, Rinehart & Winston, New York.

Varley, M., and Symmes, D. (1966). The hierarchy of dominance in a group of macaques. *Behaviour* **27**, 54–75.

Warden, C. J., and Galt, W. (1943). Study of cooperation, dominance, grooming and other social factors in monkeys. *J. Genet. Psychol.* **63**, 213–233.

Warren, J. M., and Maroney, R. J. (1958). Competitive social interaction between monkeys. *J. Soc. Psychol.* **48**, 223–233.

West, M. J. (1967). Foundress associations in Polistine wasps: Dominance hierarchies and the evolution of social behavior. *Science* **157**, 1584–1585.

Winter, P., and Ploog, D. W. (1967). Social organization and communication of squirrel monkeys in captivity. *In* "Progress in Primatology" (D. Starck, R. Schneider, and H.-J. Kuhn, eds.), pp. 263–271.

Yerkes, R. M. (1943). "Chimpanzees: A Laboratory Colony." Yale Univ. Press, New Haven, Connecticut.

Yoshiba, K. (1968). Local and intertroop variability in ecology and social behavior of common Indian langurs. *In* "Primates" (P. C. Jay, ed.), pp. 217–242. Holt, Rinehart & Winston, New York.

Zuckerman, S. (1932). "The Social Life of Monkeys and Apes." Rutledge & Kegan Paul, London.

Unlearned Responses, Differential Rearing Experiences, and the Development of Social Attachments by Rhesus Monkeys[*]

Gene P. Sackett

Regional Primate Research Center and Department of Psychology
University of Wisconsin, Madison, Wisconsin

[*] This research was supported in part by USPHS Grant MH-11894 from the National Institute of Mental Health. Portions of this paper were presented at the Symposium on Social Attachments in Man and Animal, American Psychological Association Meeting, San Francisco, September, 1968.

I. INTRODUCTION

Acquired social attachments have been frequently demonstrated for many animal species in both laboratory and natural settings. Studies of imprinting (e.g., Sluckin, 1965) reveal that attachments can be formed by many avian species toward objects having few attributes in common with more biologically appropriate stimuli. H. F. Harlow (1958) has shown that infant monkeys form strong attachments to inanimate surrogate mothers, particularly surrogates with special tactile qualities. Other observations suggest that many objects, such as diapers and metal tubes, can serve as stimuli evoking strong clinging and approach behaviors in monkeys reared without mother or peer contacts.

Such examples have suggested that sufficient conditions for forming attachments to animate and inanimate stimuli early in life include (1) contiguity conditioning, in which an animal's normal behaviors become temporally associated with the stimuli in its environment (Cairns, 1966); or (2) operant conditioning, in which stimuli take on a positive valence because they provide primary or secondary reinforcement for the developing animal's behaviors (Gewirtz, 1961). These models, based on positive learning mechanisms, suggest that acquisition through associative conditioning is not only sufficient but is also necessary for the development of social attachments. Although it seems clear that reinforcement contingencies can affect indices of social attachment (e.g., Gewirtz, 1967), it is not clear that operant mechanisms are responsible for the initial formation of attachments.

This paper presents data concerning social attachment behavior in rhesus monkeys. These data question the assumption that associative learning is necessary for attachment formation and suggest that when learning mechanisms do operate their role is not explained by simple contiguity or positive reinforcement models. Evidence is presented suggesting that (1) unlearned response biases lead to preferences that may play a large role in the normal attachment process, and (2) early rearing experiences are crucial in determining whether or not social attachments toward species members will be formed and the quality of stimuli that will serve as objects of social attachment.

II. THE SELF-SELECTION CIRCUS

Most of the studies discussed here were conducted in the self-selection circus. This free-choice situation allows monkeys to select among social stimuli by making differential approach or bodily orientation responses. These differential responses identify preferred stimuli, and serve operationally as indicators of relative attachment. Figure 1 presents a schematic drawing of the circus in several sectional views.

The latest version of the circus is constructed of aluminum channels, bolted to aluminum forms at the top and bottom, forming a hexagonal outer perimeter. Six outer *choice* compartments surround the hexagonal inner *start* compartment, and are separated from the start compartment by vertically sliding clear Plexiglas walls or by opaque Masonite walls which block all access to unused choice compartments. The front choice compartment wall is also clear Plexiglas. The floors are 0.25-inch stainless steel rods, spaced 0.75 inches apart. Each of the seven floor sections is electrically isolated and wired into contact circuits which operate clocks for recording duration of choice compartment entry by the subject. Solenoid-controlled vacuum lifters raise the doors separating start from

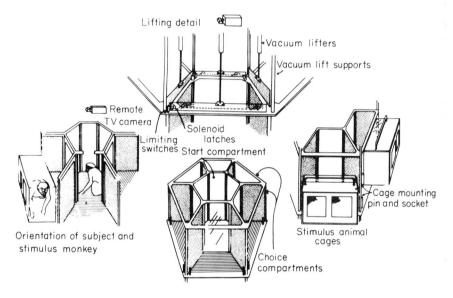

FIG. 1. The circus apparatus used to study social preferences. Lower center: View of center start compartment and choice compartments. The stippled choice compartment walls were opaque Masonite, the front and inner walls were clear Plexiglas, indicated by slashes. Upper center: View showing vacuum cylinder door lifts, position of television camera, solenoid latches which locked the walls so they could not be raised by the subject, and limit switches which operated a light indicating proper functioning of the vacuum lifters. Right: View showing mounting brackets, and aluminum stimulus animal cages with clear Plexiglas on the front and on one-half of the top section. Left: View showing the relative position of a stimulus and subject monkey during the *exposure* period.

choice compartments. A stimulus animal cage can be hooked onto the outside of each choice compartment. The side of this cage nearest the choice compartment is clear Plexiglas, allowing an unobstructed view of the stimulus animal from either the start or choice compartment. A closed-circuit television camera provides a view of the subject's behavior.

In the circus experiments reported here, different types of monkeys were randomly assigned to stimulus animal cages. The standard test method proceeded as follows. (1) The subject was placed in the center start compartment with the inner Plexiglas doors separating start from choice compartments closed. (2) During a 5-minute *exposure* period, the subject could see and hear the stimulus monkeys but could not enter the choice compartments. (3) The inner Plexiglas doors were raised for a 10-minute *choice* period during which the subject could enter and reenter any choice compartment or could remain in the center. In a choice compartment the subject had close visual and auditory contact with the stimulus animal, but physical interaction was not possible. In addition to the choice compartment entry measure, in some experiments an observer recorded a "duration of bodily orientation" score whenever the subject oriented its head and body toward a given stimulus animal, regardless of the choice compartment the subject was in.

III. SPECIES PREFERENCES OF MACAQUES

Several studies measuring preferences for own versus other species (Sackett *et al.*, 1970) suggest important relationships concerning innate and rearing condition effects on social attachment. Three types of feral-born, adult female monkeys (Fig. 2) served as stimulus animals in the circus: (1) a rhesus (*Macaca mulatta*), (2) a pigtail (*M. nemestrina*), and (3) a stumptail (*M. speciosa*). In most gross physical respects the rhesus and pigtail are more physically similar than either species is to the stumptail.

A. PREFERENCES OF FERAL-BORN ADULTS

In this study adult, feral-born rhesus (four males, four females), pigtail (four females), and stumptail (four males, four females) monkeys served as subjects. Each monkey received one standard circus trial. Figure 3 presents the choice compartment entry and bodily orientation

Fig. 2. Adult female stimulus animals used in the species preference experiment and a 30-day-old rhesus monkey. The adult pictures give a subject's eye view as they were taken in circus stimulus animal cages. Upper right, rhesus; upper left, pigtail; lower left, stumptail.

durations toward each stimulus animal by each species. One-way analyses of variance for each species, with type of stimulus animal as a repeated measure, revealed (1) significant stimulus animal differences within each species for entry duration (all $p < 0.001$), (2) significant stimulus differences for rhesus and stumptails on the orientation measure (both $p < 0.01$), but (3) no significant difference in orientation for the pigtails ($p < 0.15$) although the differences were in the same direction as the entry duration measure. Analyses of variance for rhesus and stumptails with sex and species of subject as uncorrelated variables and type of stimulus animal as a repeated measure revealed (4) a significant species $\times$ stimulus animal interaction on both measures (each $p < 0.01$),

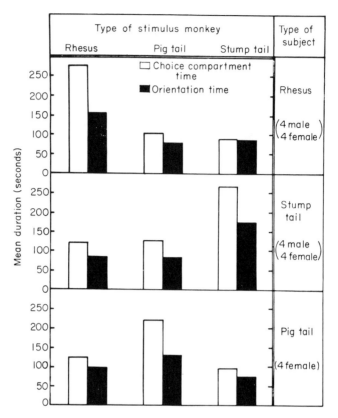

Fig. 3. Preferences by feral-born adult rhesus, stumptail, and pigtail subjects for adult females of the same and other species.

but (5) no sex of subject main effect or interactions with species or stimulus type (all $p > 0.10$).

These data show that feral-born adult macaques prefer their own species. They also show that the two preference measures, compartment entry time and orientation duration, although differing in absolute values reflect the same basic relationships. Thus adult rhesus males and females preferred the rhesus stimulus animal, adult stumptails preferred the stumptail stimulus irrespective of the subject's sex, and the female pigtail subjects preferred the pigtail stimulus alternative. Also, as can be seen in Fig. 3, there was little difference in preference for the two other-species stimulus monkeys shown by any of the three groups of subjects. In addition to identifying a basic own-species preference occurring independently of the choosing monkey's sex, these data on socially sophisticated macaques appear to validate the circus test procedure as an indicator of social preference.

B. Preferences of Laboratory-Born Partial Isolates

Rhesus monkeys ranging in age from 20–30 days through adulthood were tested under conditions identical to those in the previous experiment. Each monkey had been separated from its mother shortly after birth. For the next 9–12 months the animals lived in partial isolation (Fig. 4) in a bare wire cage from which age-mates could be seen and heard but not touched. All subjects older than 1 year had participated in physical interaction with species members during social behavior tests in playrooms and home cages. Adult subjects had spent at least some time after their first year living in gang cages with conspecifics.

Animals under 7 months, which were relatively inactive and appeared fearful, rarely entered choice compartments, so entry duration could not be used to measure preference. Therefore only the orientation duration scores are presented. The design of the experiment, sample sizes and sex distribution, and the number of animals in each age group having their highest orientation duration for a given type of stimulus animal are presented in Table I.

TABLE I

Age, Sex Distribution, Sample Size, and Number of Subjects in Each Age Group That Had Their Highest Orientation Duration Scores for the Rhesus, Pigtail, or Stumptail Stimulus Alternative[a]

Age of subject (months)	Sample size		Type of stimulus animal			p
	M	F	Rhesus	Pigtail	Stumptail	
Partial isolates						
1–2	4	4	6	2	0	0.039
3–4	4	4	7	1	0	0.005
7–9	6	4	7	2	1	0.039
1–9			20	5	1	<0.001
18	4	4	3	3	2	NS
24–36	7	5	3	8	1	0.037
>60 (adult)	4	4	2	4	2	NS
18–60+			8	15	5	0.144
Adult feral	4	4	7	0	1	0.005

[a] The table shows comparative data between the partial isolate and adult feral rhesus subjects. The p values for individual groups were calculated from two-tailed binomial tests with $p = 1/3$ and $(1 - p) = 2/3$. The two-tailed p values for the summed data were calculated from the normal approximation to the binomial tests. NS, not significant.

Fig. 4. Partial isolation cages from which the monkey could see and hear but not touch other animals.

An analysis of variance with age and sex as uncorrelated dimensions and type of stimulus animal as a repeated measure revealed (1) a significant main effect of stimulus animal type ($p < 0.001$), with rhesus and pigtail females receiving more orientation than the stumptail; (2) a significant age $\times$ stimulus animal interaction ($p < 0.05$) (Fig. 5), in which the younger, socially inexperienced subjects preferred the rhesus, while the older, socially experienced animals either had no rhesus preference or preferred the pigtail; and (3) no sex-of-subject main effect or interactions with the other variables (all $p > 0.15$).

The consistency of these effects is seen in the frequency with which subjects had their highest orientation scores toward the three alternatives (Table I). With few exceptions animals 9 months and younger significantly preferred the rhesus, while older animals showed no preference or preferred the pigtail. Further, as shown in Fig. 5 and Table I, adult ferals clearly preferred the rhesus, while partial isolates did not. Although not presented here, the same trends appeared in choice compartment entry times in comparisons between 9-month and older partial isolates, and between adult ferals and adult partial isolates.

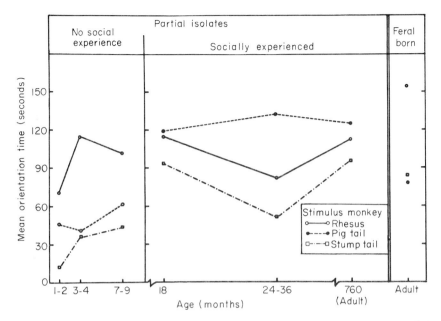

FIG. 5. Preferences of partial isolate rhesus monkeys, varying in age, for adult females of the same versus other species. The 1 to 9-month-old subjects had experienced no physical interactions with age-mates, the older partial isolates had all received physical social experience. Data for feral-born adult rhesus are presented on the far right.

There is a great difference in appearance between infant rhesus monkeys and all of the three types of adult females used as stimuli in this study (Fig. 2). It seems unlikely that rhesus neonates separated from their mothers at birth use visual and auditory contacts with other neonates to learn to discriminate an adult female rhesus from the other two species. Therefore these data suggest that this preference for their own species may be based on unlearned visual and/or auditory cues. Such unlearned, prepotent stimuli may play an important causal role in the development of attachments under normal rearing conditions.

In these circus studies no restrictions were placed on the behavior of subjects or stimulus monkeys. The preferences that were observed could therefore be based on the physical appearance of the stimulus animals, on species-specific behaviors emitted by the stimulus animals that are unrelated to the behaviors made by the subject during the choice period, or on specific behaviors of the stimulus animal which occur as interactions dependent on the subject's behaviors. Thus a rhesus infant subject might be disturbed or be making some other response which elicits species-typical responses from the rhesus adult female. These "maternal" responses might then serve as approach signals for the infant resulting in orientation toward the rhesus or movement into the choice compartment near the rhesus female. It is possible to analyze these factors in the circus by scoring the behavior of stimulus monkeys, by using motion pictures as stimuli rather than live animals, or by using one-way viewing walls through which the subject can see the stimulus monkey but the stimulus animal cannot see the subject. Although such studies have been planned, no data are available on these issues at this point. However, even if the infant rhesus preference for the rhesus female is based on specific stimulus-animal behaviors, it is unlikely that reinforcement learning processes are involved in generating the preference as the infants had experienced virtually no maternal behaviors after birth.

However, the development of this early preference into permanent own-species attachment appears to depend on the nature of the early rearing condition. Many studies have shown that the later behavior of partial isolates such as those tested in this study is grossly abnormal compared with either mother-peer laboratory-raised or feral-born animals (e.g., Cross and Harlow, 1965; Arling and Harlow, 1967; Sackett, 1967). In general, juvenile and adult partial isolates are deficient sexually and maternally, are hyperaggressive, and show low levels of play, environmental exploration, and positive affiliative behaviors, especially those involving physical contact. The presence of early visual and auditory contact is not therefore sufficient to produce adequate social development in partial isolates. These monkeys appear to develop avoidance responses to age-mates during postrearing social tests. These avoidance responses persist and generalize to other social stimuli and test situations, which may explain the loss of their initial own-species preference. The results of this study, coupled with data showing permanent behavioral anomalies following isolation rearing, suggest that maintenance of the initial own-species preference depends on learning during physical interactions with species members early in the rearing history of the monkey. Thus visual and auditory exposure to a social stimulus is not sufficient for permanent attachment formation but seemingly must be accompanied by actual physical contact as a crucial stimulus dimension.

IV. SEX PREFERENCES

Two circus studies measured preference for male versus female adults and for age-mates shown by subjects 60 days to 8 years old (Suomi *et al.*,

1970). The 48 rhesus subjects were all partial isolates separated from their mothers at birth and reared alone in wire cages for the first 9–12 months. Subjects over 1 year old had participated in some physical interaction with other monkeys before testing and were found to be socially inadequate as compared with peer- or mother-peer-raised animals (e.g., Cross and Harlow, 1965; Sackett, 1968). The subjects were individually housed prior to these studies, and no female was in estrus when tested.

A. Preferences for Adult Males and Females

Two adult, feral-reared rhesus stimulus animals, a male and an anestrous multiparous female, were placed in stimulus animal cages adjacent to opposite choice compartments in the self-selection circus. The subjects were given one standard, 5-minute exposure, 10-minute choice trial. Although entry duration and orientation duration scores were collected, only the orientation data are presented because many infant subjects failed to enter choice compartments. As in the species preference study discussed above, this appeared justified because of the high correlation between entry and orientation duration for the subjects that did enter choice compartments ($r = 0.81$, $N = 150$).

Figure 6 presents developmental trends in percentage of orientation toward own-sex stimulus animals ($n = 2$ per age for each sex) and average number of seconds orienting toward each stimulus animal by males and females in four age groups (six male and six female subjects per group). Animals under 44 months oriented more toward the adult female, regardless of the subject's sex. All 24 female subjects spent more time orienting toward the female than toward the male. Each of the 16 males 38 months old or younger spent more time orienting toward the female. Each of the 8 male subjects older than 38 months spent more time orienting toward the male, although this occurred above chance expectation only after 48 months of age. Also, in all of the four age-group breakdowns in Fig. 6, males and females showed significant sex-of-stimulus differences (all $p < 0.025$).

This analysis indicates that partial isolate monkeys do exhibit sex preferences between adult feral animals. Females of all ages prefer the adult female. Males initially prefer the female, but gradually shift to a male preference with a crossover at about 40 months of age. Of special importance are the results for neonates and infants. Relative to their total orientation time, these animals exhibit the greatest relative preference for the female stimulus of any age group.

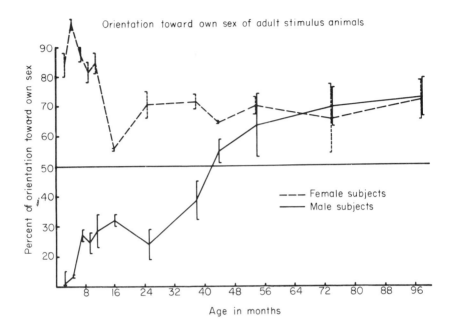

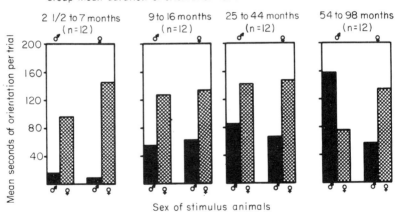

FIG. 6. The development of relative preference for feral adult male and female stimulus animals by partial isolate monkeys (top) and the mean duration (in seconds) of orientation to these adult stimuli for infants, preadolescents, juveniles, and adults (bottom).

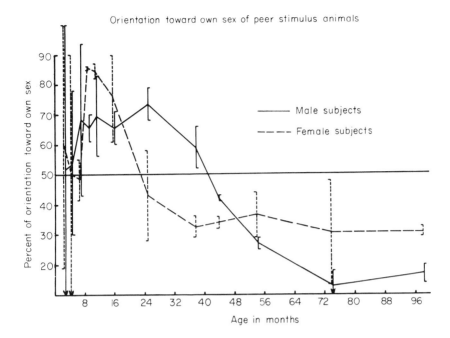

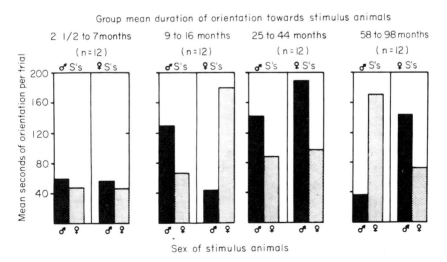

Fig. 7. The development of relative preferences for like-reared age-mate male and female stimulus animals by partial isolate monkeys (top) and the mean duration of orientation to these age-mate stimuli by infants, preadolescents, juveniles, and adults.

B. PREFERENCE FOR PEERS

In this study the same partial isolate animals tested for adult sex preferences were given one standard circus trial with a partial isolate male and female of the subjects' own age as stimuli. Subjects and stimuli were assigned so that each monkey in an age group was a stimulus for one male and one female of that group, and no two subjects had exactly the same combination of stimulus monkeys.

Figure 7 presents the percentage of total orientation by males and females ($n = 2$ per age for each sex) toward the stimulus monkey of its own sex and the average duration of orientation toward male and female stimuli for four age groupings. Infants (2–7 months) spent equal time orienting toward males and females (binomial $p > 0.50$), while pre-adolescents (9–16 months), juveniles (25–44 months), and adults (54–98 months) oriented more toward one of the two age-mate stimuli (all binomial $p < 0.001$). These and other statistical analyses (Suomi et al., 1970) can be summarized as follows. (1) Both male and female infants under 9 months old had no consistent sex preference for age-mates; (2) males and females 9–16 months old preferred their own sex; (3) females 25–44 months old shifted their preference to males, while males still preferred their own sex or showed no preference; and (4) adult subjects of either sex preferred the opposite sex. In summary, after infancy males and females spent more time orienting toward a member of their own sex, but subsequently shifted maximal orientation toward opposite sex, similarly reared peers. The male subjects shifted their preference substantially later than did females, however.

These sex preference studies have several implications concerning the necessity for learning in the determination of social approach behaviors. Although the infant subjects had no previous exposure to adult monkeys of either sex after birth, they exhibited an overwhelming preference for an adult female over a male. Contiguity notions of social attachment formation (e.g., Cairns, 1966) are inadequate to account for this preference because of the lack of any temporal exposure to adults prior to testing. Operant conditioning explanations (e.g., Gewirtz, 1967) also seem inadequate, as the baby monkeys had no prior history of reinforcement from adults. Although we do not as yet know whether the infant's preference is based on the physical or on the behavioral characteristics of the adult stimulus female, it seems clear that the baby does recognize and respond positively toward female cues in the absence of specific opportunities to learn the meaning of such cues. This appears to lend strong support to the possibility of unlearned mechanisms in the determination of infant-mother attachment formation.

In the second study, subjects of both sexes exhibited similar chronological sex preference patterns. The ages at which shifts to opposite sex preference occurred corresponded almost directly to the ages at which rhesus monkeys mature sexually. This seems to represent a specific behavioral manifestation of a known pattern of physical maturation. Studies investigating the development of rhesus peer interactions (e.g., Hansen, 1966; Rosenblum, 1961) have shown own-sex shifts in play behaviors and proximity by socially sophisticated mother-peer-raised monkeys at the ages at which such shifts were exhibited by the socially inadequate partial isolate monkeys tested in this study. Thus these shifts in sex preference patterns appear to be representative of the species, controlled by physical maturation, and largely independent of specific learning during physical interaction with mothers or with peers.

V. OTHER EVIDENCE FOR UNLEARNED SOCIAL STIMULUS EFFECTS

Further data on the existence of unlearned social responses appeared in monkeys raised in social isolation with the opportunity to respond to projected colored slides (Sackett, 1966). These eight *picture isolate* subjects were removed from their mothers and placed individually in totally enclosed cages within 24 hours after birth. Throughout the 9-month rearing period, each monkey was exposed to pictures of threatening, playing, fearful, withdrawing, exploring, and sexing monkeys, as well as infants, mothers and infants together, monkeys doing "nothing," and nonmonkey control pictures.

Three effects emerged after the first 30 days. (1) Pictures containing monkeys generally received more exploration than nonmonkey pictures. (2) Pictures of infants (but not mothers and infants) and pictures of monkeys threatening produced more exploratory and play responses, and higher motor activity, than any other pictures. (3) Until about day 80 none of the stimuli produced fear, withdrawal, or disturbance. From day 80 to day 120 the frequency of these behaviors rose markedly whenever threat pictures were presented even though these pictures had not produced fear before this time. Subsequently, the frequency of fear of threat stimuli declined but remained higher than for other pictures.

These results suggested that infant monkeys are born with prepotent responses to certain classes of visual stimuli and that at least some aspects of complex social communication may lie in innate recognition mechanisms. The data on fear suggested that this unlearned mechanism requires postnatal maturation before becoming operative.

Few other examples of complex visually evoked unlearned behaviors have been identified in primates. However, studies of human neonates and infants (e.g., Spitz, 1946; Fantz, 1965; Lewis, 1969) suggest that certain classes of visual stimuli may be prepotent in eliciting differential fixation, smiling, and distress indices. Thus human newborns seem to prefer differentially complex patterns, including natural and rearranged human "face" stimuli, and undistorted faces elicit more smiling than distorted and rearranged faces from the earliest ages at which smiling occurs.

A developmental mechanism to explain relatively simple, visually controlled, unlearned behaviors has been proposed based on differential development of neural cells responsive to specific attributes of visual input (Sackett, 1963). This feature-detection mechanism could account for the behavior of picture isolate monkeys and of human newborns, and for some of the circus data presented above. Electrophysiological microelectrode studies have shown that single nerve cells in the retina or in the visual cortex are "programmed" to encode relatively specific dimensions of visual input. For example, some cells respond with a change in firing rate primarily to specific colors, edges of figures, degree of curvature in a figure, movement in a particular direction, or particular figure-ground brightness relations. These information-processing functions appear to be unlearned, genetically determined equipment for analyzing complex visual information about the environment. Given this equipment, and the further assumption that certain of these cells require postnatal maturation before becoming functional, it is conceivable that innate recognition functions could be prewired in the primate nervous system. These functions would make certain classes of stimuli prepotent in eliciting attention, approach behavior, and possibly even more complex communicative responses.

After the initial 9-month rearing period, the picture isolates were extensively tested in a series of social and nonsocial situations. Their behavior was compared to that of partial isolates and *together-together* monkeys who had lived in cages with age-mates during the first 9 months of life. The picture isolates were behaviorally inadequate from 1.5 to 3.5 years of age, spending 70% of their time in social situations exhibiting disturbance and fear, showing little environmental exploration, and few positive affiliative responses toward other animals (Pratt, 1969). Thus, the presence during rearing of unlearned tendencies to respond positively to visual "social" stimulation, the occurrence of at least some appropriate social behaviors toward these stimuli, and the presence of spatial and temporal contiguity between these visual stimuli and the monkey's responses were not sufficient to produce attachments during social interactions subsequent to rearing. This suggests that feedback or more specific reinforcement following socially appropriate behaviors during rearing may be necessary for the long-term maintenance of positive social responses even though it is not necessary for the initiation of such behaviors.

VI. PREFERENCE FOR LIKE-REARED ANIMALS

Two-year-old picture isolates, partial isolates, and peer-reared subjects ($n = 8$ per group) were tested in the circus for preference among monkeys of their own and of the other two rearing conditions (Pratt and Sackett, 1967). Three stimulus monkeys in each circus test included a picture isolate, a partial isolate, and a peer-reared animal of the same age and sex as the subject. On one type of test the stimulus monkeys were all familiar to the subject, having been paired with it during 6 months of social testing. On a second type of test the stimulus animals were all strangers to the subject. Analysis of variance of choice compartment entry duration, with rearing condition of subject as an uncorrelated variable and rearing condition of stimulus animal and familiarity as repeated measures, revealed (1) a significant subject rearing condition $\times$ stimulus animal rearing condition interaction ($p < 0.001$, see Fig. 8), but (2) no significant effects of familiarity. As shown in Fig. 8, like-reared animals preferred each other.

These results show that a monkey can recognize and respond positively toward a like-reared animal even if that animal is a total stranger. The cues producing this discrimination may include vocalizations, body postures, facial expressions or general activity level. Thus an animal may learn some differential characteristics about monkeys of its rearing condition and employ these cues in its social approach behaviors. Such learning may occur during rearing and during subsequent social testing and could explain the preferences exhibited by partial isolates and peer-reared subjects.

Learning during rearing cannot explain the preference of total isolates for each other, however. These monkeys saw no live animals during rearing and therefore had no cues, other than self-produced stimuli, to associate with social behaviors of other isolates. Conditioning by *positive* reinforcement during subsequent social interactions is an unlikely ex-

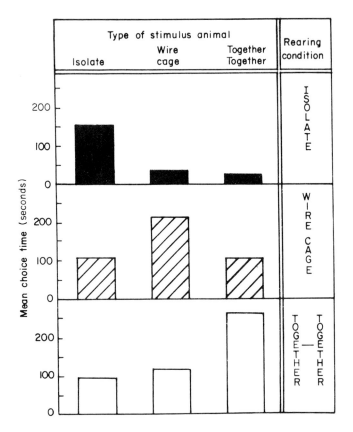

Fig. 8. Preference for like-versus unlike-reared monkeys by picture isolates, partial isolates, and together-togethers.

planation, as isolate postrearing interactions were devoid of any obvious positive experiences. One possible mechanism is conditioning by contiguity, as isolates had opportunities to view the behavior of other isolates during social tests. A second explanation is that isolates prefer each other by *default*, avoiding aversive stimulus properties of animals from the other groups by approaching the maximally distant choice compartment near the less aversive isolate. A third possibility is that during postrearing interactions the isolate forms a type of disjunctive discrimination concerning the stimulus properties of isolates. By this mechanism isolate stimulus properties take on a positive valence as a result of the dissimilarity between the behavior, facial expression, and body posture of isolates and the appearance and behavior of aversive, nonisolate animals. Thus the isolate may come to prefer a like-reared monkey on the basis of behaviors which that monkey *does not* perform, these behaviors becoming positive cues for social approach as the result of a "concept learning" process in aversive social interactions.

LIKE-REARING PREFERENCES OF ADULTS

In a recent experiment, twelve 7- to 9-year-old monkeys were grouped together in a large holding pen at the Madison Zoo. Each of four groups included one male and two females. During infancy the animals had been reared in total isolation for 6 months during year 1 (group 1), in wire cages without mothers or peers during year 1 (group 2), with mothers and peers during year 1 (group 3), or in India for the first 2 years before being brought into the laboratory (group 4). All animals had intensive social and nonsocial test experience during the 6–8 years before the start of this study, but none of the subjects had lived together for at least 4 years, or lived in groups for at least 1 year.

The group formation procedure consisted of simply putting all monkeys together and allowing dominance confrontations to determine the social structure. After 1 day of intensive physical aggression, the male mother-peer-raised monkey became the dominant animal and the feral male second. By day 2, fighting was minimal and social subgroup structure began to appear. This structure was measured by making periodic daily checks on the proximity of all subjects to all other monkeys. At least four proximity checks were made on each day. In particular, each monkey was scored for the number of times it was within 2 ft of any other monkey on either the floor, ceiling, walls, or perches of the pen. The probability that each monkey would be proximal to each of the other 11 monkeys on days 2, 3, and 4 and days 5–14 after group formation was calculated from these 2-ft distance scores.

The proximity data, averaged for the three animals in each rearing condition, are presented in Table II. Inspection reveals that isolates spent most of their time not standing within 2 ft of any other animal, but when they were proximal to another animal it was most likely a mother-peer monkey. Wire-cage-raised animals spent more time either with isolates or other cage-reared monkeys than with mother-peer or feral animals. Mother-peer and feral monkeys spent more time proximal to their own rearing group than to the other groups. Averaged for all groups, the monkeys had 29.6% proximity to a like-reared animal and 15.4% to any single other-reared group on days 2–4, and 29.0% proximity to like-reared and 13.4% to any other-reared group on days 5–14.

Although the sample sizes and sex dispositions are quite small in this

TABLE II

PROBABILITY THAT MONKEYS WILL BE PROXIMAL TO LIKE-REARED ANIMALS,
ANIMALS OF OTHER REARING CONDITIONS, OR NOT
WITHIN 2-FT PROXIMITY TO ANY ANIMAL[a]

Rearing condition	Proximal animal				
	Isolate	Wire cage	Mother-peer	Feral	None
Days 2–4					
Isolate	0.085	0.110	0.190	0.040	0.575
Wire cage	0.250	0.295	0.170	0.145	0.140
Mother-peer	0.190	0.210	0.375	0.130	0.095
Feral	0.050	0.240	0.125	0.430	0.155
Days 5–14					
Isolate	0.105	0.085	0.220	0.025	0.565
Wire cage	0.245	0.200	0.100	0.120	0.335
Mother-peer	0.210	0.205	0.410	0.110	0.065
Feral	0.040	0.100	0.150	0.410	0.300

[a] Data are given for days 2–4 and days 5–14 following formation of the 12-monkey group.

study, and there is an obvious confounding because of female estrous cycle and dominance differences, these results seem to support and partially validate the circus data presented above on like-rearing preferences. Although isolates failed to show a preference in their proximity scores for other isolates, the behavior of animals in the other three conditions suggests that they formed into social groupings compatible with their early rearing histories. This was, of course, especially true for mother-peer and feral monkeys but also showed similar trends for the socially inadequate wire-cage monkeys that spent more time proximal to each other or to the socially incompetent isolates than to the more socially sophisticated mother-peer and feral monkeys.

VII. ATTACHMENT AND VERY EARLY
SOCIAL EXPERIENCES

Imprinting data suggest that in birds early experiences before or during a critical period can influence social attachment in a relatively permanent fashion. The following experiment identified a related effect in primates (Sackett et al., 1965). Preadult monkeys (3.5–4.5 years old) were tested in the circus, with a human adult female seated in a chair next to one choice compartment and a caged monkey of the same age and sex as the subject adjacent to the opposite choice compartment. The five groups tested included (1) *partial isolates*, which were separated from their mothers at birth, given hand-feeding and intimate physical contact with human females during the first 30 days, and then housed in wire cages with visual and auditory but no physical contact with other monkeys for the rest of year 1; (2) *peer-reared* animals, which were also separated from their mothers at birth and received human care for 30 days but were then reared in wire cages with age-mates during the rest of year 1; (3) *mother-peer* animals, which were housed with their mothers and received daily peer experience during year 1; (4) *6-month total isolates*, which were kept in enclosed cages with no visual, auditory, or physical social contact from birth through 6 months; and (5) *1-year total isolates*, which were kept in total social isolation from birth through the end of year 1. After year 1 all subjects participated in extensive social interactions in cages and playrooms and lived for at least several months with other monkeys in gang cages. The circus procedure differed from the standard method in that only a 5-minute choice period was given.

Choice compartment durations for the human, monkey, and center start area are given in Table III. Analysis of variance revealed a significant rearing condition $\times$ choice compartment interaction ($p < 0.005$), produced as follows. (1) The partial isolates preferred the human. (2) Together and mother-peer subjects preferred the monkey and spent more time in the center than near the human. (3) The two isolate groups, which had received neither human nor monkey contact early in life, spent most of their time in the center, the area farthest from either social alternative. When the isolates left the center, however, they spent more time near the monkey than near the human (both $p < 0.01$).

These data show that very early experiences can affect social choice

TABLE III
MEAN CHOICE COMPARTMENT ENTRY DURATION (IN SECONDS)
IN THE HUMAN-VERSUS-MONKEY EXPERIMENT

Rearing condition	n	Choice compartment		
		Human	Monkey	Center
Partial isolate	7	226	36	38
Together-together	12	27	165	108
Mother-peer	12	11	191	98
6-Month isolate	4	4	99	197
1-Year isolate	4	0	126	174

behavior after 4 years, but whether or not this effect occurs depends on experience subsequent to the first month. The data differ from imprinting results in that the preference for a human, produced by early exposure to a human, was reversed by subsequent caging with other monkeys during the first year of life. Further, when there was no human contact and contact with monkeys did not occur within the first 6 months, an absolute preference for a monkey did not appear. This suggests that there may actually be a sensitive period in monkeys for attachment development to species members but that this period is measured in months instead of hours as in the case of birds.

A finding of importance concerns the relative preference by isolates for the monkey over the human. Other work has shown that these isolates were extremely bizarre in their social behavior as preadults (e.g., Mitchell *et al.*, 1966), rarely engaging in positive behaviors with agemates, and often receiving negative reinforcement in the form of severe maulings by test partners. Thus the relative preference of isolates for a monkey over a human seems to have occurred without a history of positive reinforcement of approach responses toward monkeys. This again suggests that positive reinforcement may not be a necessary condition for producing approach behaviors toward social stimuli.

VIII. MATERNAL EXPERIENCES AND ATTACHMENT

A series of circus tests studied preference of infants for their own versus other mothers and infant preference for adult females versus agemates (Sackett *et al.*, 1967). All subjects lived with their mothers for the first 8 months of life in a "playpen" apparatus (Harlow *et al.*, 1963). Visual and physical access between infants and between different mother-infant pairs was not available during this time as opaque Masonite partitions blocked the view between living cage units. At the end of rearing, each infant was permanently separated from its mother but continued to live in the playpen, receiving ten 30-minute peer interaction sessions during the next 2 weeks. Circus testing commenced 2 weeks after the final mother-infant separation.

Data was gathered on infants from four mothering treatments (two males and two females per condition). The mothers in the first three groups were all feral-reared (Griffin, 1966). (1) In a *multiple-mother* group, each infant was separated from a mother for 2 hours every 2 weeks and then returned to a different mother. Thus each infant lived with each mother for four 2-week periods. One of the four adult females was the infant's biological mother. (2) In a *separation* control group, each infant was separated from its real mother for 2 hours every 2 weeks and then returned to its own mother. (3) In a *normal* control group, the infants were never separated from their real mothers during the rearing period. (4) The infants in the fourth group were reared by *motherless mothers*, adult females who were born in the laboratory, taken from their own mothers at birth, and reared on surrogates or in partial isolation for 8 months (Harlow *et al.*, 1966). These last females were all maternally inadequate: either indifferent to the needs of their babies, rejecting attempts to nurse or maintain physical contact; or brutal, inflicting severe physical punishment on their offspring.

A. Preference of Infants for Own Mother

In experiment 1 each infant received four tests under the standard circus procedure. During a given test each infant subject chose between its own biological mother and the three other adult females who were in the same mothering condition. The mean choice time for own-mother versus the average for the three other mothers is presented in Table IV. These data summarize a significant mothering condition × stimulus animal interaction ($p < 0.01$), which included the following effects. (1) Infants in the multiple-mothering group did not prefer the biological mother, while those in the separation and normal control groups did prefer their own mothers. (2) The inadequately mothered offspring of the motherless-mother group also preferred their own mothers and subsequent tests indicated that their preference for their own mothers was significantly higher than that exhibited by infants in the two control conditions (both $p < 0.001$).

TABLE IV

PREFERENCE OF INFANTS FOR OWN VERSUS OTHER MOTHERS[a]

Mothering condition	Stimulus	
	Own mother	Other mothers
Multiple-mother	69	144
Separation control	187	91
Normal control	131	88
Motherless mother	328	97

[a] The data are choice compartment entry durations (in seconds) averaged over four trials for the blood relative and the average of the other three stimulus animals.

These data suggest that *positive* reinforcement by the mother is not a necessary condition for attachment of the infant to the mother. In fact, indifferent maternal behavior or active punishment produced a greater preference for the mother following a separation period than did adequate mothering. These results are in keeping with the findings of Rosenblum and Harlow (1963) in which it was shown that rhesus infants during early development spent more time in contact with a periodically rejecting cloth mother surrogate, with which they were reared, than controls of similar ages did with the usual continuously accepting cloth surrogate.

B. Preference of Infants for Adult Females versus Age-Mates

A second experiment involving these same groups of infants at 10 months of age employed a four-choice circus test. The stimuli included two 9- to 12-month-old infants and two adult females. One of the infants was familiar, having been present in the playpen during the 2-week postseparation interaction sessions; the other infant was a complete stranger. One of the adults was the subject's real mother, the other was an unfamiliar adult. Analysis of variance of the choice compartment entry durations revealed (1) a significant mothering condition × stimulus animal age interaction ($p < 0.01$), presented in Table V, but (2) no significant familiarity main effect or interactions with stimulus animal age or mothering condition (all $p > 0.10$).

The behavior of the infants in this situation is of particular importance to some views of maternal attachment. It has been shown that the initial separation of an infant from a mother induces disturbance and depression behaviors and strong motivation for responding toward the mother or, by stimulus generalization, toward a mother figure (e.g., Bowlby, 1958). An animal experiencing repeated separations from many "moth-

TABLE V
Preference of Infants for Adult Females versus Age-Mates[a]

Mothering condition	Stimulus	
	Adult female	Age-mate
Multiple-mother	123	103
Separation control	77	153
Normal control	80	133
Motherless mother	320	106

[a] The data are average choice compartment entry durations (in seconds).

ers" should be less disturbed by a final separation than an animal experiencing separation from one mother for the first time. Being less disturbed, the repeatedly separated animal should show less preference for a maternal stimulus than one experiencing the initial separation. This prediction was not fulfilled by the present data. The multiply mothered infants showed a significant ($p < 0.02$) preference for adult females, while both the separation and normal control infants preferred age-mates (both $p < 0.05$). The motherless mothered infants had the greatest preference for adult females (all $p < 0.01$).

Descriptively, rhesus monkeys at 8–12 months of age are exceptionally active in peer interactions, engaging in intense physical contacts and infantile sex behaviors. If adequate adjustment to peers depends on earlier stimulation, abnormalities in peer interactions should appear strongly at this age. In this study the adequately mothered infants in both control groups preferred age-mates over adult females, one of which was their own mother. Further, this preference appeared in equal strength in both control groups, although one group had experienced many previous separations from the mother while the other had recently experienced only one separation. However, the multiply mothered animals, which experienced repeated separations and whose maternal experiences were inadequate compared with those of the control groups (Griffin, 1966), did not prefer age-mates. This implies that past experiences are critical to the infant's readiness and ability to respond to new demands placed on its behavior by maturational changes and by changes in the environment. Thus maturational state, interacting with adequate rearing experiences, appears extremely important for attachments between infants and may override effects such as separation from the mother.

The preferences exhibited by the monkeys in this study also question the necessity for positive conditioning in forming age-mate attachments. During the first 8 months of life, adult females provided the sole source of reinforcement or contiguity for these babies. Conditioning specific to age-mates was only available for 300 minutes during the 2 weeks after maternal separation. The infants in the control groups thus appear to have formed their age-mate preference on the basis of factors other than simple reinforcement contingencies or contiguity conditioning directly related to peers.

The behavior of motherless mothered and multiply mothered offspring, groups that received atypical maternal care, suggests that inadequate maternal experiences produce strong approach responses toward maternal stimuli and retard the development of peer attachments. Thus positive reinforcement by the mother appears to produce readiness to respond positively toward peers even after minimal prior experiences with age-mates, while a lack of positive reinforcement or inconsistent reinforcement by a mother appears to produce an animal with exceptionally strong attachments to maternal stimuli.

C. Preference of Mothers for Babies

Several studies have assessed preferences of adult females for their own and other babies, and for neonates versus monkeys of other ages, in order to index maternal motivation as a function of rearing conditions. In one experiment the mothers that raised infants in the multiple-mother, separation control, normal control, and motherless mother groups discussed in the previous experiments were studied. A standard circus test for choice between their own and the other babies was given 3 weeks after mother-infant final separation (Sackett et al., 1967). The mean number of seconds spent with the mother's own baby and the average time spent with each of the other babies are presented in Table VI. Analysis of variance produced a significant group $\times$ stimulus interaction ($p < 0.001$) in which (1) multiply mothering females did not prefer their own baby, who was only one of four infants that she had reared; (2) both separation and normal control mothers preferred their own babies ($p < 0.05$); while (3) motherless mothers showed a reliable preference for babies that were not their own ($p < 0.005$). Under normal mothering conditions, feral-born females did prefer their own infants. Inadequate motherless mothers did not prefer their own infants, however, even though their infants had a very great preference for them 2 weeks after separation.

A second study assessed general effects of laboratory rearing and number of previous babies on maternal motivation. Feral- and laboratory-born females who had zero, one, or more than one previous infant were given a choice between an adult, a 3-year-old, a 1-year-old, and a 30-

TABLE VI

Mean Choice Time (in seconds) of Mothers for Their Own Infant
versus Average Time for Other Babies of Their Rearing Group

Rearing condition	Stimulus infant	
	Blood relative	Unrelated
Multiple-mother	88	121
Separation control	159	88
Normal control	146	74
Motherless mother	53	139

day-old neonate on one standard circus trial. All four stimulus monkeys were females. All laboratory females had been reared without real mothers or physical peer interaction in wire cages or with surrogate mothers. The data (Fig. 9) revealed a significant rearing condition × number of infants × stimuli interaction ($p < 0.01$). All feral groups spent more time with the neonate than the laboratory-reared subjects. The laboratory groups did not differ significantly among themselves ($F < 1.00$), preferring the adult stimulus regardless of number of babies. The feral groups did differ as a function of number of babies ($p < 0.01$), with primiparous females showing a higher neonate preference than nulliparous and multiparous females exhibiting extremely high choice time for the neonate.

Thus rearing without mothers or physical peer interaction produced a large deficiency in maternal motivation, which was not offset by the experience of raising one or more babies. These data from a single 10-minute preference test are quite consistent with the actual inadequate maternal behavior of such motherless reared females (Arling and Harlow, 1967).

IX. SUMMARY AND CONCLUSIONS

The work reviewed in this paper does not generally reinforce learning theories of social attachment acquisition, at least not for monkeys. Although it does seem clear that social attachment maintenance depends on feedback during interactions with other animals, it appears that this feedback must occur early in the animal's life to be effective. In these experiments social approach measures used to index social preference yielded the following information. (1) Neonate and infant monkeys separated from their mothers at birth and having no access to animals other than age-mates preferred adults of their own rather than other species, preferred adult females over males, and showed maturational changes in sex preferences that were not related in any obvious manner to positive reinforcement opportunities. (2) Social stimuli available during the first days and months of infancy were preferred by preadult animals 3–4 years after the specific early experiences. (3) Like-reared monkeys were preferred over animals reared under different conditions even when the like-reared social stimuli were completely unfamiliar and even though some rearing groups had not experienced positive social interactions with other animals. (4) Infants receiving adequate maternal

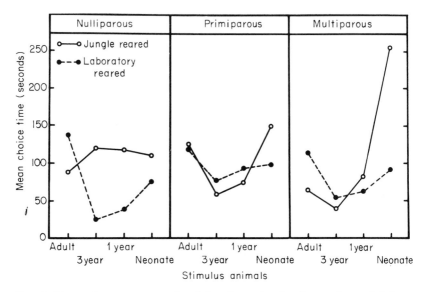

Fig. 9. Maternal motivation, indexed by choice for female monkeys of different ages by feral- and laboratory-born adult females that had reared different numbers of babies.

care preferred their own over other mothers, but infants receiving inadequate maternal care involving negative reinforcement showed a greater preference for their own mother than infants receiving species-typical mothering. (5) Within 2 weeks of separation from the mother, infants that were adequately mothered preferred age-mates over adult females even though they had almost no previous experience with age-mates. Inadequately mothered monkeys did not respond to age-mates but preferred adult females, however. (6) Adult female monkeys preferred neonates over animals of other ages only if they had been reared by a real mother. Thus these data strongly point to the existence of inborn and maturational perceptual and response mechanisms that support the acquisition of social approach behaviors in the absence of specific conditioning and reward opportunities. These results, however, suggest that the maintenance of social approach motivation depends on the quality of feedback from social stimuli experienced during the monkey's infancy.

REFERENCES

Arling, G. L., and Harlow, H. F. (1967). Effects of social deprivation on maternal behavior of rhesus monkeys. *J. Comp. Physiol. Psychol.* **64**, 371–377.

Bowlby, J. (1958). The nature of the child's ties to his mother. *Int. J. Psycho-Anal.* **39**, 350–373.

Cairns, R. B. (1966). Attachment behavior of mammals. *Psychol. Rev.* **73**, 409–426.

Cross, H. A., and Harlow, H. F. (1965). Prolonged and progressive effects of partial isolation on the behavior of macaque monkeys. *J. Exp. Res. Personality* **1**, 57–64.

Fantz, R. L. (1965). Visual perception from birth as shown by pattern selectivity. *Ann. N. Y. Acad. Sci.* **118**, 793–814.

Gewirtz, J. L. (1961). A learning analysis of the effects of normal stimulation, privation, and deprivation on the acquisition of social motivation and attachment. *In* "Determinants of Infant Behavior" (B. M. Foss, ed.), pp. 213–290. Metheun, London.

Gewirtz, J. L. (1967). Deprivation and satiation of social stimuli as determinants of their reinforcing efficacy. *In* "Minnesota Symposia on Child Psychology" (J. P. Hill, ed.), Vol. 1, pp. 3–56. Univ. of Minnesota Press, Minneapolis, Minnesota.

Griffin, G. A. (1966). The effects of multiple mothering on the infant-mother and infant-infant affectional systems. Unpublished doctoral dissertation, Univ. of Wisconsin, Madison, Wisconsin.

Hansen, E. W. (1966). The development of maternal and infant behavior in the rhesus monkey. *Behaviour* **27**, 107–149.

Harlow, H. F. (1958). The nature of love. *Amer. Psychologist* **13**, 673–685.

Harlow, H. F., Harlow, M. K., and Hansen, E. W. (1963). The maternal affectional system of rhesus monkeys. *In* "Maternal Behavior of Mammals" (H. L. Rheingold, ed.), pp. 254–281. Wiley, New York.

Harlow, H. F., Harlow, M. K., Dodsworth, R. O., and Arling, G. L. (1966). Maternal behavior of rhesus monkeys deprived of mothering and peer associations in infancy. *Proc. Amer. Phil. Soc.* **110**, 58–66.

Lewis, M. (1969). Infants' responses to facial stimuli during the first year of life. *Develop. Psychol.* **1**, 75–86.

Mitchell, G. D., Raymond, E. J., Ruppenthal, G. C., and Harlow, H. F. (1966). Long term effects of total social isolation upon the behavior of rhesus monkeys. *Psychol. Rep.* **18**, 567–580.

Pratt, C. L. (1969). The developmental consequences of variations in early social stimulation. Unpublished doctoral dissertation, Univ. of Wisconsin, Madison, Wisconsin.

Pratt, C. L., and Sackett, G. P. (1967). Selection of social partners as a function of peer contact during rearing. *Science* **155**, 1133–1135.

Rosenblum, L. A. (1961). The development of social behavior in the rhesus monkey. Unpublished doctoral dissertation, Univ. of Wisconsin, Madison, Wisconsin.

Rosenblum, L. A., and Harlow, H. F. (1963). Approach-avoidance conflict in the mother-surrogate situation. *Psychol. Rep.* **12**, 83–85.

Sackett, G. P. (1963). A neural mechanism underlying unlearned, critical period, and developmental aspects of visually controlled behavior. *Psychol. Rev.* **70**, 40–50.

Sackett, G. P. (1966). Monkeys reared in visual isolation with pictures as visual input: Evidence for an innate releasing mechanism. *Science* **154**, 1468–1472.

Sackett, G. P. (1967). Some persistent effects of different rearing conditions on preadult social behavior of monkeys. *J. Comp. Physiol. Psychol.* **64**, 363–365.

Sackett, G. P. (1968). Abnormal behavior in laboratory-reared rhesus monkeys. *In* "Abnormal Behavior in Animals" (M. W. Fox, ed.), pp. 293–331. Saunders, Philadelphia, Pennsylvania.

Sackett, G. P., Porter, M., and Holmes, H. (1965). Choice behavior in rhesus monkeys: Effect of stimulation during the first month of life. *Science* **147**, 304–306.

Sackett, G. P., Griffin, G. A., Pratt, C. L., Joslyn, W. D., and Ruppenthal, G. C. (1967). Mother-infant and adult female choice behavior in rhesus monkeys after various rearing experiences. *J. Comp. Physiol. Psychol.* **63**, 376–381.

Sackett, G. P., Suomi, S. J., and Grady, S. (1970). Species preferences by macaque monkeys. Unpublished data.

Sluckin, W. (1965). "Imprinting and Early Learning." Aldine, Chicago, Illinois.

Spitz, R. A. (1946). The smiling response: A contribution to the ontogenesis of social relations. *Genet. Psychol. Monogr.* **34**, 57–125.

Suomi, S. J., Sackett, G. P., and Harlow, H. F. (1970). Development of sex preference in rhesus monkeys. *Develop. Psychol.* (in press).

Behavior of Tree Shrews

M. W. Sorenson

Department of Zoology and Space Sciences Research Center
University of Missouri, Columbia, Missouri

I. INTRODUCTION

In Malaya tree shrews are called tupai tanah, which means ground squirrel. This is not surprising, for the similarities to the squirrel are striking. For example, the long-footed tree shrew *Tupaia longipes,* a typical member of the family, is relatively slender, with a long rostrum and a bushy tail 7½ inches long—more than two-thirds its body length.

141

The fur along the back and sides is a dull grizzled black and tawny, and the underparts are buff. A well-defined buff-colored stripe is present on each shoulder. The first written description of tree shrews also refers to them as squirrels. William Ellis, while serving as a surgeon's mate aboard the *Discover* on Captain Cook's third voyage in 1780, describes tree shrews from the island of Poulo Condore, about 100 miles south of Saigon, and classifies them simply as *Sciurus dissimilis* (Lyon, 1913).

Sir Stamford Raffles, in 1821, quoted by Lyon (1913), was first to classify the tree shrews as Insectivora and to propose the generic name *Tupaia*. Lyon (1913) describes tree shrews as "diurnal insectivorous mammals characterized by a general squirrel-like aspect, more or less arboreal habits, orbit completely encircled by bone, alisphenoid canal present, malar bone with a more or less enlarged perforation, separate radius and ulna, separate tibia and fibula, and dental formula 2/3, 1/1, 3/3, 3/3." Huxley (1872) first mentions the similarities between tree shrews and primates, and the first formal reference to tree shrews as primates is by Carlsson (1922) who refers to them as prosimiae. Overlapping characteristics continue to divide opinion among researchers as to whether tree shrews are the most primatelike insectivores or the most primitive of primates.

Today there are six living genera of tree shrews, *Tupaia, Anathana, Urogale, Dendrogale, Ptilocercus,* and *Lyonogale*. They are distributed from India on the west to the island of Mindanao on the east, and from southern China on the north down through the chain of islands near the southwest coast of Sumatra. They are not found in the Celebes or New Guinea (Lyon, 1913). The fossil record shows a broader distribution—in Germany in the Middle Paleocene, roughly 60 million years ago, and in Belgium and France in the Late Paleocene and Early Eocene, about 10 million years later.

While little is known of the habits of these animals in the wild, we have been able to make extensive observations of tree shrews maintained within a habitat designed to simulate their natural environment as closely as possible. Their advanced social organization and reproductive patterns under these conditions are indicative of early primate behavior (Sorenson and Conaway, 1964, 1966, 1968). Nevertheless, a sufficient body of conflicting evidence exists, and many researchers consider it prudent to relegate the tree shrews to an infraorder of Primates separate from the Lemuriformes or to place them in an order other than either Primates or Insectivora.

To further clarify, or confuse, the phylogenetic position of the Tupaiidae it is necessary to review criteria that either advocate or refute the primate affinities of tree shrews; however, it is beyond the scope of this article to peruse all available data. Therefore, we arbitrarily review only a limited number of studies concerned with the nervous system, the reproductive system, other anatomical features, cytology, and paleontology. Finally, the social and reproductive behavior of different species of tree shrews are examined in some detail, for behavior also expresses phyletic relationships. It should be stated that the taxonomic problem is mainly academic, for tree shrews will remain of interest to behaviorists, physiologists, anatomists, and researchers of other disciplines regardless of where man places them in his scheme of things.

II. THE NERVOUS SYSTEM

It is well known that tree shrews possess characters found in several divergent groups of mammals. As early as 1910, Gregory stated that primates are probably derived from large-brained arboreal insectivores resembling in many ways *Tupaia* and *Ptilocercus;* however, he also lists characters tree shrews share with marsupials, insectivores, and menotyphlous mammals. In his studies of the brain and skull of *Tupaia minor* and *Ptilocercus lowii*, Le Gros Clark (1924a,b, 1926, 1932) compares his findings with representative studies of insectivores and prosimians and notes that the olfactory bulb of *Tupaia* is relatively smaller than that of insectivores and that the visual cortex is expanded. He emphasizes that the visual-olfactory relationship is just the opposite of that found in insectivores. Le Gros Clark classifies *T. minor* as intermediate between *Macroscelides* and *Tarsius* and states that it is ". . . difficult to conceive two small mammalian brains which are more fundamentally different and divergent in their structure than those of *Macroscelides* and *Tupaia*." Stephan (1965) notes that the accessory olfactory bulbs of certain insectivores and primates (*Tupaia* and *Lemur*) are not similar to those of other prosimians.

Bauchot and Stephan (1966) used brain and body weights to determine an index of encephalization in their study of 33 species of insectivores and 28 species of prosimians including *Tupaia javanica, T. glis*, and *Urogale everetti*. They note that the insectivore index ranges from 80 (*Crocidura*) to 285 (*Rhynchocyon*) and the prosimian index from 240 (*Lepilemur*) to 704 (*Daubentonia*). The elephant shrews (*Elephantulus*) average 241, whereas the tupaiids average 304, well above the insectivore range. Andy and Stephan (1966) find that the percentage of septal nuclei volume to total brain volume decreases with phylogenesis, that is, insectivores 1.85%, *Tupaia* 1.08%, prosimians 0.77%, South American primitive monkeys 0.67%, African cercopithecidae 0.40%, chimpanzee 0.22%, and man 0.1%. In his excellent review article, Holloway (1968) aptly points out that the extreme variability of cranial capacity is well known and its correlation with any sort of behavioral attribute is notoriously low.

It is generally agreed that among mammals the modality of color vision is well developed only in the primates; however, color vision occurs in several of the Sciuridae, for example, in prarie dogs (Walls, 1942; Cain and Carlson, 1968), tree squirrels (Arden and Silver, 1962), ground squirrels (Arden and Tansley, 1955), and chipmunks (Tansley *et al.*, 1961), and also in certain carnivores such as the cat (Meyer and Anderson, 1965). Recent investigations by Tigges (1963, 1964) and Shriver and Noback (1967a) show that *Tupaia* possess excellent color vision and tends to remember color tasks longer than black and white tasks. *Tupaia* discriminates among red, yellow, blue, and green and as Shriver and Noback state ". . . if *Tupaia* is an insectivore, then color vision has evolved in at least one group of insectivores." The ability of *Tupaia* to discriminate red in the long wavelength of the spectrum is not characteristic of nonprimate mammals with color vision (Duke-Elder, 1958).

DeValois and Jacobs (1968) found that the macaque and the squirrel monkey differ in their color vision and in the physiology of their visual systems. The authors state that tree shrews have deuteranopic color vision. Snyder's work (1968) indicates that the geniculostriate system in tree shrews is topographically organized and shows an enormous elaboration as compared to the geniculostriate system in the hedgehog. When the geniculostriate system of *Tupaia* is completely removed, the animal retains form and pattern vision, that is, there is a secondary visual projection to the cortex parallel to the geniculostriate system (Snyder and Diamond, 1968). Tigges (1966) points out that the accessory optic system of *Tupaia* is complete, consisting of the anterior accessory optic tract, the transpeduncular tract, and its terminal nucleus. Comparisons of the retinal projections of the tree shrew and the hedgehog by Campbell *et al.* (1967) show that both animals have accessory optic systems. The differential projections to the various laminae in the tree shrew are exactly opposite those of primates, and it is suggested that the laminated pars dorsalis of *Tupaia* results from convergent evolution. Both Campbell (1966a) and Glickstein (1967) confirm that the uncrossed fibers from the retina of *Tupaia* end in the innermost and outermost laminae of the ipsilateral nucleus, whereas crossed fibers end chiefly in laminae two and four in the contralateral dorsal lateral geniculate nucleus. Campbell (1966b) states that those characteristics that indicate a tupaiid-primate affinity are really only aspects of an elaborate visual system and that none of the central nervous system evidence supports the primate classification for tree shrews.

In the examination of a large number of eyes of different primates by light and electron microscopy, Rohen (1964) found that the retina of *Tupaia* consists of cones only and that the ocular fundus is a distinct broad central area similar to that of some nocturnal prosimians. Photographs of the fundus of *T. glis* by Samorajski *et al.* (1966) reveal no specialization or differentiation of a central foveal region; however, histological, histochemical, and ultrastructural comparisons indicate only cone populations in the central region of the retina with some "rod-type" features located more peripherally. This suggests that the entire retina of *Tupaia* may function physiologically as a crude central "fovea." According to Wolin and Massopust (1967), who also photographed the ocular fundus of *Tupaia*, the morphology of the tupaiid eye is unlike any other in the primates investigated. Tigges *et al.* (1967) note that electroretinogram (ERG) recordings of the cone retina of *T. glis* indicate a spectral sensitivity curve congruent with the foveal psychophysical curve of the human and with the absorption curve of iodopsin, except for increased sensitivity in the blue region. This is in contrast to all other pure cone mammals.

A study of the cortical projections to the spinal cord in *T. glis* by the Nauta technique shows that the funicular position and caudal extent of the corticospinal tracts conform neither to primate nor insectivore patterns (Shriver and Noback, 1967b). Jane *et al.* (1965) and Verhaart (1966) confirm that the pyramidal tract of *T. glis* crosses to the dorsal funiculus and ends in the lower thoracic levels, in contrast to all other primates so far examined. The pyramidal tract is lateral in primates, ventral in insectivores, and as in tree shrews it is dorsal in monotremes, edentates, marsupials, and rodents.

III. THE REPRODUCTIVE SYSTEM

Martin (1966) reports that among tree shrews (*Tupaia belangeri*) only the male constructs a nest for the birth of the young, a fact unique among mammals. The young are born in this main nest in the course of 1 or 2 hours during which the mother disposes of the embryonic membranes and suckles the offspring. Thereafter, the young are suckled only once every 48 hours and no other tree shrew enters the main nest until the young are weaned at 4 weeks of age. Martin (1969) claims that this "absentee system" shows no similarity to that evident in living primates and is of systematic importance. He also suggests that the length of gestation and the rate of development of tree shrews reported by other authors, for example, Snedigar (1949), Wharton (1950), Hendrickson (1954), Sprankel (1961), Sorenson and Conaway (1964), Conaway and Sorenson (1966), and Hasler (1969), are possibly in error and states that there is strong circumstantial evidence for delayed implantation in *Tupaia*. Martin proposes that tree shrews be regarded as a separate order of mammals derived from menotyphlan stock. Sorenson (1964), Conaway and Sorenson (1966), and Sorenson and Conaway (1968) report quite different behavior for males and females of several species of tree shrews which suggests that perhaps members of *T. belangeri* are specialized in their reproductive mechanisms.

Aspects of the reproductive anatomy of tree shrews are often used as an indication of phyletic relationships. Wood-Jones (1917) states that the male genitalia of *Tupaia ferruginea* are definitely primate in character, with a pendulous penis and permanently scrotal testes; however, Verma (1965) notes that the testes position varies with different species, for example, they are permanently scrotal in *Tupaia*, descend only during the breeding season in *Ptilocercus*, and are abdominal in *Anathana*. Martin (1968) states that the permanent descent of the testes into a scrotum in Tupaiidae is not an indicator of specific relationship to primates, since the descensus is widespread among the Metatheria and the Eutheria.

In a study of the anatomy of the male reproductive tracts of several species of tree shrews, Alcala and Conaway (1968) found that the uterus masculinus is a constant feature in *Lyonogale tana*, *L. dorsalis*, *Tupaia montana*, *T. palawanensis*, and *T. minor* and is absent in *T. gracilis*, *T. chinensis* (*T. belangeri*), *T. glis*, *T. longipes*, and *U. everetti*. In these last-mentioned species without a uterus masculinus, a vagina masculinus is present. A uterus masculinus also occurs in males of *P. lowii* (Zuckerman and Parkes, 1935), but is absent in males of *Dendrogale frenata* (Davis, 1938) and *Anathana wroughtoni* (Verma, 1965). Alcala and Conaway (1968) state that the presence or absence of a uterus masculinus in male tree shrews may have taxonomic value. A uterus masculinus is present in the insectivore families Chrysochloridae, Erinaceidae, and Macroscelididae, whereas a vagina masculinus occurs in the higher primates. Such dichotomy in the male tract, coupled with evidence that the ovary of the female tree shrew shows similarities to soricoid insectivores, primates, and carnivores (Luckett, 1966), indicates that parts of the reproductive anatomy may have only meager taxonomic significance.

Stratz (1898) reports that female tree shrews undergo proestrous bleeding, and the possibility of menstruation is sugested by van Herwerden (1906) and by van der Horst (1949, 1954). Conaway and Sorenson (1966) describe the occurrence of menstruation in tree shrews and speculate about the possible origins of a generalized higher primate cycle from the reproductive patterns of the tupaiids. The authors point out that menstruation is often regarded as diagnostic of the primates and considered restricted to them; however, they declare that there is no valid basis for this limitation.

Mossman (1953) states that fetal membranes are so conservative that they form the best criteria for phylogentic interrelationships of recent mammals. From earlier studies of the placentation of *Tupaia javanica*, Hubrecht (1899, 1908) reports the presence of a hemochorial placenta and emphasizes consideration of the early ontogenic phenomena in the interpretation of phylogeny. Meister and Davis (1956, 1958) also report the occurrence of a labyrinthine hemochorial placenta in *Tupaia tana* (*Lyonogale tana*) and *T. minor*. Additionally, they note that the yolk sac and allantoic vesicle are nearly vestigial, similar to those of the higher primates.

Other investigators did not find evidence of hemochorial placentation among the tree shrews, rather, they report that tree shrews have endo-theliochorial placentation, a condition quite different from that of higher primates. From a reexamination of the placental material in the Hubrecht collection, van der Horst (1949) reports that the endothelium of the maternal capillaries remains intact in *T. javanica*. Hill (1965) states that *Tupaia* has a unique double placenta of the endotheliochorial type in which the fetus is connected by its umbilical vessels with two laterally placed placentas. He notes that the fetal membranes, in their early development, are arranged similarly to those of the mole (*Talpa*). A fine article by Luckett (1969) summarizes the evidence for phylogenetic relationships of tree shrews based on placentation and fetal membranes. He states that the development of fetal membrane characters of tupaiids is similar to that of soricoid insectivores and carnivores and suggests that the tree shrews be placed within the suborder Protoeutheria of Insectivora.

IV. SELECTED ANATOMICAL FEATURES

The century-old disagreement over the taxonomic status of *Tupaia* is based mainly on studies of anatomy. Cantor (1846) points out that the rudimentary unfringed sublingua of *T. ferruginea* is somewhat similar to that of *Nycticebus tardigradus,* whereas Garrod (1879) notes that the smooth surface of the brain and the presence of large olfactory lobes of *T. belangeri* resemble those of the insectivores. Chapman (1904) is concerned over the fact that some members of *Tupaia* do not have a cecum, but he regards *Tupaia* as the most lemurine of the Insectivora.

The first detailed study of the myology of *T. minor* was by Le Gros Clark (1924b). He lists 16 muscle attachments present in this tree shrew that do not occur in the Insectivora. Some of these attachments are remarkably similar to those of lemurs, galagos, and *Tarsius.* Another study by Le Gros Clark in 1926 on the anatomy of a different genus (*Ptilocercus*) confirms that the tree shrews are closely related to the primates. Lightoller's (1934) comparison of the facial musculature of certain lemurs and individuals of *T. javanica* shows a similarity between the two groups. More recent studies by Montagna *et al.* (1962) point out that the skin of *T. glis* differs from that of all other Prosimii, but contains eccrine sweat glands similar to those of the Anthropoidea. In their study of the anatomy and histochemical properties of the skin of the external genitalia of eight groups of primates including man, Machida and Giacometti (1967) found some evidence of phylogenetic ascendancies; however, there are also remarkable species differences. All prosimians except *Tupaia* and *Tarsius,* and the tamarin and the squirrel monkey, have only apocrine glands in the external genitalia, whereas higher forms have both apocrine and eccrine glands. The eccrine glands of the tree shrew, woolly, green, and pigtail monkeys, the Celebes ape, gorilla, and man are enwrapped by acetylcholinesterase-rich nerve fibers. The dermal melanocytes, absent in the lowset prosimians and highest simioids, are abundant in the Cercopithecidae.

Studies of skeletal parts of the hand of *Tupaia* by Lorenz (1927) show no relationship between tree shrews and primates. Additionally, the recent finding by Steiner (1965) of an independent fifth carpal and tarsal ossicle in the autopod of the embryo of *Tupaia* demonstrates a primitive retention in *Tupaia* not found in the primates. Studies of the development of the chondrocranium of *Tupaia* by Henckel (1928) and Roux (1947) show no evidence of primate relationships. Roux states that the primate skull could not develop from the menotyphlan-type chondrocranium exhibited by *Tupaia* and *Elephantulus.* Lyon (1913) also considers the tree shrews closely related to the family Macroscelididae but points out that the tupaiids differ from the elephant shrews in having a supraorbital foramen, an orbit surrounded by bone, a separate radius and ulna, a separate tibia and fibula, and primatelike ear bones. Evans (1942) states that the skeleton of the elephant shrew is closely related to the skeleton of the tree shrew. Romer (1966), however, from his reexamination of the elephant shrews finds no real relationship between them and the tree shrews. Martin (1968) also cites evidence against a tupaiid–elephant shrew kinship. von Spatz (1967) describes the manner of ossification of Meckel's cartilage in individuals of *T. glis* as primitive and reminiscent of marsupials.

V. CYTOLOGICAL EVIDENCE

Sarich and Wilson (1967) suggest that the albumin molecule evolves at a steady rate and therefore reflects molecular evolution. Although Hudgins *et al.* (1966) found compositional differences in myoglobin between Hominoidae, Cercopithecoidae, and Ceboidae, they agree that the myoglobin molecule also is of phylogenetic significance. Barnicot *et al.* (1967) state that molecular biology is in quite close agreement with traditional taxonomic views but emphasize that several different proteins should be studied to verify relationships among mammals.

Hafleigh and Williams (1966) consider *Tupaia* a primate, based on quantitative precipitation reactions of 22 primate serum albumins with pooled antiserums against human serum albumin. Goodman (1966) also found that the serum albumin of *Tupaia* is more similar to that of primates than to that of insectivores; however, in his 1967 study of the macromolecular specificities of primates, he showed that the Lemuroidea and Lorisoidea are closer to one another than either to Tupaioidea or Anthropoidea and closer to Anthropoidea than to Tupaioidea. Moore and Goodman (1968) now use modern computer techniques in their studies of the immunotaxonomy of primates and their data place *Tupaia* nicely among the prosimians.

Hematological studies comparing the blood of humans, monkeys, and hedgehogs with the blood of tree shrews, *L. tana* (Braun and Kloft, 1965) and *T. glis* (Hunt and Chalifoux, 1967), show similarities between the groups. The mean values for red blood cells, hemoglobin, and hematocrit are higher for the tree shrew (*T. glis*) than for man and certain monkeys, whereas the total white blood cell count is lower than in most primates. Layton (1965) showed that members of *T. glis* are unlike other prosimians (*Lemur fulvus* and *Perodicticus potto*) in that they are not sensitized by human atopic reagins such as commercial extracts of giant ragweed and mixed pollens of spring grasses. Johnson (1968) states that there is a basic difference between the hemoglobin of insectivores and primates. He points out that the migration of hemoglobin of insectivores is cathodal, whereas in *T. glis* and primates migration is anodal.

Other parameters, such as the metabolic rate of *Sorex* being eight times as high as that of tree shrews (Nelson and Asling, 1962), the biosynthesis of ascorbic acid in the liver of tree shrews and the slow loris in contrast to man and certain monkeys (Elliot *et al.*, 1966), measurements of blood sugar levels (Rabb *et al.*, 1966; Elliot and Wong, 1969), and aspects of thermal regulation among insectivores and primates (Chaffee *et al.*, 1968, 1969) point out the need for a detailed examination of the physiology of tree shrews.

Chu and Bender (1961), using chromosome cytology, neatly fitted *Urogale* into the scheme of primate chromosome evolution; however, other reports on the chromosomes of *Tupaia* do not agree with one another (Hsu and Johnson, 1963; Borgaonkar, 1967). Klinger's (1963) analysis of the somatic chromosomes of five species of primates shows that *T. glis* has 62 chromosomes. Egozcue *et al.* (1968) describe the occurrence of chromosomal polymorphism in *T. glis* and state that dimorphism is not uncommon in primates but that only in *Macaca* is chromosomal polymorphism found. In their examination of the chromosomes of seven species of tree shrews, Arrighi *et al.* (1969) found that the diploid number varies from 68 for *T. montana* to 44 for *U. everetti*. *Tupaia glis* and *T. longipes* have identical karyotypes and diploid numbers of 60. Acrocentric chromosomes with a distinct secondary constriction near the centromere are present in members of five species of tree shrews but absent in two other species. A pair of chromosomes with similar morphology is found in different species of lemurs with diploid numbers in the same range as those of tree shrews (Chu and Bender, 1961; Egozcue, 1967).

VI. PALEONTOLOGICAL EVIDENCE

Straus (1949) states that the outstanding feature of primates is that they may not be defined by any single or peculiar character. This lack of the unusual coupled with a lack of unquestioned fossil tupaiids clouds the evolutionary history of primates. Similarities between tree shrews and primates often are considered convergent although certain authors such as Buettner-Janusch (1963) point out that many tupaiids are island forms and were more likely to undergo divergence than convergence.

In his 1913 examination of the primitive tupaiid insectivores (*Ento-molestes*) and primitive lemurlike primates of the Bridger Eocene deposits, Gregory found that primates are more nearly allied to the tupaiids than to any other existing group. Simpson (1931) describes a fossil tupaioid, *Anagale*, found in the Oligocene deposits of Mongolia, which has flattened nails on the pedal digits. McKenna (1963) considers that the dentition and certain aspects of the ear region of *Anagale* are not good evidence for relating Anagalidae to Tupaioidea. Van Valen and Sloan (1965) suggest that primates arose in the late Cretaceous since primate fossils are similar in structure to those of condylarths and leptictid and erinaceoid insectivores. McKenna (1966) suggests that it is more meaningful to study fossil types than the anatomy and behavior of recent tupaiids and insectivores. He admits, however, that there is only a scanty distribution of known fossiliferous deposits from the time primates emerged in the Paleocene. McKenna believes that tupaiids are leptictid-like insectivores with special similarities to Malagasy lemurs and are the closest relatives of primates. Leptictid relationships are based on the formation of the bulla which fuses differently in tupaiids and leptictids than in primates.

Van Valen (1965) found little similarity between the middle Paleocene to early Eocene tupaiid genus *Adapisoriculus* and other Paleocene primates. He states that a primate-tupaiid relationship is possible but unlikely and that relationships between recent tupaiids and primates are probably convergences and primitive retentions. Van Valen, as does McKenna, bases certain conclusions on the formation of the bulla. Van Valen states that the bulla is formed by a single ossification (entotympanic) in *Tupaia*, whereas the bulla of lemurids and lorisids forms by an outgrowth of the medial side of the petrosal.

In his studies of the ontogenesis of the bulla of *Tupaia*, von Spatz (1964, 1966) found that the bulla tympanic develops from an independent cartilaginous entotympanicum. He suggests that the entotympanicum is a new acquisition of mammals without genetic relation to other structures. In contrast to McKenna and Van Valen, Spatz concludes that the bulla of the Malagasy prosimians develops from a tupaioid bulla by "fusio primordialis" of the entotympanicum and petrosum. Patterson (1955) postulates that the entotympanic, in the course of phylogeny, may assume different forms and that typaiid dentition may have been acquired in the post-Eocene. He states that the Paleocene and Eocene forms, not the present forms, are the "real" Prosimii and have much in common structurally. Le Gros Clark (1960) states that convergence cannot account for detailed similarities, only general features. For example, the orbitotemporal region of the skull and the

auditory bulla with its vascular foramina of the tree shrews are too
similar to the corresponding structures of lemurs to be convergent.

Szalay (1968) reports that three groups of animals may be ancestral
to primates, that is, the leptictids, the erinaceoids, and the condylarths.
He believes that Apatemyidae and Tupaiidae are derived independently from the Insectivora, not from primates, and that the insectivore-
primate transition occurred at the end of the Cretaceous and was fore-
shadowed by behavioral and physiological adaptations. Romer (1967)
more or less summarizes the available evidence by stating that the
actual ancestors of primates, 20 million or so years ago, are extinct and
that if one looks for living forms closest to the primitive stock one
chooses the tree shrews. He states that there is little in the structure of
tree shrews that prevents them from being considered parental to all
higher mammals.

VII. STUDIES OF BEHAVIOR OF TREE SHREWS

The following descriptions of the behavior of tree shrews are based
on studies of eight species of Tupaiidae (*L. tana, T. longipes, T. minor,
T. chinensis, T. glis, T. gracilis, T. montana,* and *T. palawanensis*) maintained in captivity at the University of Missouri between 1962 and the
present. Three main cages were used: one enclosure measured 16 ×
14 × 9 ft and the other two measured 8 × 14 × 9 ft. The three cages
were arranged to simulate the animals' natural habitat. Data are based
on over 1500 hours of observations made through a one-way glass window by three investigators, M. W. Sorenson, P. J. Thompson, and J. F.
Hasler. All tree shrews were individually recognized by use of fur-
clipped tail patterns. Certain results of these studies are published elsewhere (Sorenson, 1964; Sorenson and Conaway, 1964, 1966, 1968; Conaway and Sorenson, 1966; Thompson, 1969, Hasler, 1969; Williams *et al.,*
1969), whereas other data reported herein are novel. Representatives of
five species of tree shrews are shown in Fig. 1.

A. INVESTIGATIVE BEHAVIOR

Most animals of most species awaken shortly after dawn and settle
for the night by dusk. Observations at sunrise (5:00–5:20 A.M.) show
that members of *T. chinensis* are already active, looking out of windows
and eating; individuals of *T. longipes* are stretching and yawning, less

active than individuals of *T. chinensis;* other species are just awakening. If cage lights are left burning after dark, all species still become inactive and go to sleep only slightly after their normal time. During the day, the tree shrews alternate activity with rest periods. Each species displays two activity peaks, for example, *T. longipes* at 7:00 A.M. and 6:00 P.M., *T. chinensis* at 8:00 A.M. and 2:00 P.M., *T. gracilis* at 6:00 A.M. and 2:00 P.M., *T. palawanensis* at 9:00 A.M. and 4:00 P.M., and *L. tana* at 9:00 A.M. and 2:00 P.M. (Fig. 2). Sprankel (1961) reports that members of *T. glis* are diurnal and that they awake at about 6:20 A.M., after which they eat, explore, and are active until nap time at 11:30 A.M. Their major activity peak however, occurs between 5:00 and 6:00 P.M. Similarly, Vandenbergh (1963) states that individuals of *T. glis* display decreasing activity from 7:00 to 11:00 A.M. and increasing activity from 3:00 P.M. until evening. Wharton (1950) notes that members of *Urogale* move with great speed on the ground and are excellent climbers. He found that at night the animals sleep in tightly curled balls. In contrast to the daytime activity of most Tupaiidae, individuals of *Ptilocercus* are described as mainly nocturnal and sleep during the day in a pile composed of several animals (Banks, quoted by Le Gros Clark, 1926).

Temperature and humidity affect the activity levels of tree shrews. Activity declines during increased temperature and humidity; for example, members of *T. chinensis* are less active when the temperature rises above 90°F but also when it drops below 60°F. Individuals of *T. palawanensis* are less active at temperatures above 85°F and males and females of *T. montana* die at temperatures above 105°F; however, members of *T. chinensis* remain alive at a temperature of 100°F for a period of 70 days. During exposure to high temperatures (85°F+), all animals except members of *T. minor* tend to rest separately and even sprawl with their bodies flattened dosoventrally and their appendages maximally extended. This posture possibly facilitates evaporative cooling. In contrast, during cooler months all tree shrews rest in sunny spots along logs on the cage floor or on window sills. The animals sometimes puff their fur and assume an embryonic posture.

Certain species exhibit cyclic daily rest periods. Members of *T. gracilis* have periods which average 18 minutes (range 3–29 minutes) and which often are interrupted by short 8–10 minute bursts of activity. Rest periods of *T. longipes* average 75 minutes (range 35–120 minutes) and individuals of *T. chinensis* often rest for periods of 2 hours or more. Following rest periods the animals usually stretch and yawn. Stretching movements are rather stereotyped and consist of inching forward with the forelegs until they are fully extended, followed by raising the head so that the back becomes arched downward. This position is followed

Fig. 1. Representatives of five species of tree shrews: (A) *Tupaia minor,* (B) *T. gracilis,* (C) *T. longipes,* (D) *T. chinensis,* (E) *Lyonogale tana.*

Fig. 1. (*Continued*)

FIG. 1. (*Continued*)

by a forward body shift which tends to extend and then drag the rear limbs forward. The rear legs are rotated so that the plantar surfaces are dorsal. Sometimes the animals stretch their front legs individually by extending each rigidly in a straight line. When tree shrews yawn, their mouths are widely open and frequently their tongues extend ¼ to ½ inch beyond the end of their rostrums.

Members of *T. minor* are the most secretive of the species studied, are less active, and participate in fewer social interactions. In contrast, *T. gracilis* males become tame after 1 year in captivity and allow themselves to be held and stroked in a fashion similar to the behavior of individuals of *Urogale* in captivity (Polyak, 1957). Both *T. minor* and *T. gracilis* are less active during periods when the larger animals are hyperactive, and individuals of *T. minor* even show an activity decline in the presence of the slightly larger members of *T. gracilis*. The larger species, however, do not modify their activity in relation to each other except when new animals are introduced into their cages. Introductions always result in hyperactivity among the resident animals. As a general rule, males are more active than females and members of *L. tana* and *T. palawanesis* are far less active and excitable than are individuals of other species.

Activity involves various movements associated with sexual, aggressive, and exploratory behavior. Additionally, often for no apparent reason several animals spontaneously race around the cage three or four times. During these races, the tree shrews utilize any vertical surface from which they either spring off in long jumps or use as stops to facilitate quick turns. The animals run by means of consecutive leaps of 2 or 3 ft. During slower travel, members of *T. gracilis* still more or less hop whereas individuals of *T. longipes* actually walk, placing one foot in front of the other in stiff, jerky strides. Individuals of *T. chinensis* move more slowly than do members of *T. longipes* and are less nervous and more ratlike in form. Individuals of *L. tana, T. chinensis,* and *T. pala-wanensis* jump less than members of other species and are more terrestrial in captivity. All species are capable of walking inverted along tree branches, but the smaller animals do this most frequently.

When climbing on wire mesh, members of *T. minor,* and *T. gracilis* flatten their bodies by extending their limbs laterally with their feet pointed diagonally away from their body. They take short steps by grasping the wire with their digits. This dorsoventral flattening is not common among the other species. During descent, the forefeet are pointed forward or slightly outward, whereas the rear feet are turned around pointing posteriorly and allowing the claws to act as a brake. Neither members of *T. chinensis* nor *L. tana* reverse their rear feet to the degree that the other species do, which may indicate a more ter-restrial habitat. Also, newly introduced hand-reared individuals of *T. longipes* (raised in small cages) display poor arboreal ability, fre-quently requiring weeks before they can move with agility along tree limbs. Perhaps arboreal ability depends on structural modifications and learning; thus, naturally existing terrestrial or low bush forms would not display this behavior in captivity.

B. Sensory and Learning Abilities

Each species has good daylight vision and depth perception, adequate crepuscular vision, but very poor night vision. When the animals are disturbed after dark, they either remain stationary or panic and crash into cage objects. Their pure cone retina is described earlier in this chapter. Members of *T. longipes* and *T. chinensis* often stand on their hind legs and look over logs, out of windows, and into nest boxes. All species are very dependent on sight when catching live crickets, grass-hoppers, and mice. It is apparent that olfaction is not relied upon, for hungry animals walk within inches of motionless crickets or grasshoppers without finding them; however, young *T. longipes* make continuous sniffing noises when exploring new areas and all species utilize olfaction

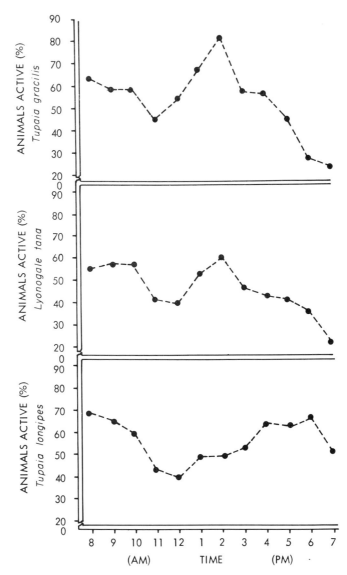

Fig. 2. Activity periods of five species of tree shrews in captivity.

during sexual stimulation. Auditory reception is acute, although the animals adapt to noise much more rapidly than to visual disturbance.

Some evidence of problem solving appears when food is placed through a small hole into a hollow log. An animal looks through the hole at the food, attempts to reach it with its forefeet and, failing, goes to the end of the log, enters and obtains the food. If a test tube containing

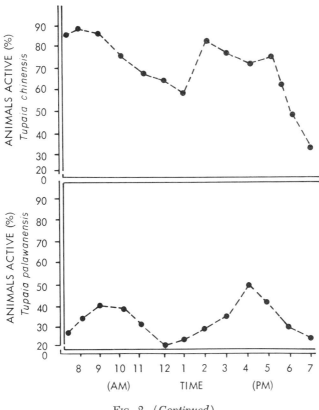

FIG. 2. (*Continued*)

something of interest is placed in front of young tame *T. longipes*, they try to put either their noses or tongues into the tube and, again failing to reach far enough, immediately thrust a foreleg deep into the tube and then lick their digits. Additionally, when a Kleenex tissue is waved in front of tame animals they quickly attack and rip it. When the tissue is placed in a pocket of the investigator, the tree shrews try time and again to pull it out. The animals recall over a period of several days which pocket contained the Kleenex tissue and each morning climb up the pant leg of the investigator and examine that pocket. In contrast, Leonard *et al.* (1966) found that members of *T. glis* show no interproblem improvement on 800 six-trial object discrimination problems. The authors state that on both learning-set and delayed-response tasks, performance is inferior to that previously reported for primates and cats. It is possible, as Tinbergen (1951, 1953) has freely suggested, that the senses are used preferentially and therefore certain psychological tests do not apply to broad groups of animals. Often it appears that tree shrews fail to examine a certain situation thoroughly, relying only upon sight or touch. For example, when a rubber bottle stopper is placed in a hole in the wall of the cage, the animals display immediate alert postures, whereas if the same stopper is placed on the floor of the cage they bite it and play with it.

C. Sociality

Although all species clearly are social, members of *T. minor, T. chinensis,* and *T. palawanensis* are less so than the other species. Sociality is evidenced in the heterosexual grouping of two to five species maintained for nearly 2 years without aggressive fatalities. The males of most species display linear dominance hierarchies based mainly on aggressive interactions, and intraspecies and interspecies recognition of dominance expressions is pronounced. The most common aggressive displays are (1) threat calls, (2) threat postures, (3) lunges, (4) chases, and (5) fights. Members of *T. montana* engage in "boxing" wherein two animals stand upright on their hind legs near one another and thrust their forefeet toward the head of the other animal. This behavior is not common among the other species.

Dominance is more subtly expressed by the displacement of subordinate animals from rest boards or food trays, by the mounting of subordinate by dominant animals, and by sexually related anal nuzzling, following, and copulation.

Females of *T. chinensis* display linear hierarchies, but among the males of this species a single dominant animal assumes the role of a despot. The rank of this male is rarely reversed; he harrasses all subordinate males; he is occasionally groomed by females; and his presence alone is enough to disrupt all sexual behavior among the animals of other species. A similar situation is reported for individuals of *T. glis* by Hendrickson (1954), Sprankel (1961), and Vandenbergh (1963).

Among members of *T. montana,* social rankings include both males and females with the number one- and the number two-ranked males being for the most part codominant or mutually tolerant. Individuals of *T. montana* are the most social of the species studied and exhibit fewer aggressive interactions but more threat calls and displacement activities.

On occasions following serious male-male aggression, the defeated animal becomes asocial. He waits until the others have eaten, rests alone for excessive periods in out-of-the-way places, sleeps alone, is consistently intimidated by the ranking male and often by females as well; his social position remains precarious for 2–3 days prior to slowly renewed interactions. During these periods, the next-ranked male often shows an increase in activity.

Dominance among captive tree shrews is influenced by the amount and arrangement of cage space, the number of animals, and the ratio of males to females. Fewer members of *T. chinensis* can be maintained in a limited area than members of *T. montana* or *T. longipes*. Individuals of *T. palawanensis* are particularly affected by population density. In this species the most dominant and the most subordinate animals are the first to die when confined in high-density populations (15 or more animals kept in a cage measuring 16 × 10 × 8 feet). The most dominant tree shrews (males) display a high degree of activity and aggressive behavior while the most subordinate animals (mainly females) are the least active and the most timid in the population (Thompson, 1969). The dominant tree shrews must defend their position in the linear hierarchy against a large segment of the high-density population, and therefore they are committed to an increased number of aggressive interactions. The subordinate group, unable to assert itself, is the object of threats and aggressive behavior. The middle-ranked part of the population enters into fewer interactions and is not subjected to constant threat and aggression. Individuals of *T. palawanensis* in this middle group retain their linear ranking when the most dominant and the most subordinate individuals die; however, the level of activity of all animals in the enclosure is less than that of the initial population. This situation partially confirms Myers' (1966) studies of density-stressed rabbit populations.

A recent study of social stress among captive members of *T. glis* by Autrum and von Holst (1968) shows that tree shrews ruffle the hair of their tails in response to social and nonsocial stimuli in their environment. The percent ruffling is related to the degree of subordination and the number of individuals in the cage. When ruffling is above 50%, the males retract their testes and the females become infertile; if above 80% for 10 days, the captive animals die. It has been shown that tail hair ruffling is under the control of endocrine organs.

As a general rule, males and females of each species share rest areas, nest boxes, and food. Actual male-female consorts, however, occur only for 1 hour or less during the height of female receptivity. At night the dominant male *T. chinensis* sleeps in a nest box with as many as 12 females. These females are not in estrus, however, and no lasting consorts are formed with the male. Female-female consorts do exist and sometimes last for months. Consort females lie one on the other or next to each other during rest periods and even sleep together at night. They groom each other daily although most grooming is not reciprocal. Most often female-female consorts are restricted to wild-wild or tame-tame pairs and are most prevalent among members of *T. longipes*.

Members of *T. glis* do not form large social groups, rather, only short-lived family units occur in captivity (Vandenbergh, 1963; Kaufmann, 1965; Sprankel, 1961). Martin (1968) reports that specimens of *T. belangeri* are not social. Manley (1966) summarizes most of these studies. In contrast to laboratory studies of *T. glis*, individuals of *T. montana* form social units of up to 12 animals of mixed sexes under captive and natural conditions (Sorenson and Conaway, 1968). In captivity, all species we have observed are social although members of *T. minor* are the least so. It is possible that environmental changes cause shifts in social systems and that local populations of the same species have different social systems (Crook, 1967). Crook and Gartlan (1966) suggest that the small size of social units among forest frugivores is related to the limited conditions of food supply occasioned by the stable conditions of the rain forest. Also, the concept of social dominance as a structuring mechanism is questioned by Gartlan (1968), who also suggests that learning plays a significant part in social roles and that genetic influences are minimal. He notes that a hierarchy among animals is often considered to reduce aggression; however, when hierarchies are most rigid, for example, under captive conditions, aggression is most common. Gartlan suggests that the habitat determines the social responses among animals and that advanced and primitive labels on social evolution are not warranted. Chance (1962) expresses a somewhat different view. He believes that the majority of contemporary primates display competition for social position in which success is rewarded by a breeding premium and conflict resolved by social spacing.

D. Play Behavior

Many instances of chasing among tree shrews are interpreted as "play." Such episodes are not accompanied by biting, vocalization, and the aggressive chasing usually associated with dominant-subordinate interactions. Sometimes animals chase each other around and around the cage using the same pathway, as though playing "follow-the-leader." As many as eight animals, mostly females, are involved in these chases along terrestrial and arboreal routes. The animals run at high speeds and are spaced at about 1- or 2-ft intervals. Most chases start without obvious cause and last for 1–3 minutes. Individuals of *T. longipes* chase objects rolled along the floor of their cage, catch the objects, and wait for them to be rolled again. Tree shrews also take strips of cloth in their mouths and drag these around the cage, often eliciting interest from other animals, then they exchange short chases following thefts of cloths from one another.

On occasion an animal approaches another from behind and surprises him. The surprised animal jumps into the air and is then pursued by the other animal. After running a short distance, the chased animal may reverse direction and chase the other. Aggression does not result from these changes although the animals become very hyperactive and seemingly excited. On occasion two females of *T. longipes* may peer at each other through a small hole in the side of a hollow log. With one animal inside the log, the other animal repeatedly pokes her nose into the hole and then runs away. Moments later she enters the log and the second animal then races out and looks through the hole several times.

E. Imprinting

Imprinting occurs among all hand-reared members of *T. longipes*. Young animals taken from their mothers less than 2 hours after birth and maintained in an area isolated from other adult animals for a period of 60 days become imprinted. The offspring are fed two or three times daily until they are able to eat solid foods. The eyes of newborn tree shrews do not open for approximately 2 weeks and therefore their first visual contacts are with the feeder's hand.

During the first week or two after the animals' eyes open, the offspring begin to chin and slide their bodies along the hand of their feeder. Chinning and rubbing become a greeting expressed on nearly all occasions as the animals mature. These movements are not simple forms of play since they develop into adult sexual postures and are expressed each time the human hand is encountered. Such early human contact remains a dominating factor in the behavior of the tree shrews for months after they are allowed association with various wild tree shrews. The basic behavior patterns of the tame animals are similar to those of the wild animals except for sexual expressions. Most hand-reared females fail to copulate with either tame or wild males, and young tame males and females seem to prefer human association to that of their own species. After sexual maturity, hand-reared animals may leave a consort during coitus and come to a human hand and chin and rub on it. This activity is not merely an artifact of captivity, for animals reared in cages by their parents behave normally with conspecifics.

F. Ingestive Behavior

Tree shrews begin eating immediately after fresh food is offered each morning and continue eating intermittently throughout the day. The ani-

mals eat a variety of fresh fruits and vegetables (melons, grapes, bananas, apples, oranges, lettuce, carrots) and dry and canned dog foods.
The diet is supplemented with live or freshly frozen mice, and with
crickets, grasshoppers, cockroaches, peanuts, and sunflower seeds. The
tree shrews demonstrate a marked preference for meats and soft juicy
fruits such as melons and grapes.

The animals approach food trays by indirect routes and often stop and
peer about before selecting pieces of food. They use either their mouths
or forefeet to grasp food from food trays. Most of the time the food is
carried a foot or so to an adjacent log before it is eaten. All species may
eat side by side without aggression; however, individuals of *T. minor*
and *T. gracilis* usually eat alone while sitting on raised rest boards or
tree limbs. When very juicy fruits are consumed, the animals raise their
muzzles to facilitate swallowing of the juice. The lower incisors are used
as a scoop. Slices of apple or banana are held by one or both forefeet
and often this food is shoved into the mouth. Although the tree shrews
often take only a bite or two of a piece of fruit and then drop it, there is
little waste. The original animal either returns later to eat the food or
another animal eats it. Chunks of dry dog food become widely scattered
in the wood shavings on the cage floor, and the animals spend a considerable amount of time foraging through these shavings looking for
pieces of food.

In the most frequently seen eating posture, an animal sits upright with
its rear feet pointing out laterally and its tail extended directly back-
ward. The shoulders are slumped forward slightly and the forefeet are
held with the palms upward. Food is held by the claws and the palms of
the forefeet; opposability of the thumbs is minimal (Fig. 3). Pieces of
meat are firmly held by the forefeet and then pulled over the teeth. The
shredded meat is masticated slowly and thoroughly.

Adult tree shrews of all species are cannibalistic and eat both newborn animals and other adults that die. Many of the young animals born
in captivity are consumed within 48 hours of birth. An adult begins eating the offspring at the head end while holding it firmly with the forefeet. Other animals try to take the freshly killed young away from the
first animal by creeping near it and grabbing the young, or by chasing
the first animal around and around the cage until it drops the offspring.
Under these circumstances, aggression often occurs and several animals
become involved. Much of the flesh of the heads and forequarters of the
adults that die is also eaten.

When a live mouse is introduced into the cage, most of the tree shrews
converge on the mouse and kill it in seconds by biting it on the head
and throat. Sometimes the mouse is eaten simultaneously by two or three

FIG. 3. Typical eating posture of *T. gracilis*.

tree shrews without aggression, and at other times the animals fight for possession. Sometimes a mouse runs directly into a nest box and several members of *T. longipes* and *T. chinensis* run after it, look into the box, start to enter it, and then back out when the mouse begins to squeak. An individual *T. gracilis* may then enter the nest box and both it and the mouse squeak; after a moment the tree shrew comes out of the box without the mouse. A female *T. longipes* may then repeat the sequence. The mouse remains inside the box for about 15 minutes while the tree shrews just mill around in front of the opening. Finally, a tree shrew enters the box and ejects the mouse, which runs nearly the length of the cage prior to being caught and killed. On occasion mice escape from the cages without injury and some mice are found dead in the cages but not eaten.

The tree shrews often spend several minutes at a time trying to catch flies outside the windows. Flies inside the cage are closely watched and occasionally are caught, always with the mouth. Individuals of *T. minor* are adept at catching ants on hollow logs, and members of *Lyonogale* spend many minutes "rooting" off loose bark from the logs in search of insects. Crickets and grasshoppers are easily caught by tree shrews. The insects are trapped beneath the forefeet of the tree shrew and then either eaten while held in this position or placed in the shrew's mouth with the forefeet. Only rarely does a tree shrew catch these insects directly with its mouth.

Location of prey is almost entirely visual and insects that remain stationary frequently are by-passed. The fact that crickets, which are choice food items, live inside and under logs of the cage for months at a time implies that olfaction is infrequently used to find prey and also demonstrates the lack of nocturnal activity by the tree shrews.

During morning feeding periods, all species may share food with each other and steal food from one another without aggression. The shrews sit side by side and simultaneously eat on the same piece of food or pass it back and forth between bites. Sometimes a tree shrew with food raises his forefoot and gently pushes another animal away. This shoving is performed most often with the left foreleg. Two or three female *T. longipes* may band together and successfully steal food from a male. The females approach the male at about a 45° angle from the rear, and when he turns to face one of them another female grabs his food and runs away. The male rarely gives chase, but after two or three thefts he emits threat calls or makes short lunges when the females approach him. Food sharing may be an adaptive behavior which brings animals together at major food sources, prevents waste, and allows heterosexual contact.

Often members of *T. gracilis* display a peculiar behavior which consists of taking pieces of food in their mouths and then rubbing the food against the hollow logs or the floor in a series of sweeping arcs. The animals stand firmly on all four feet and pivot from the shoulders. These arcs are fairly rapid and travel 2–4 inches. There seems to be no fixed number of sweeps, ranging from 2 to 96 consecutive swings with a single piece of food. This behavior is seen when the animals eat fruits as well as live prey. Observations show that these arcs are severe enough to tear legs and wings from grasshoppers and it is assumed that this is a method of disabling prey. Alternatively, this activity may serve to clean foodstuffs. Similar behavior occurs rarely among the other species and is always at a reduced intensity.

Vomiting is a rare occurrence and the regurgitated matter is quickly eaten by all tree shrews present. Coprophagy is quite common, especially among newly captured tree shrews and during the first 2 months among hand-reared young. Individuals of *Lyonogale* and *T. chinensis* continue to eat feces throughout long periods in captivity but at a reduced rate. The animals eat fresh as well as old feces; sometimes they watch others defecate and then run over and consume the droppings.

There is no evidence of food hoarding, and excess food is not found inside nest boxes prior to births.

Water is supplied to the animals ad libitum in suspended bottles with lick tubes. Water consumption increases with increased temperatures and following periods of hyperactivity. Injured animals also drink more than noninjured animals. Water offered in a flat dish is lapped with the tongue, but the animals do not bathe.

G. ELIMINATIVE BEHAVIOR

Defecation nearly always involves a fixed posture. The tree shrews extend their hind quarters over a limb, a log, or a rest board with their tails held up or straight out and their heads held horizontally. In the case of the male, abdominal contractions cause the penis and scrotum to move upward and forward.

Except for members of *T. gracilis* and *T. minor,* two or three definite toilet areas are maintained. Newly introduced tame *T. longipes* fail to use established toilet areas, however; they either defecate at random or start new areas. All adult members of *T. longipes* orient themselves toward the window of their cage while defecating.

Following defecation, an animal drags his anus 4–10 inches along a limb or a hollow log in what appears to be a wiping function. Since other animals are rarely seen to either sniff or avoid freshly wiped areas, it is assumed that anus dragging is not a means of territorial marking and differs from the placing of glandular secretions on objects as noted by Sprankel (1961). Feces are seldom observed in nest boxes of adults or young, and females of *L. tana* most probably remove feces from the nests of their young.

All species merely squat when urinating, although a male *T. longipes* sometimes slowly walks along a rest board while dribbling urine and his rear feet become impregnated with it. This seems to be a method of territorial marking although other animals do not respond.

H. AGGRESSIVE BEHAVIOR

All species display linear dominance hierarchies although the number and sex of animals involved and the stability of the hierarchies vary among species. Several methods of dominance expression occur, and intra- and interspecies recognition of these expressions is pronounced. Subordinate animals react at distances of up to 10 ft and quite clearly recognize individual members of their own and other species. The most common aggressive displays are (1) shrill staccato threat calls emitted by nonreceptive females during male sexual advances and by subordinate animals in precarious situations; (2) threat postures in which the animal either sits or assumes a crouched stance with its neck stretched forward, its head lifted upward slightly, its mouth open, and its body held rigid; (3) lunges in which an animal moves forward no more than 10 inches; (4) rapid, determined chases that often cover several circuits of the cage, through and over hollow logs, up and down trees, and bounding off vertical surfaces; (5) fights in which the animals "face off" at about 18 inches, crouch in a threat posture, move side to side or circle each other, and then lunge forward to bite one another. Most bites are superficial and occur as the animals roll over and over on the floor of the cage; however, tails, thighs, digits, and ears are injured. Members of *T. montana* sometimes stand upright on their hind legs and thrust their forefeet toward the head of the other animal; striking of the animal seldom occurs but aggression is reduced. Dominance is more subtly expressed by the displacement of subordinate animals from rest boards or food trays, by the social mounting of subordinates, and by visual staring.

Aggression among species varies. Members of *T. minor* and *T. gracilis* seldom display overt aggression; chases are of low intensity and rarely evolve into fights. Individuals of *T. minor* use threat postures to ward off aggression by animals of larger species and rely on arboreal agility for escape. Members of *T. gracilis* seldom fight among themselves and are tolerated by members of larger species under almost all conditions. They steal food from tree shrews of the other species, sleep with members of all species, and mount females of the larger *T. longipes* without causing aggression.

Individuals of *T. longipes* are dominant over all species except *T. chinensis*. Both males and females of *T. longipes* form dominance hierarchies and reversals rarely occur. Most aggression among members of *T. longipes* is initiated by the second-ranked male and centers around sexual activity involving estrous females. Chases may start a chain reaction with the first-ranked male chasing the second-ranked male and the latter chasing a lower-ranked male. Some chases are identical to "follow-the-leader" activities although conflict develops when one animal catches another. The active animals often jump over passive animals without incident. The victor of these encounters rubs his chin and chest areas on the hollow logs and the subordinate tree shrews escape into nest boxes from which the dominant animal rarely ejects them.

Tupaia chinensis is the most dominant and aggressive species. Both males and females attack members of other species without apparent reason. The dominant male is a despot and often roams about the cage disrupting social activities of the other species. He violently attacks males of *T. longipes* whenever they attempt copulation with females of their own species and chases these males until fatigue prevents further action. The larger individuals of *L. tana* fare no better and are consistently intimidated. As noted, members of *T. gracilis* are an exception in that males of *T. chinensis* do not chase them and allow them to share food and attempt copulations with females of the larger species without being driven away.

Fights between males or females of *T. chinensis* in which two animals "stand their ground" are rare. Fights occur when strange males are introduced into the colony and challenge the dominant male. Before a fight, both animals usually face one another in a crouch, about 1 or 2 ft apart, for a period of up to 2 minutes. During this motionless time the animals stare at one another and then one animal lunges forward and the other animal either meets the attack or runs away. During a fight the animals roll on the floor, biting and emitting "squealing" calls. Finally one animal flees and the other animal chases him and utters "snort" calls. An open-mouth stance apparently is a submissive posture, for it usually stops aggression.

Both males and females of *T. palawanensis* form dominance hierarchies; however, in a population of 25 animals (5 males, 20 females) 12 females may have equal standing and threaten and chase each other. One male and three females are midway in the hierarchy and six animals (one male and five females) in the population never assert themselves and are always the object of threats and charges by the three high-ranked males. Among members of *T. palawanensis*, the highest- and the lowest-ranked animals in the population die after only a few months in captivity. The vacated positions are filled by animals next of rank and these in turn die after short periods. The result of the continued deaths of high- and low-ranked animals is a stable population composed of the original center portion of the hierarchy. In captivity, aggression among members of the stable population is minimal and directed mainly toward animals of the opposite sex. Whenever new animals are added to one of the stable groups, the existing hierarchy deteriorates and a new ranking takes place which includes several of the new tree shrews. Following establishment of this new ranking, the most dominant and most subordinate animals again die after a period of 1 or 2 months (Thompson, 1969).

The introduction of strange animals into cages containing established groups of animals has different effects depending on the resident and introduced species. When males of *L. tana* are placed in a cage containing members of *T. longipes*, the males touch noses and sniff each others' anal areas and then more or less ignore each other. The resident animals do not defend their food, water, or rest areas. After a day or two, the dominant male *T. longipes* starts to chase the males of *L. tana;* within hours he establishes his position and the males of *Lyonogale* become subordinate animals. Introductions of males of *T. minor* and *T. gracilis* into cages containing members of *T. longipes* are uneventful, each species accepting the other without aggression and their individual hierarchies being maintained.

The introduction of four females of *T. chinensis* into a cage containing other *T. chinensis* and members of *T. minor, T. gracilis, T. longipes,* and *L. tana* has a profound effect on the social structure of the resident animals. Moments after females are placed in the cage, males of *T. gracilis* attempt to mount them; there is no aggression. Next, a female *T. longipes* and a female *T. chinensis* touch noses and then separate. A large female *T. chinensis* then nuzzles the anal area of a male *Lyonogale* who immediately runs away; however, she continues to follow and molest him. She nuzzles his ears, neck, and testes and then mounts him. He remains passive. She leaves him and drags her anus along the surface of a hollow log; she begins stamping up and down on the same spot, then dashes at the male *T. chinensis* in the cage who runs away. The male *T. chinensis* returns to her a moment later and nuzzles her and a male *L. tana*. The male *T. chinensis* then tries to mount each of them in turn, but the female resists and the male *Lyonogale* runs away. The male *T. chinensis* then displaces both members of *T. minor* and *T. gracilis* by forcing them up a tree. During the next few days the male *T. chinensis* becomes increasingly aggressive and begins chasing all other animals. He soon completely dominates the group. From this time onward, the male *T. chinensis* remains a despot and the *T. chinensis* females also start to dominate members of the other species so that a new interspecies hierarchy develops with individuals of *T. chinensis* dominant over members of the other species.

Presentation occurs among males and females of *L. tana* and *T. longipes* during the first month or so of captivity. The animal presenting lifts its tail directly overhead and positions its anal area 2–3 inches in front of the other animal who assumes the posture of an open-mouthed aggressor. Usually the aggressor then either licks the exposed rear of the animal presenting or ignores it. Sometimes two animals present and then turn face to face and display threat postures. One of these tree shrews then again presents and the other mounts him. When a female of

T. longipes grooms the rump of a male, he responds by presenting to her. Presenting nearly always expresses subordinance, and after separation of animals into species groups and the establishment of a dominance hierarchy presenting rapidly declines.

There are few indications of territorial behavior among captive tree shrews. Dominant males of *T. chinensis* seem to prefer certain rest areas and often eject one or more resting animals from these areas and then defend the position. Occasionally a male shares his rest area with certain females and responds to other males that approach within about 2½ ft. Members of *T. minor*, maintained with individuals of *T. montana*, use only a restricted part of their cage and defend that area against the other species by emitting shrill vocalizations.

Under natural conditions members of *T. montana* occur as discrete groups of about 12 animals separated from other small groups by distances of approximately ½ mile. The defense of the boundary between groups has not been observed (Sorenson and Conaway, 1968).

I. Vocal Behavior

Members of *T. minor* are the least vocal of the species observed. They emit only two basic calls, interpreted as *warning* and *threat* calls. The warning call is a very rapid, high-pitched chatter made with an open mouth while the animal crouches in a semithreat posture. This call is made following intense chasing or immediately after a major disturbance such as our entering the cage. The threat call is made with an open mouth and sounds like a muffled bark. It is emitted during stressful moments when an animal is cornered by another. After a year in captivity, members of *T. minor* call far less and become nearly mute.

Tupaia gracilis also has two basic calls somewhat similar to those given by *T. minor* except that the warning call is more rhythmic and melodic and is made with a nearly closed mouth. This call is triggered by the warning calls of members of *T. minor* and continues for 2–3 minutes. Such calls also occur following aggression and cage disturbance. Characteristic of their passive nature, members of *T. gracilis* emit few *threat* calls. Those given are made with an open mouth and consist of two or three quick, barking sounds. Members of *T. longipes* have six distinct vocalizations which are interpreted as *warning* calls, *threat* calls, *rage* calls, *fear* or *pain* calls, *excitement* calls, and *pleasure* calls. Warning calls are the most frequent and are made with an almost closed mouth and minimal abdominal contraction. They consist of a series of medium-pitched warbles which slowly decrease in volume. The calls are in response to sudden noises or other disturbances and follow aggressive chases. Often the subordinate males call while the dominant male is chasing another animal. Unlike individuals of *T. minor*, members of *T. longipes* never call immediately after we enter the cage. It is difficult to ascertain which animal makes warning calls because of a ventriloquistic quality of the call; however, the calling animal normally flicks his tail back and forth and remains in elevated areas of the cage. The animals seldom call while in motion, commonly waiting until they are out of immediate danger. This behavior may be adaptive in that tree shrews do not advertise their positions until they are safe.

Threat calls are delivered rapidly with an open mouth and sound like an explosive bark or hiss. Often they are followed by short, forward lunges. These calls are mostly in response to overt aggression by other tree shrews. Nonreceptive females give threat calls when males approach them. Both sexes emit threat calls in area defense and also when we disturb them at night by flashing light in their eyes or by handling them. When threat calls fail to thwart an aggressor, these calls shift to shrill continuous screams indicative of rage calls. The animals squat low on the floor with their open mouths inches apart and their eyes fixed on each other. After 30 or 40 seconds, one or both animals stop squealing but they remain staring at each other.

Fear or pain calls are similar in intensity to rage calls, although not continuous. Instead, they are shrill, chopped-off cries emitted following a bite or other injury. Excitement calls consist of sharp, clear whistles often accompanied by tail flicks and nervous pacing; sometimes the animals stand up on their hind legs and look in the direction of the disturbance. These calls occur during moments of curiosity associated with investigating new objects or watching movements outside the cage area. The animals also frequently emit these calls during play chases.

Pleasure calls (*chirps* and *clicks*) are restricted to the young, particularly during the first 2 or 3 months following birth. Even before their eyes open, young tree shrews are vocal and when awakened they often give a typical threat call of low intensity. Then they emit recognition chirps, single birdlike calls, upon being picked up. Immediately after being fed the young make soft, continuous, clicking calls reminiscent of distant castanets. These sounds are made with the mouth closed and last for 1 or 2 minutes. More mature animals give variations of these clicking sounds when we pick them up and play with them. Loud, quick, heavy breathing by the juveniles occurs whenever they are allowed to explore our hair or insert their noses into our pockets or ears.

Members of *L. tana* rarely emit warning calls; when disturbed they run inside hollow logs where they remain silent. They have distinct *threat* calls which are low-pitched guttural barks made with an open mouth during aggressive encounters. These calls also become shrill, staccato cries of *rage* similar to those of members of *T. longipes*. Additionally, males of *Lyonogale* emit very low *quacking* calls which are quite rapid and rhythmical and made with an almost closed mouth during courtship dances. These calls slowly increase in volume as the males become more sexually excited; however, at no time are the calls easily audible.

Tape recordings of vocalizations of members of *T. chinensis* are made periodically without disturbance of the animals. Calls are printed on a Kay Electric Sonagraph 6061A and analyzed by Dr. H. W. Williams, Westminster College, Fulton, Missouri. Five distinct calls are identified: a *quack* call, a *snort* call, a *cluck* call, a *twitter* call, and a *squeal* call (Hasler, 1969). Sound spectrographs of representative calls appear in Fig. 4. The quack call is the most commonly emitted of all calls and is similar to the quacking of a duck. It is made by both males and females with the mouth nearly closed and with obvious abdominal contractions. It occurs rhythmically for periods ranging from a few minutes to several hours. At times, a series of two or three quacks alternate with periods of silence. Quacking seems to indicate a state of excitement or alertness, and the dominant male usually quacks before and after reproductive episodes. Quacking is often accompanied by tail flicking. The snort call is a deep, explosive, coughlike sound delivered with an open mouth and sometimes uttered repeatedly after aggressive encounters. The cluck call, which is just audible, is emitted with a closed mouth by estrous females. The twitter call is a series of rapid, high-pitched sounds given by females not completely receptive. The squeal call is emitted with an open mouth by subordinate males and females during aggression.

Individuals of *T. palawanensis* have four basic vocalizations, each elicited by a different stimulus and given under different situations. There is a low-volume *chatter* call associated with aggression, a *rage* call given by both sexes when hand held, a *whistle* call emitted as a *warning* call and given as a pure note in the frequency range of 6.2–8.2 kcps [this call is particularly interesting because of its antiphonal character (Williams *et al.*, 1969)], and a *chirp* call which is characteristic of the dominant animal and which occurs regularly with the whistle call. Individuals of *T. palawanensis* are not too vocal if not disturbed and therefore interpretation of their calls is difficult; however, the chatter call is heard most commonly and is given by dominant animals during aggression. Whistle calls are only emitted following a disturbance or at dusk, and each call lasts for approximately 1 second. There is a pause of several seconds between each whistle call unless another animal answers antiphonally. When another animal does answer, the answer call occurs

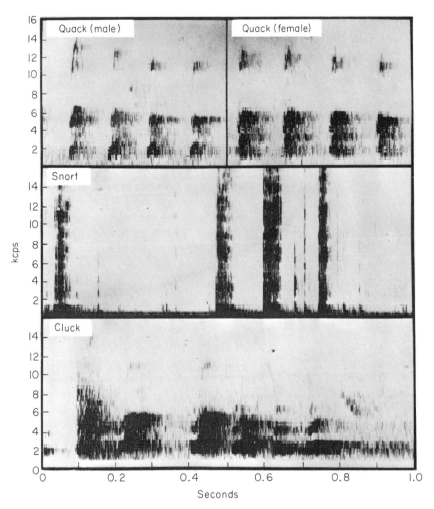

Fig. 4. Sound spectrograms of calls of *T. chinensis*.

at an interval of about 1 msec, or without a time lapse, or the answer call overlaps the first call. Marler (1957), states that bird calls of high frequency and long duration without any sharp beginning or ending are difficult for predators to locate. The whistle calls of members of *T. palawanensis* are very similar to such calls and possibly also confuse predators of tree shrews in the wild.

Mountain tree shrews (*T. montana*) are the most vocal of all species studied in captivity although the number of different calls is less than that found in most other species. *Warning* calls are very pronounced, and after an initial call by one animal up to four or five animals begin calling and continue to call for as long as the disturbance lasts. During these vocalizations, the animals hold their tails upright and forward over their backs and flick their tails back and forth. The calls are forced from the abdomen and the mouth is open. *Threat* and *ejaculation* calls also are given. Ejaculation calls are short, shrill calls emitted at the moment of ejaculation and accompanied by a pronounced forward thrust of the male that often jars the female. Such calls are not part of the vocal behavior of other species and may serve to advertise the presence of a receptive female (Sorenson and Conaway, 1968).

Each species recognizes the discrete warning, threat, and rage calls of other species and frequently the calls of animals from one cage elicit similar calls from animals of another cage. Sometimes the warning calls of members of *T. chinensis* cause all other vocalizations to cease.

J. CONTACTUAL BEHAVIOR

During daily rest periods two to five tree shrews lie one on top of the other in piles. These groups are composed of specific females as a general rule although both sexes and even different species sometimes are present. This piling phenomenon is less frequent when flat rest boards are provided and also declines during periods of increased temperature. As indicated previously, above 85°F all animals except individuals of *T. minor* tend to rest separately and even sprawl with their bodies flattened dorsoventrally and their appendages maximally extended. This posture presumably facilitates evaporative cooling. During cooler periods all tree shrews rest in sunny spots along logs or on window sills. Sometimes the animals puff their fur.

Female-female consorts are common, but male-male pairs seldom rest together although males of *T. gracilis* and *T. longipes* do lie one on another. At night members of most species sleep in pairs and in mixed groups inside various nest boxes. Often five tame individuals of *T. longipes* utilize a single nest box ($4 \times 4 \times 6$ inches) and as many as 12 animals of *T. chinensis* use a single box which measures only $6 \times 8 \times 12$ inches. Individuals of *T. minor* and *T. gracilis* have three basic rest postures: (1) They position themselves head up on high, oblique tree limbs, clasping the branch with their forefeet while their tails extend out behind them or form an S-shaped curve along the branch. This posture is more typical of *T. minor* than of *T. gracilis*. (2) They sit on flat rest boards with their tails curled around their bodies and their heads resting on their tails. All four feet are tucked in close to the body. (3) They sit on flat surfaces with their tails curled up and over their necks, their heads between their forelegs and either lying flat or tucked under their abdomens. This posture is more typical of *T. gracilis* than of *T. minor* and, with the head under the abdomen, is similar to an embryonic position. Sometimes specimens of *T. gracilis* turn around and around in tight circles when wrapping their tails around themselves.

Individuals of *L. tana* rest on flat surfaces with their tails curled around their bodies. Their tails either hide their eyes or lie under their noses as in Fig. 5. Sometimes their noses are tucked into their abdomens. On occasion, the animals lie on their backs with their feet extending up in the air although more normally they sprawl out on their stomachs with their heads resting between their forelegs.

Members of *T. longipes* may also rest on their backs with their feet in the air, generally when several of them are resting in a pile; however, they usually rest on elevated flat boards in a stretched-out posture with their tails hanging limply over the edge and their heads tilted downward. They also rest in a sitting position similar to that of members of *T. gracilis*, with their tails wrapped around their bodies and their heads lying between their forelegs. Sometimes a female assumes a semimount posture on another female and the two animals rest this way for extended periods.

Self-grooming or cleansing occurs after all major feeding periods. All species thoroughly clean their forefeet by licking their palms and digits. They spread their digits and either place them one at a time into their mouths or extend their tongues between each digit. The lower incisors are used to scrape material from the claws. Tree shrews wet their forefeet with their tongues and vigorously rub their muzzles from nose to ear. At the bottom of each stroke, they relick their palms and as their forefeet pass over their eyes, they momentarily blink. At other times, the animals just rub the sides of their mouths along tree branches or hollow logs.

Occasionally, a member of *T. gracilis* or *T. longipes* moves its tail around in front, picks it up with its forefeet, and then licks it until it is very wet. Then the animal grasps the distal end of the tail with the forefeet and rubs it alternatively back and forth along each side of its muzzle as though using a towel.

During regular fur cleansing, an animal holds its tail with its forefeet and licks it along its length. Tail dressing by all species is similar. Often a tree shrew stretches out a rear leg, holds it with its forefoot, and dresses the inner and outer sides with its tongue or lower incisors. The tree shrew's thumb is quite opposable when it holds its rear leg. Individuals of *L. tana*, *T. gracilis*, and *T. longipes* definitely use their lower incisors to comb and scrape their fur in a manner similar to the behavior of members of *T. glis* (Buettner-Janusch and Andrew, 1962).

Mutual grooming between males and females, males and males, and

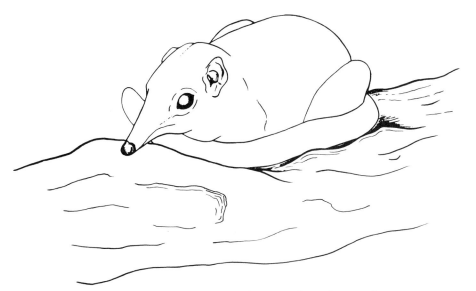

FIG. 5. Rest posture of *L. tana*. Illustration from photograph.

females and females occurs, although it is rarely reciprocal. The most typical posture involves one animal sitting behind another and grasping the latter's rump with its forefeet. The tree shrew licks and combs the dorsal fur of the other until it appears ruffed and wet (Fig. 6). Sometimes a female forces another female to lift her head and then grooms her ventrally along her throat or thrusts her muzzle into the other female's abdomen. Generally the passive animal is groomed and although grooming appears pleasurable, active solicitation does not occur.

Male-female grooming is sex-directed and males of all species except

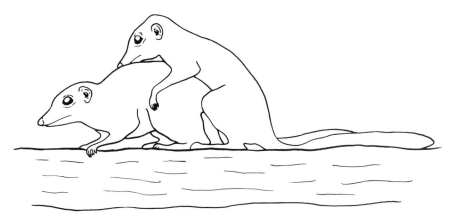

FIG. 6. Grooming posture of *T. longipes*. Illustration from photograph.

T. minor attempt to groom females whenever the females near estrus. Grooming may evolve into bites by the male when he tries to hold the female prior to mounting her.

Often males and females of *T. longipes* rub and slide their bodies against one another. Following this behavior, the females usually race around the cage in an excited manner. Conceivably, such contactual association may replace reciprocal grooming and function as a social bond which may evolve into solicitation among higher primates.

K. Sexual Behavior

Courtship behavior varies among the different species, and copulation does not occur unless females are fully receptive. Males of *T. gracilis* often follow and anal-nuzzle females of other species even though these females are not in estrus; however, the males are more stimulated during female receptivity. Males of *T. gracilis* attempt mounts of both resting and active females of *T. longipes*. The male clasps the rump of the female with his forefeet, and vigorously thrusts back and forth along her dorsal fur (Fig. 7). Sometimes he grooms her flank and bites her ears and forelegs. When a female of another species fails to remain stationary, the male *T. gracilis* jumps onto her back and holds onto her fur with his four feet. In this manner he rides the female along the cage floor for 15–20 ft. Often, *T. gracilis* males interrupt copulation between pairs of *T. longipes* without causing aggression. Interspecies copulation does not occur because of a disparity in size and different reproductive postures. When females of *T. longipes* are in estrus they display pro-

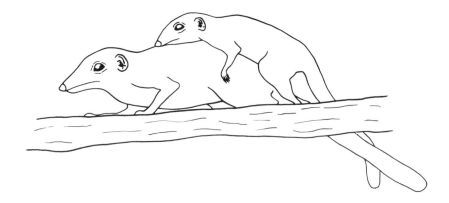

Fig. 7. Interspecies mounting posture of *T. gracilis* and *T. longipes*. Illustration from photograph.

nounced lordosis and the smaller males of *T. gracilis* simply do not achieve intromission.

Tupaia longipes males do not attempt sexual interactions with females of other species. Each day the males anal-nuzzle *T. longipes* females and as these females near estrus the frequency of nuzzling, following, and mounting by the males increases. Sometimes females wiggle and push their rumps into the face of a male or another female and then rapidly turn away. Sexually excited females mount each other or rub their bodies against one another. Such mounts are of short duration and are usually without thrusts. By contrast, female-male mounts nearly always involve pelvic thrusts by the female and sometimes the male shakes his hindquarters or thrashs his tail; more often he remains passive. Anestrous females lie down during attempted mounting by males thereby blocking copulation.

Following of estrous females by males is intense, and the males keep their noses as close as possible to the perineum of the females (Fig. 8). Whenever a female stops, a male either tries to mount her or pushes her rump upward, forcing her into lordosis. After several follows, the male semimounts the female and begins to groom her middorsally. As grooming intensifies, the male bites the female along her nape and shoulder. At this point, if the female fails to display lordosis and accept the male, he becomes aggressive and tries to force copulation. Often the two animals fight momentarily, but nearly always the female suddenly becomes

Fɪɢ. 8. Anal nuzzling of female by male *T. longipes*. Illustration from photograph.

passive and allows intromission by the male. During estrus the female sometimes remains in lordosis in between mounts.

Copulation is ventrodorsal, except for individuals of *L. tana,* with the male standing behind the female and clasping her along her flanks with his forelegs. The male's rear legs are fully extended and he assumes a hunched-over posture. His rear feet are outside of and anterior to the female's rear feet. Sometimes he moves her tail with his foreleg although usually she holds her tail to one side. After intromission, the male of *T. longipes* often uses his forelimbs to "quiver" the female (Fig. 9). A pronounced lordosis by the female is essential for successful copulation. This fact is demonstrated when experienced males attempt copulation with virgin hand-reared females and suggests that sexual competence requires learning as well as innate ability.

During copulation, males make a series of three to four thrusts with 2- to 4-second pauses occurring between each thrust and a longer lapse of 1 or 2 minutes between each series. Intromission time averages 5 minutes for members of *T. longipes,* and following intromissions the male dismounts and both he and the female lick their respective genital areas. Mounts are often disrupted by subordinate males, causing the dominant male to give chase. On rare occasions during these chases, another subordinate male attempts to copulate with the receptive female; however, the dominant male is the most reproductively active, and copulations by subordinates only occur after the dominant male has been satiated. In contrast, among members of *T. montana* two or three males may achieve ejaculation with the same estrous female.

When males or females are changed from one cage to another, sexual interactions increase. Females that fail to cycle in one cage often be-

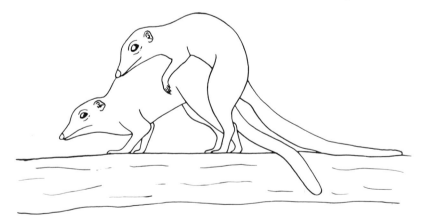

Fig. 9. Mounting posture of *T. longipes.* Illustration from photograph.

came regular breeders in another cage. When animals in one cage are copulating, males in adjacent cages become sexually stimulated. Hand-reared individuals of *T. longipes* become sexually excited when taken from their cage and stroked along their backs. When they are returned to their cage they attempt to mount each other. Such behavior suggests a communication of sexual status by pheromones or other means (Michael and Keverne, 1968).

Hand-reared young of *T. longipes* display the same sequence of repro-ductive patterns as adults, for example, the males start by grooming and biting the females along their napes, then try to mount while holding the females with their teeth, and then become aggressive when the females attempt to run away. When these tame animals are placed in cages containing wild adults, the tame males try to mount nonspecific females.

Lyonogale tana males attempt copulation with all captive females and display a stereotyped courtship dance, female following, and anal nuz-zling. The dance consists of five sequential steps: (1) The male turns around and around in small circles in front of a female with his tail held vertically and flicking; (2) he starts to stamp up and down on all four feet; (3) as stamping increases in intensity, he begins to emit soft quack-ing calls; (4) he suddenly stops dancing and backs up toward the fe-male, placing his rear in her face; (5) he rapidly stands up on his rear legs and tries to mount her. If the female fails to turn around as he stands up, he sometimes mounts her head (Conaway and Sorenson, 1966).

Mounts by males of *L. tana* are different from those of other males in that the animals always stand up on their rear legs and walk bipedal toward the female. The male often performs pelvic thrusts as he walks. In a mount position, he stands immediately behind the female, nearly erect, with his shoulders hunched slightly forward, and his forelegs held limp. His forefeet rarely touch the female (Fig. 10). This posture re-quires extreme passiveness by the female if copulation is to be successful. Such behavior reproductively isolates this species from other groups of tree shrews; however, male *Lyonogale* frequently attempt mounts of both *T. longipes* and *T. chinensis* females and often lick the pudendum of *T. chinensis* females for 2–3 minutes at a time. This results in maximal male stimulation and the males run around the cage trying to mount any available animal.

Males of *T. chinensis* only attempt mounts of specific females. Their courtship is similar to that of males of *T. longipes*, involving lengthy follows and anal nuzzling prior to mounting. In contrast to *T. longipes* females, *T. chinensis* females sometimes appear to solicit copulation.

FIG. 10. Mounting posture of male *L. tana*. Illustration from photograph.

They walk slowly with their rear legs spread apart and their tails held well to one side exposing their vaginal areas. They even back up when males lick their vulval regions. During these periods, their posture and disposition is such that males of *L. tana* almost achieve intromission. Male *T. chinensis* are very aggressive and immediately chase other large males away from receptive females.

Tupaia longipes males do not copulate with *T. chinensis* females although size and behavior are somewhat compatible. Possibly a hormonal rhythm influences this behavior since most sex activity of *T. longipes* occurs during the morning, whereas *T. chinensis* is more active during the late afternoon.

Contrary to the views of Martin (1968), a reproductive cycle exists among tree shrews in captivity. Copulations occur at intervals of 8–16 days with a mode at 11 days. This cycle represents a basic estrous cycle in which either ovulation does not occur or a functional corpus luteum does not develop. Most of these short cycles occur during the spring and fall months when parturitions are minimal. A second cycle occurs at intervals of 18–28 days with a mode at 21 days. This cycle is one of pseudopregnancy and is terminated by a postpseudopregnancy estrus and menstruation, phenomena verified histologically. Additionally, pseudopregnant cycles are demonstrated repeatedly by females maintained with vasectomized males.

Copulations and high-intensity male-female interactions at intervals of 30–40 days denote possible embryo resorptions, and on occasion uterine bleeding is evident.

Births take place between 41 and 50 days in most species and are followed by postpartum heat, ovulation, copulation, and possible conception. Litter sizes range from one to three with two being normal. As expected, individual cycles vary and some females are far more regular than others. Most females display several short estrous cycles in captivity, whereas females better adapted to captivity, such as females of *T. montana*, have fewer short cycles, fewer pseudopregnancies, and more births per individual (Sorenson and Conaway, 1968).

Postpseudopregnancy estrus is as pronounced as regular estrus. The females' behavior is identical with that of short-cycle breeders. On some occasions copulation fails to occur even though interactions are intense for 1 hour or more. Although pseudopregnancy and menstruation are luteal phenomena, external vaginal bleeding is seldom seen. However, an examination of uterine horns removed 48–72 hours after a postpseudo-pregnancy copulation shows pronounced uterine desquamation and blood laking along the pad areas. Moreover, red blood cells and tissue flakes occur within the uterine lumen. Estrogen plus progesterone withdrawal produces identical effects in experimental *T. chinensis* females. Quite possibly, under natural conditions, uterine bleeding is less intense and tree shrews may fail to menstruate although they retain the capacity to do so. The amount of uterine bleeding is always microscopic, similar to that of New World monkeys (Conaway and Sorenson, 1966).

A comparison of all reproductive interactions indicates that members of *T. longipes* copulate throughout the year in captivity but that copulations are less frequent during the summer months. The short estrous cycles are minimal during May, June, and July, and the pseudopregnancy cycles are minimal during May, June, July, and August. This implies that summer copulations are postpartum induced. These data suggest that individuals of *T. longipes* are seasonal breeders with minimal summer copulations balanced by maximal summer births. We expect that under natural conditions, favoring increased environmental stimuli, the seasonal aspect would be enhanced.

L. MATERNAL BEHAVIOR

Although numerous litters are born in captivity, cannibalism or parental desertion usually occurs. In the former case, the offspring are eaten within minutes of birth by either the mother or other adult animals. Frequently, members of *T. chinensis* eat *T. longipes* newborn without any display of protective behavior by the parental species. Sometimes young are cared for over a 2- or 3-day span and then consumed.

In the latter cases, the mothers cleanse and nurse their offspring immediately after birth. Further nursing by females of *T. longipes* takes place twice daily, once in the early morning hours and again toward evening. This behavior is in contrast to that observed by Martin (1968) for members of *T. belangeri*. At all other times the mothers avoid the nest area and do not rest or sleep with their young. Sometimes mothers place their young in nest boxes with their mouths.

In contrast to most females of *T. longipes*, females of *L. tana* usually rear their young in captivity. On the day before birth, the mother makes numerous trips in and out of a ground-level nest box, constantly arranging the nest area with cloth strips and wood shavings. Sometimes she remains inside the box for 1 hour or more at a time. Although males frequently enter the nest area, they do not aid in nest preparation. Immediately following parturition, the young are thoroughly cleansed of blood and abundantly nursed. During the next 7 days, the young are nursed every evening and every other morning. Following this period they are nursed twice daily. The mother does not enter the nest box during the daytime and sleeps inside an adjacent nest box at night; however, the mother and other adults often look into the nest box. On the eighth day after birth, the young display the characteristic dorsal stripe of their species, and by day 16 their eye sutures begin to split and their eyes open.

During this time the mother remains secretive and stays inside hollow logs for hours at a time. She avoids adult males whenever possible. Feces are never seen inside the nest box; either the young eat them or their mother removes them. The offspring nurse for about 27 days and by 30 days they leave the nest area for the first time. They tend to remain close to their mother, who is now more bold and participates in normal daily activities.

By 5 weeks of age, the young display typical adult behavior. They frequently dress themselves, eat a variety of fresh fruits, chase each other, nuzzle their mother, and even rest lying against adult males. With the exception of an occasional threat call by the males, the young are tolerated. The offspring are quite agile when climbing but very clumsy while descending. Overall, their arboreal abilities are more advanced than those of hand-reared individuals of *T. longipes* at a comparable age. Also, juvenile *Lyonogale* appear to grow faster than do hand-reared members of *T. longipes*.

Adult males attempt to mount juvenile females of *L. tana* when the latter are about 50 days old but most mounts are repelled by use of threat calls. During the following days, male sexual activity intensifies and males begin to display courtship dancing. It is during this period that the mother *L. tana* actively defends her young against male advances.

VIII. DISCUSSION

An attempt to relate the data in the preceding review in a manner that clarifies the phylogenetic position of the tree shrews is more of a task than meets the eye. There are always problems inherent in the laboratory maintenance of wild species, namely, providing a habitat that simulates natural conditions, assuring an adequate diet, and fulfilling the psychobiological requirements of the species. We contend, however, that when conditions of captivity are suitable behavioral displays are valid expressions of the basic wild patterns. Captive animals brought into the laboratory are forced into social groups in which a reconstruction of their behavioral patterns is readily observable. A comparison of behavioral traits of eight species of tree shrews in captivity is given in Table I. Whether these views are realistic will be shown only by field observations and, so far, only an initial study of tree shrews in the wild has been made by Morris *et al.* (1967).

Therefore we are left with the choice of either supporting or refuting the primate affinities of the Tupaiidae by using evidence gained from studies of animals in captivity and from data gleaned from the literature. Since our position during the past few years has not changed, although it has been dampened somewhat, we shall continue to believe that tree shrews are indeed the bottom rung of the primate ladder. At the same

TABLE I

A COMPARISON OF BEHAVIORAL TRAITS OF EIGHT SPECIES
OF TREE SHREWS IN CAPTIVITY[a]

Genus and species	Most active	Most arboreal	Most vocal	Most aggressive	Most complex social structure	Most adaptable to the laboratory
Tupaia palawanensis	5	4	4	4	4	5
Tupaia glis	3	3	2	1	4	4
Tupaia chinensis	4	4	2	1	4	3
Tupaia montana	1	3	1	5	2	1
Tupaia longipes	1	2	1	2	2	2
Tupaia gracilis	2	1	3	3	3	4
Tupaia minor	3	1	4	4	5	4
Lyonogale tana	5	5	5	5	1	3

[a] Numbers indicate ranking.

time, we again plead for a series of sound observational studies of tree shrews under natural conditions.

Our reasons for supporting the Tupaiidae in the suborder Prosimii are open to question and undoubtedly other investigators have criteria just as valid for removal of the tree shrews from the primate stock. We are certain that tree shrews are in no way interested in these semantics and that they will continue to prosper and initiate new problems of affinity for years to come; however, we see no reason other than timidness to remain on the fence and avoid the issue by placing the tupaiids in a separate order by themselves.

The fact that tree shrews display remarkable visual acuity with a reduction in olfaction is indicative of primate evolution and is certainly not typical of the Insectivora. The generalized, pure cone retina of *Tupaia* and the ability of the tree shrews to discriminate red in the long wavelength of the spectrum is also significantly different from the corresponding traits of insectivores. Although the geniculostriate system of *Tupaia* is far more complex than the comparable system in the hedgehog, the fact remains that the retinal projections to the various laminae in the tree shrew are opposite those of primates. Perusal of the literature therefore suggests that opinions on tupaiid-primate affinities depend on those structures one wishes to evaluate. In this respect, we find it difficult to discount the excellent anatomical evidence put forth in several studies by Le Gros Clark (1924a,b, 1925, 1926, 1932, 1960). Certainly he is not an amateur and we are surprised at the number of scientists who apparently have not or do not recognize his abilities. Few, if any, researchers have made a more thorough study of the overall anatomy of *Tupaia*, and it is the complete picture that is required.

Evidence for and against primate affinities is also available from studies of the reproductive morphology and behavior of tree shrews. Martin's (1968) suggestion that tree shrews may have delayed implantation, unknown in primates, seems to us to be naive and lacks any experimental support. On the contrary, known-age embryos indicate a normal 41 to 50-day gestation period as reported by several investigators. The maternal "absentee system" reported by Martin (1966) for members of *T. belangeri* is not only different from that of primates but also from that of several other species of tree shrews. Again, a single criterion or the study of a single species of animals rarely is of phylogenetic significance.

Whereas Luckett (1969) and Hill (1965) state that there is no valid reason for considering tupaiids to be related to primates based on evidence of placentation and fetal membranes, Alcala and Conaway (1968) point out that certain species of *Tupaia* have a vagina masculinus similar to that of primates, and Sorenson (1964) and Conaway and Sorenson

(1966) have shown how the menstrual cycle of primates may have evolved from the more generalized reproductive cycle of the tree shrews.

Studies by Goodman (1966), Moore and Goodman (1968), and Hafleigh and Williams (1966) using immunotaxonomy tend to place *Tupaia* among the prosimians. Johnson (1968) reports that the migration of hemoglobin of insectivores is cathodal whereas it is anodal in *Tupaia* and primates. Although only a single molecule is considered, and it is obvious that several molecules need examination, the anodal migration of hemoglobin seems to us to be a basic difference of phyletic importance.

Investigations by Hopkins (1949) and Dunn (1966) consider phylogenetic and ecological interpretations of host-parasite associations and point out that parasites that infest closely related animals show taxonomic relationships, that is, *Docophthirus* of the tree shrews is related to *Lemurphthirus* characteristic of African lemurs.

Nearly all studies of the behavior of tree shrews point out the fact that the tupaiids have a more complex social and reproductive structure than do the insectivores with which they are compared. If tree shrews are indeed insectivores, then at least one group of insectivores has not only developed color vision but has also developed an advanced social order and a reproductive cycle approaching the primate plan.

The entire question would be easily settled if a fossil series were available; however, as we know, there is no such link between tree shrews and primates. In fact, there is only a meager scattering of fossiliferous deposits from the time primates emerged in the Paleocene. This plight leads to comparisons between living forms and those few documented fossils—a hazardous path to explore. Recent studies of the ontogenesis of the bulla of *Tupaia* by McKenna (1963), Van Valen (1965), and von Spatz (1964, 1966) are not in agreement and raise the question whether or not single structures can assume different forms during phylogeny. There is also the possibility that many of the tupaiid-primate structures are acquired by means of convergence and, just as surely, as Buettner-Janusch (1963) points out, most tupaiids are island forms and therefore are more likely to have undergone divergence rather than convergence.

It is again apparent that if the criteria are cleverly chosen the tree shrews can be assigned to one of several phyletic positions. We wish that we were able beyond doubt to classify the tree shrews among the primates but in all honesty we cannot do so. Tree shrews are, and will remain, a transitional group in the eyes of most taxonomists, and nothing is really gained by shifting the tupaiids in and out of various orders or by placing them into an order of their own. Romer (1967) states our view best when he declares that if one looks for living forms closest to the primitive stock of perhaps all mammals one chooses the tree shrews—and we suspect that Romer, like Le Gros Clark, knows more about phylogeny than most people are willing to admit.

REFERENCES

Alcala, J. R., and Conaway, C. H. (1968). The gross and microscopic anatomy of the uterus masculinus of tree shrews. *Folia Primatol.* 9, 216–245.

Andy, O. J., and Stephan, H. (1966). Primate septum in phylogeny. *Anat. Rec.* 154, 310.

Arden, G. B., and Silver, P. H. (1962). Visual thresholds and spectral sensitivities of the grey squirrel (*Sciurus carolinensis leucotis*). *J. Physiol.* (*London*) 163, 540–557.

Arden, G. B., and Tansley, K. (1955). The spectral sensitivity of the pure-cone retina of the souslik (*Citellus citellus*). *J. Physiol.* (*London*) 130, 225–232.

Arrighi, F. E., Sorenson, M. W., and Shirley, L. R. (1969). Chromosomes of the tree shrews (Tupaiidae). *Cytogenetics* 8, 199–208.

Autrum, H., and von Holst, D. (1968). Sozialer "Stress" bei Tupajas (*Tupaia glis*) und seine Wirkung auf Wachstum, Körpergewicht und Fortpflanzung. *Z. vergl. Physiol.* 58, 347–355.

Barnicot, N. A., Jolly, C. J., and Wade, P. T. (1967). Protein variations and primatology. *Amer. J. Phys. Anthropol.* 27, 343–356.

Bauchot, R., and Stephan, H. (1966). Données nouvelles sur l'encephalization des insectivores et des prosimians. *Mammalia* 30, 160–196.

Borgaonkar, D. S. (1967). Additions to the lists of chromosome numbers in the insectivores and primates. *J. Hered.* 58, 211–213.

Braun, H., and Kloft, W. (1965). Hämatologische Untersuchungen am Spitzhornchen *Tupaia tana* (Raffles 1821) (Tupaiidae, Prosimiae). *Experientia* 21, 663–664.

Buettner-Janusch, J. (1963). An introduction to the primates. *In* "Evolutionary and Genetic Biology of Primates," (J. Buettner-Janusch, ed.), Vol. 1, pp. 1–64. Academic Press, New York.

Buettner-Janusch, J., and Andrew, R. J. (1962). The use of the incisors by primates in grooming. *Amer. J. Phys. Anthropol.* 20, 127–129.

Cain, R. E., and Carlson, R. H. (1968). Evidence for color vision in the prairie dog (*Cynomys ludovicianus*). *Psychonomic Sci.* 13, 185–186.

Campbell, C. B. G. (1966a). Taxonomic status of tree shrews. *Science* 153, 436.

Campbell, C. B. G. (1966b). The relationships of the tree shrews: the evidence of the nervous system. *Evolution* 20, 276–281.

Campbell, C. B. G., Yashon, D., and Jane, J. A. (1967). The retinal projections of the tree shrew and hedgehog. *Brain Res.* 5, 406–418.

Cantor, T. (1846). Catalogue of Mammalia inhabiting the Malayan Peninsula and Islands. *J. Asiat. Soc. Bengal* 15, 189.

Carlsson, A. (1922). Ueber die Tupaiidae und ihre Beziehungen zu den Insectivora und den Prosimiae. *Acta Zool.* (*Stockholm*) 3, 227–270.

Chaffee, R. R. J., Sorenson, M. W., and Conaway, C. H. (1968). Studies on the effects of cold on brown fat in two species of insectivores compared with *Tupaia chinensis*. *J. Amer. Oil Chem. Soc.* 45, 111.

Chaffee, R. R. J., Kaufman, W. C., Kratochvil, C. H., Sorenson, M. W., Conaway, C. H., and Middleton, C. C. (1969). Comparative chemical thermoregulation in cold- and heat-acclimated rodents, insectivores, protoprimates, and primates. *Fed. Proc. Fed. Amer. Soc. Exp. Biol.* **28,** 1029–1034.

Chance, M. R. A. (1962). Social behaviour and primate evolution. *In* "Culture and the Evolution of Man" (M. F. Ashley Montagu, ed.), pp. 84–130. Oxford Univ. Press, New York.

Chapman, H. C. (1904). Observations on *Tupaia* with reflection on the origin of Primates. *Proc. Nat. Acad. Sci. U. S.* **56,** 148–156.

Chu, E. H. Y., and Bender, M. R. (1961). Chromosome cytology and evolution in primates. *Science* **133,** 1399–1405.

Conaway, C. H., and Sorenson, M. W. (1966). Reproduction in tree shrews. *In* "Comparative Biology of Reproduction in Mammals" (I. W. Rowlands, ed.), Symp. Zool. Soc. London No. 15, pp. 471–492. Academic Press, New York.

Crook, J. H. (1967). Evolutionary changes in primate societies. *Sci. J.* **3,** 66–72.

Crook, J. H., and Gartlan, J. S. (1966). Evolution of primate societies. *Nature* (*London*) **210,** 1200–1203.

Davis, D. D. (1938). Notes on the anatomy of the tree shrew *Dendrogale*. *Fieldiana* **20,** 383–404.

DeValois, R. L., and Jacobs, G. H. (1968). Primate color vision. *Science* **162,** 533–540.

Duke-Elder, S. (1958). "The Eye in Evolution." Kimpton, London.

Dunn, F. L. (1966). Patterns of parasitism in primates: Phylogenetic and ecological interpretations, with particular reference to the Hominoidea. *Folia Primatol.* **4,** 329–345.

Egozcue, J. (1967). Position of the centromere in the marked acrocentric chromosomes of Primates. *Folia Primatol.* **7,** 238–242.

Egozcue, J., Chiarelli, B., Sarti-Chiarelli, M., and Hagemenas, F. (1968). Chromosome polymorphism in the tree shrew (*Tupaia glis*). *Folia Primatol.* **8,** 150–158.

Elliot, O., and Wong, M. (1969). Blood sugar of Malayan tree shrews. *J. Mammal.* **50,** 361–362.

Elliot, O., Yess, N. J., and Hegsted, D. M. (1966). Biosynthesis of ascorbic acid in the tree shrew and slow loris. *Nature* (*London*) **212,** 739–740.

Evans, F. G. (1942). The osteology and relationships of the elephant shrews (Macroscelididae). *Bull. Amer. Mus. Natur. Hist.* **80,** 83–125.

Garrod, A. H. (1879). Notes on the visceral anatomy of the *Tupaia* of Burma (*T. belangeri*). *Proc. Zool. Soc. London* pp. 301–305.

Gartlan, J. S. (1968). Structure and function in primate society. *Folia Primatol.* **8,** 89–120.

Glickstein, M. (1967). Laminar structure of the dorsal lateral geniculate nucleus in the tree shrew (*Tupaia glis*). *J. Comp. Neurol.* **131,** 93–102.

Goodman, M. (1966). Phyletic position of tree shrews. *Science* **153,** 1550.

Goodman, M. (1967). Deciphering primate phylogeny from macromolecular specificities. *Amer. J. Phys. Anthropol.* **26,** 255–275.

Gregory, W. K. (1910). The orders of mammals. *Bull. Amer. Mus. Natur. Hist.* **27,** 268–285, 321–322.

Gregory, W. K. (1913). Relationship of the Tupaiidae and of Eocene lemurs, especially *Notharctus*. *Geol. Soc. Amer. Bull.* **24,** 247–252.

Hafleigh, A. S., and Williams, C. A. (1966). Antigenic correspondence of serum albumins among the primates. *Science* **151,** 1530–1535.

Hasler, J. F. (1969). Behavior of *Tupaia chinensis* in captivity. M.A. Thesis, Univ. of Missouri, Columbia, Missouri.

Henckel, K. O. (1928). Das Primordialcranium von *Tupaja* und der Ursprung der Primaten. *Z. Anat. Entwicklungsgesch.* **84,** 204–227.

Hendrickson, J. R. (1954). Breeding of the tree shrew. *Nature* (*London*) **174,** 794–795.

van Herwerden, M. A. (1906). Die puerperalen Vorgänge in der Mucosa Uteri von *Tupaia javanica. Ergeb. Anat. Entwicklungsgesch.* **32,** 157–169.

Hill, J. P. (1965). On the placentation of *Tupaia. J. Zool.* **146,** 278–304.

Holloway, R. L. (1968). The evolution of the primate brain: Some aspects of quantitative relations. *Brain Res.* **7,** 121–172.

Hopkins, G. H. E. (1949). Host associations of the lice of mammals. *Proc. Zool. Soc. London* **119,** 387–604.

Hsu, T. C., and Johnson, M. L. (1963). Karyotypes of two mammals from Malaya. *Amer. Natur.* **97,** 127–129.

Hubrecht, A. A. W. (1899). Über die Entwicklung der Placenta von *Tarsius* und *Tupaia*, nebst Bemerkungen über deren Bedeutung als haemopoeitisches Organ. *Proc. 4th Int. Congr. Zool., Cambridge, Eng.* pp. 345–411.

Hubrecht, A. A. W. (1908). Early ontogenic phenomena in mammals and their bearing on our interpretation of the phylogeny of vertebrates. *Quart. J. Microsc. Sci.* **53,** 1–181.

Hudgins, P. C., Whorton, C. M., Tomoyoshi, T., and Riopelle, A. J. (1966). Comparison of the molecular structure of myoglobin of fourteen primate species. *Nature* (*London*) **212,** 693–695.

Hunt, R. D., and Chalifoux, L. (1967). The hemogram of the tree shrew (*Tupaia glis*). *Folia Primatol.* **7,** 34–36.

Huxley, T. H. (1872). "A Manual of the Anatomy of Vertebrated Animals." Appleton, New York.

Jane, J. A., Campbell, C. B. G., and Yashon, D. (1965). Pyramidal tract: A comparison of two prosimian primates. *Science* **147,** 153–155.

Johnson, M. L. (1968). Application of blood protein electrophoretic studies to problems in mammalian taxonomy. *Syst. Zool.* **17,** 23–30.

Kaufmann, J. H. (1965). Studies on the behavior of captive tree shrews (*Tupaia glis*). *Folia Primatol.* **3,** 50–74.

Klinger, H. P. (1963). The somatic chromosomes of some primates (*Tupaia glis, Nycticebus coucang, Tarsius bancanus, Cercocebus aterrimus, Symphalangus syndactylus*). *Cytogenetics* **2,** 140–151.

Layton, L. (1965). Passive transfer of human atopic allergies. *J. Allergy* **36,** 523–531.

Le Gros Clark, W. E. (1924a). The myology of the tree shrew (*Tupaia minor*). *Proc. Zool. Soc. London* pp. 461–497.

Le Gros Clark, W. E. (1924b). On the brain of the tree shrew (*Tupaia minor*). *Proc. Zool. Soc. London* pp. 1053–1074.

Le Gros Clark, W. E. (1925). On the skull of *Tupaia. Proc. Zool. Soc. London* pp. 559–567.

Le Gros Clark, W. E. (1926). On the anatomy of the pen-tailed tree shrew (*Ptilocercus lowii*). *Proc. Zool. Soc. London* pp. 1179–1309.

Le Gros Clark, W. E. (1932). The brain of the Insectivora. *Proc. Zool. Soc. London* pp. 975–1013.

Le Gros Clark, W. E. (1960). "The Antecedents of Man." Quadrangle Books, Chicago, Illinois.

Leonard, C., Schneider, G. E., and Gross, C. G. (1966). Performance on learning set and delayed-response tasks by tree shrews (*Tupaia glis*). *J. Comp. Physiol. Psychol.* **62,** 501–504.

Lightoller, G. S. (1934). The facial musculature of some lesser primates and *Tupaia*. *Proc. Zool. Soc. London* 2, 259–309.

Lorenz, G. F. (1927). Ueber Ontogenese und Phylogenese der Tupajahand. *Morphol. Jahrb.* 58, 431–439.

Luckett, W. P. (1966). The ovarian cycle of the tree shrews (Family Tupaiidae), with reference to the phylogenetic relationship of the tupaiids, *Amer. Zool.* 6, 574.

Luckett, W. P. (1969). Evidence for the phylogenetic relationships of tree shrews (Family Tupaiidae) based on the placenta and foetal membranes. *J. Reprod. Fert. Suppl.* 6, 419–433.

Lyon, M. W. (1913). Treeshrews: An account of the mammalian family Tupaiidae. *Proc. U. S. Nat. Mus.* 45, 1–188.

Machida, H., and Giacometti, L. (1967). The anatomical and histochemical properties of the skin of the external genitalia of the primates (*Tupaia, Galago, Lemur, Tarsius, Lagothrix, Cercopithecus, Gorilla*, man). *Folia Primatol.* 6, 48–69.

McKenna, M. C. (1963). New evidence against tupaioid affinities of the mammalian family Anagalidae. *Amer. Mus. Nov.* 2158, 1–16.

McKenna, M. C. (1966). Paleontology and the origin of the primates. *Folia Primatol.* 4, 1–25.

Manley, G. H. (1966). Prosimians as laboratory animals. *Symp. Zool. Soc. London* 17, 11–39.

Marler, P. (1957). Specific distinctiveness in the communication signals of birds. *Behaviour* 11, 13–29.

Martin, R. D. (1966). Tree shrews: unique reproductive mechanism of systematic importance. *Science* 152, 1402–1404.

Martin, R. D. (1968). Reproduction and ontogeny in tree-shrews (*Tupaia belangeri*), with reference to their general behavior and taxonomic relationships. *Z. Tierpsychol.* 4, 409–529.

Martin, R. D. (1969). The evolution of reproductive mechanisms in primates. *J. Reprod. Fert. Suppl.* 6, 49–66.

Meister, W., and Davis, D. D. (1956). Placentation of the pigmy tree shrew (*Tupaia minor*). *Fieldiana* 35, 71–84.

Meister, W., and Davis, D. D. (1958). Placentation of the terrestrial tree shrew (*Tupaia tana*). *Anat. Rec.* 132, 541–546.

Meyer, D. R., and Anderson, R. A. (1965). Colour discrimination in cats. *In* "Colour Vision—Physiology and Experimental Psychology, Ciba Foundation Symp." pp. 325–339. Little, Brown, Boston, Massachusetts.

Michael, R. P., and Keverne, E. B. (1968). Pheromones in the communication of sexual status in primates. *Nature* (*London*) 218, 746–749.

Montagna, W., Yun, J. S., Silver, A. F., and Quevedo, W. C. (1962). The skin of primates. XIII. The skin of the tree shrew (*Tupaia glis*). *Amer. J. Phys. Anthropol.* 20, 431–439.

Moore, W. G., and Goodman, M. (1968). A set theoretical approach to immunotaxonomy: Analysis of species comparisons in modified ouchterlony plates. *Bull. Math. Biophys.* 30, 279–289.

Morris, J. H., Negus, N. C., and Spertzel, R. O. (1967). Colonization of the tree shrew (*Tupaia glis*). *Lab. Anim. Care* 17, 514–520.

Mossman, H. W. (1953). The genital system and the fetal membranes as criteria for mammalian phylogeny and taxonomy. *J. Mammal.* 34, 289–298.

Myers, K. (1966). The effects of density on sociality and health in mammals. *Proc. Ecol. Soc. Aust.* 1, 40–64.

Nelson, L. E., and Asling, C. W. (1962). Metabolic rate of tree-shrews (*Urogale everetti*). *Proc. Soc. Exp. Biol. Med.* **109**, 602–604.

Patterson, B. (1955). The geologic history of non-humanoid primates in the old world. *In* "The Non-human Primates and Human Evolution," pp. 13–31. Wayne Univ. Press, Detroit, Michigan.

Polyak, S. (1957). "The Vertebrate Visual System." Univ. of Chicago Press, Chicago, Illinois.

Rabb, G. B., Getty, R. E., Williamson, W. M., and Lombard, L. S. (1966). Spontaneous diabetes mellitus in tree shrews, *Urogale everetti*. *Diabetes* **15**, 327–330.

Rohen, J. W. (1964). Neuere Ergebnisse einer funktionellen Anatomie des Sehorgans unter besonderer Berücksichtigung der Elektronenmikroskopie. *Arch. Biol. Suppl.* **75**, 975–1001.

Romer, A. S. (1966). "Vertebrate Paleonotology." Univ. of Chicago Press, Chicago, Illinois.

Romer, A. S. (1967). Major steps in vertebrate evolution. *Science* **158**, 1629–1637.

Roux, G. H. (1947). The cranial development of certain Ethiopian "insectivores," and its bearing on the mutual affinities of the group. *Acta Zool.* (*Stockholm*) **28**, 165–397.

Samorajski, T., Ordy, J. M., and Keefe, J. R. (1966). Structural organization of the retina in the tree shrew (*Tupaia glis*). *J. Cell Biol.* **28**, 489–504.

Sarich,, V. M., and Wilson, A. C. (1967). Immunological time scale for hominid evolution. *Science* **158**, 1200–1203.

Shriver, J. E., and Noback, C. R. (1967a). Color vision in the tree shrew (*Tupaia glis*). *Folia Primatol.* **6**, 161–169.

Shriver, J. E., and Noback, C. R. (1967b). Cortical projections to the lower brain stem and spinal cord in the tree shrew (*Tupaia glis*). *J. Comp. Neurol.* **130**, 25–54.

Simpson, G. G. (1931). A new insectivore from the Oligocene, Ulan Gochu horizon, of Mongolia. *Amer. Mus. Nov.* **505**, 1–22.

Snedigar, R. (1949). Breeding of the Philippine tree shrew, *Urogale everetti* Thomas. *J. Mammal.* **30**, 194–195.

Snyder, M. (1968). Neuroanatomical and behavioral studies of vision in the tree shrew, *Tupaia glis*. *Diss. Abstr. B* **29**, 796–797.

Snyder, M., and Diamond, I. T. (1968). The organization and function of the visual cortex in the tree shrew. *Brain, Behav. Evol.* **1**, 244–288.

Sorenson, M. W. (1964). The behavior of tree shrews in captivity. Ph.D. Thesis, Univ. of Missouri, Columbia, Missouri.

Sorenson, M. W., and Conaway, C. H. (1964). Observations of tree shrews in capitvity. *Sabah Soc. J.* **2**, 77–91.

Sorenson, M. W., and Conaway, C. H. (1966). Observations on the social behavior of tree shrews in captivity. *Folia Primatol.* **4**, 124–145.

Sorenson, M. W., and Conaway, C. H. (1968). The social and reproductive behavior of *Tupaia montana* in captivity. *J. Mammal.* **49**, 502–512.

Sprankel, H. (1961). Über Verhaltensweisen und Zucht von *Tupaia glis* Diard 1820 in Gefangenschaft. *Z. Wiss. Zool., Abt. A* **165**, 186–220.

Steiner, H. (1965). Die vergleichend-anatomische und oekologische Bedeutung bei rudimentare Anlage eines selbständigen fünften Carpale bei *Tupaia*. Betrachtugen zum Homologieproblem. *Israel J. Zool.* **14**, 221–233.

Stephan, H. (1965). Der Bulbus olfactorius accessorius bei Insektivoren und Primaten. *Acta Anat.* **62**, 215–253.

Stratz, K. H. (1898). "Der geschlechtsreife Saugethiereierstock." Den Haag, Holland.

Straus, W. L. (1949). The riddle of man's ancestry. *Quart. Rev. Biol.* **24**, 200–223.

Szalay, F. S. (1968). The beginnings of primates. *Evolution* **22**, 19–36.

Tansley, K., Copenhaver, R. M., and Gunkel, R. D. (1961). Spectral sensitivity curves of diurnal squirrels. *Vision Res.* **1**, 154–165.

Thompson, P. J. (1969). Behavior of tree shrews (*Tupaia palawanensis*) in captivity. M.A. Thesis, Univ. of Missouri, Columbia, Missouri.

Tigges, J. (1963). Untersuchungen über den Farbensinn von *Tupaia glis* Diard 1820. *Z. Anthropol. Morphol.* **53**, 109–123.

Tigges, J. (1964). On visual learning capacity, retention and memory in *Tupaia glis* Diard 1820. *Folia Primatol.* **2**, 232–245.

Tigges, J. (1966). Ein experimenteller Beitrag zum subkortikalen optischen System von *Tupaia glis. Folia Primatol.* **4**, 103–123.

Tigges, J., Brooks, B. A., and Klee, M. R. (1967). ERG recordings of a primate pure cone retina (*Tupaia glis*). *Vision Res.* **7**, 553–563.

Tinbergen, N. (1951). "The Study of Instinct." Oxford Univ. Press (Clarendon), London and New York.

Tinbergen, N. (1953). "Social Behaviour in Animals." Methuen, London.

Vandenbergh, J. G. (1963). Feeding, activity and social behavior of the tree shrew, *Tupaia glis*, in a large outdoor enclosure. *Folia Primatol.* **1**, 199–207.

van der Horst, C. J. (1949). The placentation of *Tupaia javanica. Proc. Kon. Ned. Akad. Wetensch.* **52**, 1205–1213.

van der Horst, C. J. (1954). *Elephantulus* going into anoestrus; menstruation and abortion. *Phil. Trans. Roy. Soc. London, Ser. B* **238**, 27–61.

Van Valen, L. (1965). Treeshrews, primates and fossils. *Evolution* **19**, 137–151.

Van Valen, L., and Sloan, R. E. (1965). The earliest primates. *Science* **150**, 743–745.

Verhaart, W. J. C. (1966). The pyramidal tract of *Tupaia*, compared to that in other primates. *J. Comp. Neurol.* **126**, 43–50.

Verma, K. (1965). Notes on the biology and anatomy of the Indian tree-shrew, *Anathana wroughtoni. Mammalia* **29**, 289–330.

von Spatz, W. B. (1964). Beitrag zur Kenntnis der Ontogenese der cranium von *Tupaia glis* Diard 1820. *Morphol. Jahrb.* **106**, 321–416.

von Spatz, W. B. (1966). Zur Ontogenese der Bulla tympanica von *Tupaia glis* Diard 1820. *Folia Primatol.* **4**, 26–50.

von Spatz, W. B. (1967). Die Ontogenese der Cartilago Meckeli und der Symphysis Mandibulair bei *Tupaia glis* (Diard, 1820). *Folia Primatol.* **6**, 180–203.

Walls, G. L. (1942). "The Vertebrate Eye." Cranbrook Press, Bloomfield Hills, New Jersey.

Wharton, C. H. (1950). Notes on the Philippine tree-shrew *Urogale everetti* Thomas. *J. Mammal.* **31**, 352–354.

Williams, H. W., Sorenson, M. W., and Thompson, P. J. (1969). Antiphonal calling of the tree shrew *Tupaia palawanensis. Folia Primatol.* **11**, 200–205.

Wolin, L. R., and Massopust, L. C. (1967). Characteristics of the ocular fundus in primates. *J. Anat.* **101**, 693–699.

Wood-Jones, F. (1917). The genitalia of *Tupaia. J. Anat.* **51**, 118–126.

Zuckerman, S., and Parkes, A. S. (1935). Observations on the structure of the uterus masculinus in various primates. *J. Anat.* **69**, 484–496.

Abnormal Behavior in Primates*

G. MITCHELL

Department of Behavioral Biology and Department of Psychology
University of California, Davis, California

* Supported by National Institute of Health Grant Numbers FR00169, HD04335-01, and MH17425-01.

I. INTRODUCTION

A. THE RESEARCH AREA

Since the well-publicized research by Harlow and Zimmerman (1959) on the development of affectional responses in the rhesus monkey, cloth and wire surrogate mothers have become well known to almost every behaviorist and to many people outside the behavioral area, particularly those in primate research. Although outstanding research on primate psychological, social, and emotional development had been reported prior to the appearance of Harlow's classic project, none of it had captured the fancy of both the public and the scientific communities as much as the Harlow research. Harlow's key word was *love,* and no word has ever been more contemporary or captivating. The writer loves love, but this chapter is a rather rigid review of the scientific literature on abnormal behavior in primates and from this point on the reader should be prepared to tolerate the little tedium that is inevitable in comprehensive reviews.

Recently, there have been several reviews of portions of the research area covered here. Berkson (1967) has written a comprehensive chapter on abnormal stereotyped motor acts which reviews an area very basic to the present chapter. Anyone interested in abnormal behavior in primates cannot ignore it. Harlow and Harlow (1969) and Sackett (1968) have written reviews on the effects of differential early rearing in the Wisconsin laboratories, and William A. Mason (1968) has recently described some more general trends in primate behavioral development. Mason has also presented some of his own recent findings, placed them in a comparative perspective, and considered their implications for human behavior. The present chapter draws heavily from the efforts of the Harlows, Berkson, Sackett and, especially, Mason. It is intended to be a comprehensive review of the area but not an exhaustive one. For example, many kinds of abnormalities reported in field studies have been omitted.

B. Definitions

Sackett (1968) has pointed out some problems that arise in discussing abnormal behavior. One of these concerns the definition of normal behavior, and the other the persistence of the abnormality. This chapter employs the term "abnormal" in the same manner as Sackett, and we quote him directly:

> The term abnormal will be applied here only with respect to (a) a particular control condition as a reference—and not necessarily a "normal control" group—and (b) a test situation that has ecological validity. It seems reasonable to collect normative data on animals reared in any condition and to compare the quality and quantity of their behavior with that of animals reared under different conditions. This approach yields statements about the degree to which a particular set of rearing variables produces, more or less, some type of behavior (which is different) from the arbitrary normative condition. The terms "normal" or "abnormal" might then be applied or not applied, depending on the views of the investigator.
>
> Ecological validity implies that questions of behavioral adequacy depend on the test environment. For example, feral rearing can be a poor control for assessing some behavior of laboratory animals because many types of feral behavior in the laboratory many be as inappropriate as would be that of laboratory-born monkeys transferred to the feral environment. It seems likely that many behavioral "abnormalities" are of this type, i.e., dependent on processes in the environment rather than on permanent anomalies within the animal itself.
>
> Another issue in studying the development of abnormal behavior involves permanence of response modification. Many researchers have measured effects of early experiences an early behavior, drawing inappropriate conclusions about long-term effects because the subjects were not tested as adults. It seems important to distinguish between variables producing only transient effects early in life and those producing pervasive and long-lasting behavioral deficiencies. (Sackett, 1968, p. 305.)

In addition to reference control groups, ecological validity, and persistence of the "abnormality," several other factors related to treatment and testing should be carefully scrutinized. In the case of abnormalities produced by differential early experiences, the following treatment factors can be very important: (1) the age of the animal when the treatment experience is given, (2) the age of the animal at the time of post-treatment testing, (3) the duration or quantity of the treatment experience, (4) the type or quality of the treatment experience, (5) the relation of the treatment experience to the genetic background of the animal, and (6) the sex of the animal (cf. King, 1958).

Factors related to post-treatment testing are also important determinants of abnormality relative to a control group. Post-treatment testing factors include: (1) differences in the type of general nonsocial environment employed by testing (e.g., novel versus familiar; large versus small), (2) differences in the length of the test sessions (e.g., the longer the test sessions the more likely it is that affiliative behaviors will be seen), (3) the type of scoring system utilized, (4) reliability of observers or experimenters, (5) type of data employed (e.g., frequency, duration, latency, and so on), (6) number of social partners provided, (7) species of social partners provided, (8) age and sex of social partners and, (9) familiarity of social partners. All of these post-treatment testing factors have measurable effects on the behaviors of both experimental and control animals. In some cases there may be differential effects which can produce artifactual differences between groups.

II. CAPTIVITY

A. ABSENCE OF STIMULATION

Before one can understand the effects of adverse early experiences on the behaviors of nonhuman primates, one must have an appreciation of the general effects of captivity on behavior. Changing the environment from one that allows considerable freedom of movement and variety of environmental exposure to one that restricts movement and exposure of itself can produce abnormalities in the most ontogenetically and phylogenetically advanced individuals. Many extreme examples of such a change are seen in the effects produced by social and sensory deprivation in adult humans (Wheaton, 1959). While the effects of captivity upon nonhuman primates may not be as extreme as those reported for human sensory deprivation, there are nevertheless many symptoms related to restrictions of sensory and motor function in primate captivity. Boredom is often mentioned as a possible cause of aberrant behavior in zoo animals (Hill, 1966).

Draper (1965) has found a relationship between an increased quantity of sensory stimulation, particularly light, and systematic increases in activity level in *Macaca mulatta*. Once an animal has become accustomed to low levels of stimulation (e.g., sitting in an individual cage for months), any novel stimulus tends to have greater stress effects than if the animal has been exposed to changing stimulation all along (Sackett, 1965). The act of removing an animal from his cage for behavioral testing can be extremely stressful to an animal not habituated to being moved. Rowell and Hinde (1963) have shown that as little as 6 hours of isolation can increase the stressful nature of a novel test situation for a rhesus monkey.

B. Other Factors Associated with Captivity

Not only the quantity but also the kinds of stimuli available in the laboratory are important factors in the behaviors of captive primates. A monkey or an ape that has been reared in the wild has been habituated to many or most of the salient stimuli in that environment. The fact that the animal has spent his formative years in the wild is of particular importance because at least some of his later environmental preferences are very strongly controlled by his earlier "exposure learning" (Sackett, 1966a and Sluckin, 1965). Such preferences can range from very specific ones akin to "imprinting" to very general differences in responsiveness to the environment as a whole. For example, Singh (1968) compared the responsiveness of two groups of rhesus monkeys, one of which had been captured in Indian cities and the other in jungle areas. The two groups were exposed to several visual stimulus displays of varying complexity. The urban monkeys showed greater responsiveness to the stimulus displays of higher complexity value and Singh attributed this difference to more varied visual and motor experiences acquired in the urban area of India. In a previous study, Singh (1966) compared six urban and six forest adult rhesus monkeys with regard to timidity in the presence of human beings. The urban monkeys, in contrast to the forest subjects, were found to be active, manipulative, and not at all timid of people. The habitat or ecology of the wild-reared primate may produce differences in behavior which go beyond their responsiveness to people; however, Rowell (1966) has found very great differences between the behavior of her forest living baboons (*Papio anubis*) and the savannah-dwelling baboons of the same species studied by DeVore (1963). Low levels of intragroup tension, dominance, and agonistic displays characterizing the forest-living baboons were in sharp contrast to the high levels of such behaviors in the same species on the savannah. So-called "typical" species behaviors in the olive baboon evidently do not exist unless very specific qualifications are made concerning environment, ecology, and early experience.

1. Experimenter Effects

Since urban-reared rhesus monkeys respond to humans in a different way than forest-reared rhesus monkeys, the definition of "control" groups in early experience studies as well as in all behavioral studies in the laboratory should include more than the word "feral." If control groups are not better specified, such factors as the experimenter himself might well account for many of the presumably abnormal behaviors that are observed.

Sackett et al. (1965) used an apparatus for preference testing consisting of a six-sided frame of aluminum channels having six outer chambers, each adjoining a central chamber (see Sackett, this volume). Materials forming the walls, floor, and ceiling of each chamber were inserted into the channel. The walls of the central chamber formed guillotine-type doors raised by a cable. Only two exactly opposed outer chambers were used in the experiment to be described. One stimulus was a human female laboratory attendant and the second stimulus was a nonaggressive monkey of the same size as the subjects. The central chamber was the start chamber. After 2 minutes exposure to the start chamber, the subject was free to enter and reenter the human, the monkey, or the start chamber. Monkeys (M. mulatta) reared from birth away from other monkeys and handled by humans during the first month of life preferred humans to monkeys when tested at the age of 2–3 years. Rhesus monkeys having both early human handling and physical contact with other monkeys, or physical contact with other monkeys and no human handling during the first month, preferred monkeys at the age of 2–3 years. Rhesus monkeys reared in complete isolation from both humans and monkeys spent less time with either the human or the monkey choice stimulus but also preferred monkeys to humans (Sackett et al., 1965).

Physiological data concerning the effects of the presence of humans on the subjects support behavioral data such as those reported by Singh (1966) and Sackett et al. (1965); and, in addition, these data show that specific humans may come to evoke entirely different physiological and behavioral responses from an animal. Heart rate is a particularly sensitive index of such specific preferences. Strange persons may accelerate heart rate, while the presence of a familiar or preferred person may lead to a significant decrease (Gantt et al., 1966).

2. Preference for Animals with Similar Behavior

Pratt and Sackett (1967) found still another factor of early experience that affected selection of social partners. As stated above, monkeys raised with humans prefer humans to monkeys and monkeys raised apart from humans prefer monkeys to humans (Sacket et al., 1965). Pratt and Sackett (1967) also found, however, that the kind of monkey preferred also depended upon early experience. They reared three groups of rhesus monkeys with different degrees of contact with their peers. One group was allowed no contact, another only visual and auditory contact, and a third group was allowed complete and normal contact with peers. When the animals were 18 months old they were tested for their preference for monkeys raised under the same conditions or for monkeys raised under different conditions. Monkeys raised under the same conditions preferred each other, even when the stimulus animals were completely strange to the test monkey (Pratt and Sackett, 1967). Thus a principal of like-prefer-like apparently prevails when all choice animals are strangers. But what if the choice is between a stranger and a friend?

3. Effects of Friendly Ties

Strange animals of the same species as well as strange humans have different effects on a monkey than do familiar animals of the same species. For example, the presence of familiar cage mates in a group raises an animal's dominance with respect to strangers (Masserman et al., 1968). Interactions involving familiar subjects are often found to be less negative than those involving unfamiliar subjects; for example, disturbance behaviors (rocking and cooing) are more frequent in the presence of strangers (Hansen et al., 1966). Primate social behavior in the wild has been found to depend on early social attachments, kinships, and friendships, and many behavioral phenomena seen in the laboratory can be explained in terms of an absence of such friendly ties. Several strange monkeys placed together in a pen do not constitute a stable group. Increases in the frequency and intensity of agonistic behavior may occur when strange monkeys are introduced into a pen (Southwick, 1966). Sugiyama (1966) has observed that the appearance of a new alpha male may increase aggressive behavior in a very specific way. New leaders in the wild Hanuman langur troops he studied, for example, immediately kill or chase away all of the young infants (Sugiyama, 1966). In a captive situation therefore it seems sagacious to introduce an adult male to a langur group prior to the birth of infants. Infanticide has also been observed in the laboratory in *Macaca fascicularis* (Thompson, 1967), but it is possible that such behavior by new adult male leaders could also be "normal" behavior accompanying macaque social changes in the wild as well as langur social changes. Its appearance in the laboratory should not immediately be attributed to an aberration produced by captivity.

4. Ecological Factors

Other behavioral abnormalities in captivity might be related to differences between troop and individual travel in the wild and similar movements in captivity. Alexander and Bowers (1968) have noted that in captivity a troop cannot be nomadic and the function of leaders in initiating and guiding troop movements is therefore reduced. They see this lack of travel as a factor in producing the increased aggressiveness seen in captivity, but increased aggressiveness may also be related to absence of external threat. In the wild the leader male or males must often protect the troop from sources of danger originating outside the troop itself. As Alexander and Bowers point out, captivity also prevents the process of solitarization and precludes movement of individual monkeys from one troop to another. In the wild, males often become solitary or change troops, but in captivity this is not possible.

Although social changes in group composition, changes in leadership, introduction of strangers, and changes in sex ratio are said to have a more pronounced effect on levels of intragroup aggression than environmental changes in food or space (Southwick, 1966), the latter are also important in determining the stability of social organization. Southwick (1966) reported a significant increase in agonistic behavior in a balanced captive group of 17 rhesus monkeys after a space reduction from 1000 ft² to 500 ft². Agonistic behavior has also been said to increase with "provisionization" or with the restriction of food distribution (Southwick, 1966).

Rowell (1967) has recently made direct quantitative comparisons between the behaviors of a wild and a caged baboon group. She found that interactions were nearly four times more frequent in the caged group than in the wild group. Kollar et al. (1968) have reported nearly the same thing for captive versus wild chimpanzees. Approach-retreat interactions were particularly more common in Rowell's caged baboons and this was consistent with the idea of there being a higher level of stress in the caged group. The caged group also showed appeasement gestures more often. Rowell had previously noted an inverse relationship between grooming and stress, and her wild group of baboons did more grooming than the caged group. Mitchell and Stevens (1969) also noted a significant decrease in grooming in single-caged mother-infant pairs when they were compared with mother-infant pairs in a larger captive colony. Apparently, stress interferes with the most basic social stabilizing behaviors. (See also Lindburg, 1970.)

5. Crowding

Gartlan (1968) pointed out that captivity often means crowding and that crowding leads to changes in physiological states as well as behavioral states. Aggression often increases with captivity, evidently because captivity does not permit escape from crowded conditions. Hill et al. (1967) reported elevated serum cortisol levels in a group of vervet monkeys (Cercopithecus aethiops) when the animals were exposed to irregular noise, light, and vertical movement in the laboratory. Many newly caught animals also show such signs and many of the deaths occurring in the first few months of capture might be attributed to crowding (Fiennes, 1968). Autrum and von Holst (1968) have used the ruffle of the hair on the tail of the tree shrew (Tupaia glis) as an index of "social stress" (e.g., crowding) and have determined the effects of crowding on growth, body weight, and reproduction. Tail ruffling increased as crowding increased. The hairs on the tree shrew's tail are erected by muscles that are innervated by the sympathetic nervous system. The ruffling measurement was expressed as percentage of total observation time, and as this percentage of ruffling time increased, Autrum and von Holst saw the following phenomena appear: (1) retardation of growth and loss of weight in adults at relatively low ruffling percentages (below 20% ruffling), (2) male copulatory behavior appearing in adult females at above 20% ruffling, (3) lack of protective marking of the newborn by the mother with the consequence that the young were eaten by members of the group (above 20% ruffling), (4) infertility of the females and retraction of the testes in the male (above 50% ruffling), and (5) death (above 80% ruffling for a period of about 10 days). Ruffling of the hair was also induced by such social stimuli as prolonged and intense subordination to another member of the same species as well as by crowding.

6. Importance of Both Field and Laboratory Data

Some conditions in captivity have been correlated with the appearance of behaviors rarely seen in the wild. The study of groups of animals in their natural habitat often goes far toward alleviating many of the problems of keeping animals in captivity. Those who have studied the same species both in its native habitat and in the laboratory (e.g., Rowell, 1967; Bernstein, 1967; Hall and Mayer, 1967; Kummer and Kurt, 1965) emphasize the similarities between laboratory and feral behavior, but such comparative studies develop a more complete understanding of the nature of the animal than rigid adherence to traditional absolute control in experimental psychology. In fact it could be argued very persuasively that by ignoring field data and observation the comparative social psychologist of the past, who was perhaps properly trained in theory, controlled very little in practice. (Cf. Menzel, 1968.)

III. AGE AND SEX

A. INFANTILIZATION

Behavior may often inappropriately be labeled abnormal when the ontogenetic development of behavior has not been properly examined in the wild or in normative laboratory studies. Various elements of behavior normally change in frequency and form as an animal matures. Bodily postures, vocalizations, and facial expressions change with age (Møller et al., 1968). It is an interesting fact that many forms of early deprivation in the laboratory tend to "infantilize" animals. The primate that behaves "abnormally" not only exhibits abnormal frequencies of a given behavior in abnormal contexts but in many cases such abnormality represents the infantile form of behavior (such as digit sucking) long beyond the appropriate age (Mitchell, 1968a).

B. AGE DIFFERENCES

Draper (1966) has attempted to characterize changes with age in rhesus monkey behavior in the wild and has found that both form and frequency of activity are influenced strongly by age. Younger animals "assumed quadrupedal and bipedal postures more frequently and also walked, ran, climbed, jumped and changed from one posture to another more." Older monkeys, however, were less mobile but visually scanned the environment more often than younger animals.

In the laboratory, Bernstein and Mason (1962) have measured emotional responses to complex stimuli in *M. mulatta* as a function of age. A total emotion score, including such behaviors as barks, convulsive jerks, crouches, fear grimaces, lipsmacks, stereotypy, screeches, threats, and other vocalizations, increased in frequency with the complexity of the stimulus presented regardless of the age of the animals. The form of the emotional response, however, *did* vary with the age of the subject. From birth to 3 months of age, emotional responses consisted of vocalizations, rocking, crouching, and sucking. The animals withdrew from the complex stimuli very infrequently before 3 months, but up to at least 2 years there was an increase in directed responses such as barks, lipsmacks and fear grimaces and, in addition, an increase in withdrawal. All of these immature monkeys ($N = 47$) had been separated from their mothers on the first day of life and housed in wire cages individually. The authors utilized no reference "control" group. The results reported therefore refer to changes with age in wire-cage-reared rhesus monkeys.

These findings indicate that there are not only age differences in the frequency and form of behaviors in the wild but that there are also quite predictable age changes in the behaviors of monkeys reared in cages. A wire-cage isolate 3 months of age does not exhibit directed responses toward complex stimuli and does not withdraw from them, whereas a 2-year-old social isolate directs responses toward complex stimuli and withdraws from a disturbing stimulus. Other studies have also reported changes in both normal and abnormal behavior with age (Green, 1965; Mitchell, 1968a; Møller *et al.*, 1968).

C. Sex Differences

Just as the age of the laboratory animal strongly influences the frequency and form of a behavioral abnormality, so does the sex of the subject. One frequent speculation made by students of developmental psychology is that males are more severely affected by adverse ontogenetic conditions than are females. Some feel that this is a general law. Many male vertebrates appear to be more sensitive than their female counterparts to environmental changes early in life.

Mitchell (1968c) and Jensen *et al.* (1968) have reported normative sex differences in mother-infant attachment in *M. mulatta* and *M. nemestrina*, respectively. In general, the results of both studies suggested that males become independent of their mothers earlier than females. The mothers, the male infants themselves, and perhaps even other animals in the troop appear to promote this sex difference. Later in life, males are more active and more aggressive (Hansen, 1966; Seay, 1966). Although females vocalize more frequently than males and are less dominant, early isolation seems to increase dominance in females, decrease dominance in males, and in general decrease the sexual dichotomy (Mitchell, 1968a). Other references to sex differences in *M. mulatta* may be found in Altmann (1968), Chamove *et al.* (1967), Angermeier *et al.* (1968), Møller *et al.* (1968), and Mason *et al.* (1960). The point to be made here, however, is simply that the sex of the experimental subject may have profound effects on the frequency and form of the dependent variable serving as an index of normality or abnormality. For example, little has been made of sex differences in studies of social isolation; yet such differences, when considered along with normative sex differences in development, could very well clarify several factors related to isolation which are at present difficult to explain.

D. Longitudinal Approach

It is only good common sense in experimental and statistical design to account for as much variability related to extraexperimental factors as possible. In this regard, it is rapidly becoming our opinion that detailed longitudinal life histories of individual subjects may tell us more about the effects of differential early experience than hit-or-miss cross-sectional statistically dictated data on large samples of subjects. As long as the longitudinal data are not merely "baby biographies" or anthropomorphic analogies of post hoc human clinical psychology, extensive, reliable, and very specific hypotheses concerning social stimulation in early life may be generated. All conditions in the laboratory that might affect an individual animal might be measured or controlled, perhaps even 24 hours a day. This may not be practical or even possible with $N = 16$, but it is definitely within bounds for $N = 1$.

IV. "ABNORMALITIES" ASSOCIATED WITH BIRTH

A. Asphyxia Neonatorum

Several of the specific factors that should be adequately controlled or measured when longitudinal individual life histories are recorded are related to the birth process. As Windle has pointed out: "The role of asphyxia neonatorum in organic forms of mental retardation is so large that we cannot much longer hide it behind a facade of lesser or contributing causes . . ." (Windle, 1967, p. 140). Asphyxia neonatorium is a condition in which a newborn is not breathing but is alive after some disorder or mishap. According to Windle (1967), monkey (*M. mulatta*) fetuses at full term can withstand 7 minutes of asphyxia during birth without displaying symptoms of neurological deficit, but when the asphyxiation lasts for more than 7 minutes, permanent structural brain damage is demonstrated in *all cases.* The longer the duration of asphyxia the greater the neurological deficit. Saxon (1961) reported that many infant rhesus monkeys asphyxiated during birth lack appropriate emotional outbursts that characterize normal monkey infants. In the human being many emotional and intellectual aberrations have been correlated with clinical histories of asphyxia neonatorum.

B. Mode of Birth

Studies involving manipulations of early social experience certainly cannot ignore behavioral effects associated with the birth process. Even mode of delivery has a measurable effect on the behavior of the neonate. Rhesus monkeys delivered by cesarean section are less active and vocalize less than monkeys delivered vaginally (Meier and Garcia-Rodriguez, 1966). (Low activity and low frequency of vocalization are also characteristic of monkeys raised in social isolation; therefore it is vital in isolation studies to control both mode of delivery and the likelihood of birth anomalies.)

V. MATERNAL EXPERIENCE IN CAPTIVE BUT FERAL-RAISED MONKEYS

A. Parturition

Not only the mode of delivery of an infant but also the length of the labor and the degree of difficulty of the labor are factors that could affect the later behavior of the offspring. Female rhesus monkeys that have never delivered a previous infant generally have a longer labor and a more difficult labor than do multiparous females (personal observation). This difference may be intensified by bringing feral-raised pregnant mothers into the laboratory (Van Wagenen, 1966). Maternal parity therefore becomes still another extraneous factor requiring control in early experience studies.

B. Postnatal Behavior

Following delivery the maternal behavior of primiparous and that of multiparous mothers in restricted laboratory settings differ slightly. The primiparous rhesus monkey mother normally gives adequate care to her infant but appears to show higher "anxiety" and perhaps greater protectiveness than the multipara (Seay, 1966; Mitchell and Stevens, 1969). Seay (1966) reported that primiparous mothers displayed retrieval grimaces more often than multiparous mothers. The retrieval grimace, according to Seay, "is similar to the fear grimace frequently seen in disturbed or frightened monkeys." The "retrieval grimace tends to occur when the infant is in a situation which is perceived to be threatening by the mother." Mother-infant grooming and maternal punishment occur most frequently in multiparous mothers which suggests that the multipara are less protective and less stressed (Seay, 1966; Rowell, 1967).

Mitchell and Stevens (1969) have also reported data suggesting that the primiparous rhesus mother is more "anxious" than the multiparous mother. Primiparous mothers in another recent study stroked their infants frequently and threatened, lipsmacked, and fear-grimaced to various social stimuli in their immediate environment more frequently than multiparous mothers. In the second 3 months, multiparous mothers rejected, mouthed, cuffed, and clasp-pulled their infants earlier and more frequently than the inexperienced mothers (Mitchell, 1969a).

C. BEHAVIOR OF THE OFFSPRING

The *infants* raised by primiparous mothers do not seem to display grossly inappropriate social behavior, although some differences from multiparous-raised infants have been reported. Long after weaning, primiparous-raised juveniles in one study appeared to be somewhat more subordinate, less assertive, less playful, and emitted more distress vocalizations ("coo") than multiparous-raised juveniles. In addition, the primiparous-raised juveniles paced more often and showed more pilo-erection than the multiparous-raised juveniles (Mitchell et al., 1966b). Recent research at the National Center for Primate Biology in Davis, California is attempting to replicate these findings. No differences between two groups of eight infants have been detected as of this writing; however, the infants are only 10 months old.

The ecological validity of these maternal experience studies is open to question. Differences related to maternal experience may be either intensified or decreased by captivity. In the wild an entire troop is present and several members of the troop may either contribute to the adequate socialization of the infant or provide a source of added stress on the mother. Differences between primiparous and multiparous mothering detected in a restricted laboratory environment may thus be either naturally compensated for in the wild or naturally increased. Such possibilities have not been adequately determined in the laboratory since most laboratory studies involving manipulations of early environments do not involve the entire troop. In many cases these compensating or augmenting factors (i.e., the rest of the troop) are purposely left out in laboratory manipulations in order to restrict measurement to the effects related to differential mothering or in order to make quantification and statistical analysis easier. As will be seen in the next section, however, there is a price the laboratory investigator must pay for such methodological cleanliness.

VI. PEER DEPRIVATION

A. MOTHER-ONLY REARING

A monkey mother and a monkey infant maintained by themselves in an individual home cage are both socially deprived. The mother may be deprived of the "help" and "hindrances" in rearing the infant that she would normally receive in the wild. The infant, however, is deprived of a wide range of social experiences, most notably those included in peer play.

Dr. Bruce Alexander, formerly at the University of Wisconsin Primate Laboratory, made the first systematic measurements on the effects of peer privation on later social adjustment (Alexander, 1966). Eight infant monkeys experiencing the "mother-only" treatment for the first 4 or 8 months of life were initially hostile toward other infants when they were exposed to age-mates following their early treatment. The amount of bodily contact between Alexander's mother-only infants at 8 months was initially lower than that between control infants that had the benefit of peer experience throughout the first 8 months of life. The mothers were present when the infants were first allowed to see each other.

In a study by Griffin (1966), 8-month mother-only infants showed neither an increase in aggression nor a decrease in contact with peers upon emerging from their treatment. Griffin, however, removed the mothers from the scene of initial confrontation with peers. In addition, many of Griffin's animals also had experienced repeated separations which produced subordination and other effects to be discussed in Section VII, C. It is possible that: (1) either presence of the mother at the time of first confrontation with peers promotes aggression or (2) a history of repeated separation overrides the effects of peer deprivation.

A study involving a mother-only situation which attempted to evaluate the effects of two different environments throws some light on the nature of peer deprivation. Jensen *et al.* (1968) housed eight mother-infant pairs of *M. nemestrina* in bare cages located in soundproof rooms and four more pairs in an open laboratory in cages containing toys. Behavior was recorded for the first 15 weeks of the infants' lives. Pairs in the quiet bare cages spent more time in physical contact than pairs in the rich environment. The infants in the quiet bare environment oriented more behavior toward themselves and less toward the environment and were less active than the infants in the richer environment. Jensen *et al.*, concluded that: (1) a stimulus-poor environment intensifies the physical relationship between a mother and her infant, (2) the environment does not affect the basic nature of the mother's role, and (3) a stimulus-poor environment produces some retardation in infant development. There were no differences in maternal protection or maternal punishment between the two environments, thus the infants in the rich environment showed greater independence even though the mothers' role was the same in both environments. However, none of the infants in either of the environments were severely emotionally disturbed. They displayed none of the stereotyped or "autistic" behaviors that generally appear in macaque infants raised without mothers. The Jensen mother-only infants were exposed to peers at the conclusion of their 15-week sojourn and at 6 months of age were paired with each other and with controls. The mother-only infants were subordinate to monkey infants that had been reared with both mothers and peers (Jensen and Bobbitt, 1968).

Alexander's mother-only animals, referred to previously, were also followed-up several times in many different test situations as they matured. With age it became more and more obvious that they were not behaving like mother-and-peer animals. Levels of aggression and disturbance became higher than in mother-and-peer monkeys and the most recent report received from Wisconsin indicates that they were definitely socially deficient at 6 years of age (Ruppenthal, personal communication), although not by any means as deficient as social isolates.

B. PEER-ONLY REARING

While absence of peers but presence of mother is not an optimal rearing condition, neither is presence of peers with absence of mother. Peer interaction partially compensates for maternal absence, but Chamove (1966) has found that peer-only rearing, producing compensating clinging and huddling with the cage mate or cage mates, depresses the appearance of play. Monkeys reared with more than one peer were more "adequate" in play and affiliation than those reared with a single cage mate, yet even these animals developed prolonged and excessive cling-ing behavior (cf. Harlow and Harlow, 1962a).

Rosenblum (1961) separated eight rhesus monkeys from their mothers 6–12 hours after birth and raised them in individual cages with artificial mother surrogates. These monkeys received peer experience in the pres-ence of artificial mothers once daily throughout the first 6 months of life. The peer experience was provided in a playroom and in two separate sex-balanced groups of four.

In the first month of life, the Rosenblum infants responded markedly to the surrogate mothers as well as to their peers. Threat responses and rough-and-tumble play behaviors appeared more often in male infants than in females. Rigidity, withdrawals, and early attention to surrogate mother occurred more frequently in females. Immature sexual behavior occurred in both sexes and dominance relationships did not develop until after 2 months of age.

Rosenblum's overall impression was that the behavior of his subjects, at least during the first 6 months, was quite similar to behavior in the wild. The artificial mother received a lot of the clinging the infants would have directed toward each other in its absence. Thus play often appeared in Rosenblum's infants instead of the persistent clinging that appeared in Chamove's (1966) infants.

C. Toward Optimal Laboratory Rearing

Clearly, a mother is better than nothing, and so is a peer, but some-thing more than a mother and a peer is closer to the optimal social rear-ing situation. As the laboratory data accumulate and as laboratory tech-niques become more sophisticated, we are beginning to find deficiencies in the social behaviors of even mother-with-peer-reared animals. Mother and peer monkeys are more adequately socialized than mother-only and peer-only monkeys, but clearly "aunts," juveniles, and adult and subadult males (Mitchell, 1969b) play a role in the socialization of the infant as well as in stimulating "normal" maternal behavior in the infant's mother. Careful control of each of these complex social stimuli in turn provides a more complete picture of "natural" monkey behavior.

VII. MOTHER-INFANT SEPARATION

Frequently, standard laboratory maintenance and wise animal hus-bandry practices make it necessary to separate an infant from its mother.

The mother of course represents a significant portion of the infant's environment. The infant becomes attached to most of the stimuli associated with sensory contact with the mother and to a lesser extent the mother becomes attached to the infant. It surprises no one therefore that both mother and infant protest when they are separated from one another. Their behaviors change in a regular way at separation, while separated, and a reunion.

A. THE SEAY STUDIES

Seay et al. (1962) separated four 6-month-old infants from their mothers (M. mulatta) for a 3-week period. During separation the mother and infant were allowed auditory, visual, and olfactory intercommunication. The infants showed increased "crying" (coo vocalization) and mother viewing, and decreased peer-directed social behavior. Following reunion there was increased mother-infant cradling, clinging, and ventral contact. One of the male infants showed a considerable delay in his reattachment to the mother. The authors attributed this to a more intense infant-mother tie in this animal.

In a later study, Seay and Harlow (1965) separated eight rhesus monkey mother-infant pairs for a period of 2 weeks when the infants were 7 months old. In this study they removed the mother from visual and auditory contact as well as from physical contact. All eight infants showed emotional disturbance in response to separation with accompanying drastic decreases in play while separated. The infants initially responded to the separation by violently running about, climbing, screeching, and crying; "they then passed into a stage characterized by low activity, little or no play, and occasional crying."

In this last Seay study, aggression toward peers increased after reunion with the mother. Seay and Harlow (1965) explain this appearance of aggression as follows. "Dr. Leonard Berkowitz (1964) has suggested . . . that the mother . . . may have become an aggression-evoking stimulus, but because of her intolerance of personal aggression, the aggressive behavior was displaced to peers" (Seay and Harlow, 1965, p. 440; also see Berkowitz, 1964).

Unlike clinical reports of human mother-infant separation, the rhesus infants in the Seay studies did not exhibit what Bowlby (1961) labeled "detachment," although the first two stages of the Bowlby separation syndrome—protest and despair—were evident. In the human detachment stage, the child initially refuses to go to the mother upon reunion and may even run from her or direct aggression toward her. No infant monkey ran from its mother upon reunion in the Seay et al. studies and only one animal displayed a delay in reattachment. Positive responsiveness to the mother was increased markedly by separation. There was, however, a possibility of a displaced detachment from peers. As noted in Section VII, F, however, detachment from the mother has also been demonstrated.

B. Degree of Attachment and Long-Term Effects

Two or three weeks is a long time for an infant to be separated from its mother and most routine laboratory procedures do not require a period of separation of this length. Hinde *et al.* (1966) separated four rhesus infants from their respective mothers at 7 months of age for a 6-day period. They found that coo vocalizations increased and that the behavior of the infants became depressed; but, two additional contributions to the understanding of separation effects were provided by the Hinde *et al.* study. First, the more an infant had been off the mother prior to separation, the less clinging it displayed after return to its mother after separation, that is, the severity of the separation effect varied directly with the degree of preseparation attachment. Second, all four of the Hinde infants showed some effects of the separation during the rest of the first year of life, that is, even brief (6-day) separations from the mother had long-term effects on behavior (Spencer-Booth and Hinde, 1966).

C. Brief but Repeated Separations

It may be argued that even a 6-day maternal separation is not necessary for routine care of the monkeys. Are there data available on the effects of extremely brief but repeated separations? Griffin (1966), ran a study at the Wisconsin Primate Laboratory which is pertinent to this question. He raised three sex-balanced groups of four rhesus infants in a mother-only situation for the first 8 months of life. One group was reared by four different mothers (rotated mother group). These infants were repetitively separated from one mother for 2 hours every 2 weeks and then returned to one of three other mothers. Another four infants were separated but returned to the same mother after each separation (repeated separation group). The final group was never separated.

1. Sensitization to Separation

Following each brief separation, the first two stages of the separation syndrome were witnessed: increased vocalizations and depression in the infants. There were also predictable overprotection responses in the mother upon return but only when the infant was returned to the same mother each time. *Previous separations did not reduce but rather increased the disturbing effects of subsequent separations from the same mother.* With return to a strange mother there was often an initial rejection of the infant by the mother.

At the end of 8 months, the infants were weaned from their mothers and allowed to interact with peers. The mothers were *not* present when the separated infants were first exposed to peers. The contact behavior with peers was immediately high in the repeatedly separated infants. The contact was not hostile but quite calm. Griffin often observed mutual ventral-ventral clinging and fellatio in the repeatedly separated males.

2. Long-Term Effects

In a follow-up study at 19 months of age, 11 months after the final separation (weaning), all 12 of the Griffin animals were paired several times with a huge hostile adult female and with a 1-month old infant (Mitchell et al., 1967b). The repeatedly separated monkeys were more disturbed in the test cage at 19 months of age than were those not separated. The separated animals screeched significantly more often than the controls, and those that had been repeatedly separated and returned to the same mother were the only subjects seen screeching in the presence of an infant stimulus animal. Fear grimaces and social submissions were high and threats were low in the repeatedly separated group, but hostility was high in the rotated mother group. Both groups of separated animals emitted many coo vocalizations, but those that had been returned to the same mother during the first 8 months emitted nearly seven times as many coos at 19 months than did the control animals. The separated infants never adjusted to being separated and, as we have seen, an exaggerated "separation response" persisted a full year after the final maternal separation.

D. Separation in a Group

The Griffin separation data were recorded in a mother-only situation, while the Seay data were collected in a mother-and-peer playpen situation. It has already been noted that optimal socialization in early life involves more than a mother and peers. What happens when a infant is separated from its mother but is allowed to remain in a group of monkeys approximating feral composition? Rosenblum and Kaufman have found that what happens in the group situation depends on the species observed.

1. Species Differences

According to Rosenblum and Kaufman (1968), sustained physical contact between animals characterizes bonnet (*Macaca radiata*) but not pigtail (*M. nemestrina*) monkey groups. The bonnet pattern encourages maternal permissiveness and enhances social orientation in bonnet infants. Separated bonnet infants, if housed in a social group, are often adopted by other animals in the group and hence "suffer minimal affective and behavioral changes." Pigtail infants, however, show "depression and behavioral debilitation after maternal loss." The same separation syndrome described for rhesus monkey in both a colony (Spencer-Booth and Hinde, 1966) and a mother-only (Griffin, 1966) situation is apparently also characteristic of the pigtail in a group situation (Rosenblum and Kaufman, 1968 and Rosenblum *et al.*, 1964).

2. Species Differences in the Playpen Situation

Recently, Schlottman and Seay (1968) separated four infant Java monkeys (*M. fascicularis*) for 3 weeks between 7 and 8 months of age. The test situation involved a mother-and-peer playpen situation similar to that used by Seay in his earlier separation studies on rhesus monkeys. The results of the Java monkey separations were "generally in accord with previous studies of maternal separation in monkeys." Both the protest and despair stages of the Bowlby syndrome were seen in *M. fascicularis* but, as was true of all previous monkey separation studies, a clear detachment phase was not seen. Seay and his collaborators at the Louisiana State University Primate Laboratory are presently examining the effects of the presence of a strange adult female during separation. The Louisiana researchers are also studying separation in the patas monkey (*Erythrocebus patas*), and initial reports from their laboratory suggest that the infant patas does not show a strong separation syndrome (Seay, personal communication).

E. Effects of Separation on the Mother

1. Immediate Effects

Most of the separation studies have been concerned with the effects of mother-infant separation on the infant. There have been, however, a few studies that have indicated that the behavior of the mother is also changed by separation from her infant. Mother monkeys (pigtails) separated from their infants were always more active and vocal than either nonmother female monkeys or males, and their activity and response vocalizations increased during presentations of infant calls via tape recorder (Simons *et al.,* 1968). "When five pigtailed monkey mothers (*M. nemestrina*) were separated from their six-month-old infants, they reacted initially with agitation. Eighteen days after separation several behavioral measures still suggested depression in the mothers. Two months after separation all of the measures suggesting depression returned to preseparation levels and showed no change after the infants were reunited with their mothers. The depression was mild and more subtle than the depression described for infants" (Jensen, 1968).

2. Long-Term Effects

It has already been noted that there are long-term effects of maternal separation on the behavior of macaque infants (Mitchell *et al.,* 1967b). Jensen's (1968) study, discussed above, suggests that there might also be quite a lengthy period of recovery for the mother. Is future maternal behavior affected by a history of separation from a previous infant? There have been no direct tests of this possibility, but Sackett *et al.* (1966) compared the social preferences of rhesus mothers that have experienced repeated separations with the social preferences of mothers that have experienced only a terminal weaning. The test unit used was the so-called "self-selection circus" in which six outer choice compartments surrounded a hexagonal inner start chamber. A stmulus animal was placed in each choice compartment and the experimental animal in the start chamber could enter any of the choice compartments but could not physically interact with a stimulus animal. The number of seconds spent in each choice compartment during a 10-minute test trial was used as an index of preference for the stimuli available. Mothers that had experienced brief, repeated separations throughout the first 8 months of their infants' lives exhibited a greater preference for their own infants than mothers normally reared without a history of periodic separation. "Forcible separation during rearing may, therefore, produce exceptionally strong ties for own infant by mother monkeys." If given a choice between infants and other adult females, the repeatedly separated mothers preferred adult females. Control females, however, "had a slight, nonsignificant, preference for infants." Sackett *et al.* (1966) were unwilling to speculate about the theoretical importance of this last rather surprising finding. It appears, however, that repeatedly separated mothers are less likely to prefer to interact with infants following repeated separation crises unless the infant from whom she has been repeatedly separated is one of the choice objects. Since these preference tests were conducted almost a year after the final separation from the infant, there is at least suggestive evidence that the mother suffers long-term separation effects as well as the infant, although not necessarily of the same quantity or form.

F. Age of Infant as a Factor

1. Separations after the Second Year of Life

Separations from the mother at about 6 or 7 months clearly produce protest and depression, and repeated but brief separations throughout the first 8 months result in long-term separation effects in both mother and infant. Do separations at later ages produce the same effects? Joslyn (1967) separated six rhesus monkeys from their mothers at 26 months of age. These juvenile or preadolescent offspring were overtly disturbed for 2 weeks and exhibited depressions in play for at least 6 weeks. As was true in the Spencer-Booth and Hinde (1966) separations, which took place at an earlier age, the closeness of the maternal relationships prior to separation was directly related (Spearman's rho = +0.87) to the amount of disturbance displayed by the juvenile following separation. When returned to their mothers for 1 week at the age of 30 months, all of Joslyn's subjects resumed an infantile filial relationship even though they were almost sexually mature. Some of these subjects were carried ventrally by their mothers while their heads scraped the floor of the cage. After 1 week with their mothers at 30 months, Joslyn's subjects were again separated from their mothers. At this time, however, they were denied physical contact with their mothers for 1 week but allowed visual, auditory, and olfactory communication. This partial separation, as Joslyn calls it, led to peer-directed fear and aggression. "When *completely* separated from their mothers the offspring were disturbed—but were not aggressive toward each other." Thus when an animal is able to see the mother but not touch her during separation, his peer-directed agonistic behavior increases at both 7 months of age (Seay *et al.,* 1962) and at 30 months (Joslyn, 1967). In addition, total separation produces protest and depression at least up to sexual maturity.

2. Separations prior to 6 Months of Age

It is obvious from the data presented above that single, prolonged separations have telling effects on the macaque infant and mother at least *after* the first 6 months of the infant's life. Recent research at the National Center for Primate Biology at Davis, California has shown that separation also has similar effects at as early as 2 months of age in *M. mulatta.* Mrs. Patricia Abrams, working in collaboration with Dr. Donald Lindburg and the writer, has studied the effects of separation in relation to the age of the infant. Her study is particularly important because of the number of monkeys involved. Twenty-four rhesus monkey infants were separated from their mothers for 48 hours each. The 24 infants, 12 males and 12 females, were divided into three groups of eight, and each group was separated at a different age: eight at 2 months, eight at 3½ months, and eight at 5 months of age.

In the Abrams (1969) study, the mothers as well as the infants were observed and both mothers and infants of all three age groups initially protested the separation by moving rapidly about the cage and by vocalizing frequently. Many of the infants became depressed on the second day of the separation phase, during which they moved about and cooed infrequently. Thus, as in previous studies, both protest and despair were observed.

3. The Bowlby Syndrome

As mentioned above, John Bowlby (1961) has formulated a three-phase separation syndrome for human infants. The first phase, called *protest,* involves crying and extreme agitation when the mother first leaves. The second stage, called *despair* or *depression,* is accompanied by marked inactivity and withdrawal from social contact. The third phase, called *detachment* or *denial,* is characterized by initial withdrawal from the mother upon reunion. In most previous separation studies on simians, the first two phases of the Bowlby (1961) separation syndrome for human children were observed, but only one example of displaced aggression and one case of delayed reattachment at reunion were reported. In the Abrams study, 6 of the 24 infants showed clear signs of detachment. The mother entered the cage at reunion and retrieved the infant as usual, but the infant screeched as she retrieved it; and, after a brief period of ventral contact with the mother, the infant broke contact and withdrew from her. The duration of withdrawal lasted from a few seconds to an hour during which the mother made repeated attempts to retrieve the infant who attempted to evade her. When the mother was successful in retrieving the infant, the infant soon withdrew again and a chase about the cage ensued with the infant cooing loudly. Gradually, by grooming and by increasing incidental contact, the mother was able to establish ventral contact and the infant closed its eyes and relaxed. Of the six infants displaying detachment or denial, four were males and two were females. Three were 8 weeks old, two were 14 weeks old, and one was 20 weeks old. Abrams felt that younger infants and male infants seemed to be more sensitive to separation than older and female infants, and it seemed to be the younger males that displayed detachment. Although age and sex appeared to be important factors in the infant's reaction to separation, they were not important factors in the mother's reaction to separation. Separation occurring within the first 5 months and particularly within the first 4 months were more likely to result in complete Bowlby syndromes than were later separations. Even before 5 months of age, however, Abrams detected detachment or denial in only 6 out of 24 infants.

4. Effects of Separation on Nonseparated Neighbors

Another interesting facet of the extensive Abrams separation study was the measurement of the behavior of nonseparated neighbors during a separation period. Mother-infant pairs that observed a separation but were not themselves separated showed an increase in mother-infant contact, and the mother showed an increase in cooing. Thus the macaque mother and infant apparently display a separation syndrome to only the sight and sound of a separation. Such results suggest that mother-infant separation is indeed a serious matter.

VIII. REARING IN SOCIAL ISOLATION

A. History of Primate Isolation

In 1928 Tinklepaugh described the appearance of self-mutilation in "Cupid," a male rhesus monkey. Little was known about the early life history of Cupid because Tinklepaugh (1928) did not obtain him until he was between 2 and 3 years of age. At first he did not display self-directed aggression. However, an adult, Cupid, although capable of normal copulation, developed the habit of self-mutilation, usually through biting his hind feet and arms. Recent studies have reported self-mutilation in isolation-raised adult males but adequate sexual behavior has not been seen in such isolation-raised animals (Mitchell, 1968a).

The behavioral development of a rhesus monkey reared in wire-cage isolation from its own kind was described by Foley (1934, 1935). Foley followed the development of the animal through the second year of its life. In his two articles he referred to what he called "neurotic" symptoms. Oscillating or swaying movements were present from the third day of life in the isolated rhesus monkey. Thumb sucking and toe sucking were noted by Foley on the tenth day, and a phenomenon Foley labeled a "habit residual" developed from the toe sucking: "Lastly, we may mention examples of behavior not unlike that classed as 'habit residuals' in the abnormal human subject. . . . The infant was observed to begin the upward motion of the hind foot as if to place the great toe in the mouth, and then to cease before the behavior segment was completed" (Foley, 1934, p. 39). Foley (1934) observed almost no sexual behavior in his isolation-raised infant macaque during the first year, not even an erection. Lashley and Watson (1913), however, observed an erection in a mother-raised rhesus infant on the sixty-first day.

In the second year of life, Foley's isolation-raised rhesus monkey continued to show habit residuals:

> A further example of habit residual is illustrated by frequent clutching of the right eyebrow with the corresponding hand. No foreign particle was detected in the eye itself, and the mannerism was probably due to the extremely long and projecting eyebrows which once undoubtedly partially obstructed the infant's visual field. The reaching response persisted, however, although the longer eyebrows were carefully cut. (Foley, 1935, p. 90.)

Sexual behavior began to occur in Foley's monkey in the second year. In week 59, penile erection was first noted but there was a minimum of sexual behavior during the period of isolation (18 months). After 18 months of age, the examples of sexual behavior described by Foley appeared to be aberrant:

> After having been permitted closer association with other monkeys from the age of 18 months, however, Kras began to exhibit such sexual behavior. The earliest reactions consisted of olfactorily exploring the body of the other monkey, clutching some part of the animal's mody, usually the leg or back, and making a few sporadic thrusting movements of the pelvis. Erection was not always present on such occasions, and no preference was evidenced for female over male animals nor for rhesus over the cebus or capuchin types. In fact such reactions were on a few occasions observed in response to a rag or other soft object. By the end of the second year, the sexual behavior had become more or less localized with respect to the genital regions, and there was a general although by no means exclusive preference for female animals. (Foley, 1935, p. 91.)

According to Foley (1935), aggressiveness characterized the behavior of his subject throughout the second year, and neither self-grooming nor social grooming was exhibited by his subject during the first 2 years of life.

McCulloch and Haslerud (1939) described the affective responses of an infant chimpanzee that had been isolated from other chimpanzees during early development. At 7 months of age the subject avoided moving objects and showed disturbance behavior. Very little aggression toward the stimulus object was observed. When tested again at 15 months, the chimpanzee exhibited disturbance to many objects regardless of whether or not they were moving; in addition, he exhibited much more aggression toward the objects at 15 months than at 7 months. In a series of articles, McCulloch (1939a,b) also stressed the importance of clasping a soft object in decreasing excitement. The role of "contact comfort" in early socialization became evident to psychologists at this time, although Alfred Russel Wallace emphasized its importance as early as 1869 (cf. Mason, 1968).

The importance of physical contact was emphasized still more when Nissen *et al.* (1951) raised a chimpanzee with its hands and arms covered. One of the effects of this restricted opportunity for tactile, kinesthetic, and manipulative experience was a decrease in the ability to groom.[*] Since grooming is an extremely important means of social communication in the primate (cf. Lindburg, in press), it seems apparent that even isolation of the limbs from physical contact can have deleterious effects on later social behavior. Not only grooming but also clinging and other social behaviors were affected by the limb restriction. The restricted chimpanzee did not cling to the attendant who carried him, and the lip movements and sounds that are generally a part of grooming were absent. There was some improvement in manipulation when the tubes covering the limbs were removed at age 31 months and a tendency to cling to the attendant developed, although very slowly. Grooming did not appear. There is some recent evidence from monkeys that eye-hand coordination interference, resulting from *visual* isolation of the limbs, might also have been involved in the Nissen study (Held and Bauer, 1967), but the effects of such visual isolation alone are transient.

B. The Role of "Physical Contact"

All of the so-called "symptoms of abnormality" discussed in previous sections of this chapter are mild in comparison with the symptoms seen in social isolates. A social isolate is an animal that has been removed from its mother at or soon after birth and reared in an environment in which no "physical contact" with another member of its species is allowed. Usually this rearing environment is a bare wire cage which permits visual, auditory, and olfactory contact with other animals but which allows no direct physical contact.

When Harlow (1958) and Harlow and Zimmerman (1959) published their now famous articles concerning the development of strong and persistent attachment of orphaned baby monkeys to inanimate surrogate mothers, the importance of physical contact to the infant monkey became well-known to all psychologists. But the Harlow and Zimmerman

[*] Dr. L. A. Rosenblum has pointed out that the Nissen chimp may have also been raised in relative social isolation from species mates. If this is true it may have played a role in deficient social grooming.

surrogate technique also demonstrated that movement (rocking surrogates versus stationary), clinging (cylindrical versus flat-plane surrogates), sucking, and warmth were also important factors in the formation
of attachment in primates (Harlow and Zimmerman, 1959). Since all
of these are involved in physical contact, one must be careful with termi-
nology. When the term physical contact is used throughout the rest of
this chapter, it refers to an animal's receiving a complex combination of
contact, clinging, movement, oral contact, and warmth from another animal. The social isolate is isolated from receiving this complex combination of physical contacts although he may be allowed visual, auditory,
or olfactory contact with another primate.

C. Original Mason Studies

In the early 1960's Mason published detailed measurements on subadult monkeys (*M. mulatta*) that had been maintained in bare wire
cages from birth. Such socially restricted monkeys, particularly males,
showed markedly deficient sexual behavior near maturity. They were
responsive to receptive females but their attempts at copulation were
inefficient at best. They also showed less grooming and more fighting and
assertive play than feral monkeys (Mason, 1960a). In tests of gregariousness, pairs of wire-cage-raised monkeys "made fewer social choices and
fought more frequently following a social choice than did pairs of feral
monkeys" (Mason, 1961a). Their dominance relations were also unstable
as compared to the dominance relations of ferals (Mason, 1961a). In
tests of the responses of wire-cage isolates to an alien species (albino
rats), Mason and Green (1962) found that restricted (wire-cage) monkeys were not as gentle with the rats and made fewer contacts with the
rats in the living cage than did ferals. In an unfamiliar room, restricted
monkeys crouched, sucked their thumbs or toes, clasped themselves, and
engaged in rocking or other stereotyped behaviors. The ferals, however,
responded to the novel environment with more locomotion and more
gross motor activities than the wire-cage isolates.

It might be asked whether or not visual access to other monkeys during rearing provides any advantages over enclosed cage rearing. Mason
(1960b) has investigated this problem in twelve 9-month-old rhesus
monkey infants that had been wire-cage reared from 18 hours after birth.
The presence of visually familiar stimuli, whether social or nonsocial,
decreased disturbance in such animals. For example, in the wire-cage
monkeys, responses indicative of disturbance in the presence of a familiar
animal, or even in the presence of an unfamiliar animal of the same age,

were less frequent than they were in an empty detention cage. Obviously, such animals preferred *seeing* other monkeys over *not seeing* other monkeys. The nature of the monkey's previous visual social experience was important. Adult monkeys, which were never seen in adjacent cages during rearing, were a source of disturbance.

Mason (1963b) allowed six pairs of wire-cage isolates the opportunity to contact each other in brief test sessions between the ages of 25 and 85 days and again between the ages of 120 and 150 days. Initially, clasping at each other was the dominant social response and this was sometimes accompanied by mouthing or thrusting. As the infants matured between 25 and 85 days, the clasping response decreased and play became the dominant response. After a return to isolation between 85 and 120 days of age, clasping again became the dominant response but was again surpassed by play during the social sessions before 150 days of age. Aggression, grooming, and sexual behavior were never seen in these brief sessions which terminated at the age of 5 months, but the addition of physical contact for brief periods of time during the first year facilitated social maturation from clasping to play.

Mason (1961c) also reared two rhesus monkeys in enclosed cages so that they were visually as well as physically isolated from other animals. During the period from birth to 90 days of age, movable, flashing, and stationary stimulus objects were presented to the isolates in their isolation cages. Manipulatory responses appeared in both monkeys on the second day of life and the movable and flashing stimuli reduced more manipulation than the stationary stimuli. Mason concluded that "the early onset and rapid growth of manipulatory behavior is not noticeably retarded by this reduction in the amount and variety of environmental stimulation."

Mason and Sponholz (1963) described the social behavior of the same two rhesus monkeys discussed above after they had been raised in enclosed isolation cages from birth until early adolescence (16 months). These isolates appeared to be traumatized by the extracage environment. Crouching was their characteristic posture throughout the experiment. Few responses were directed toward other animals or toward the physical environment, and the most common reactions to social contact were submission and flight. Mason and Sponholz described rocking, head banging, crouching, and self-clasping in the infant isolates and reported that wire-cage-raised controls also displayed such disturbances but not as often as the enclosed isolates. The difference between the two degrees of debilitation was not in the presence or form of disturbance behaviors but in the lack of any other responses in the more severely isolated group. As long as the isolates were in their own enclosed isolation cages, however, they did not appear to behave any differently than the wire-cage monkeys.

In general, Mason (1963a) concluded from all his studies that the socially deprived animals had troubles with response integration and communication. Although the socially deprived animals exhibited most of the components of normal social behavior, these components were not combined into integrated patterns and effectively applied in social interaction. Mason (1963a) believed this to be a deficiency in sensory-motor learning or "shaping." Although all the basic postures, gestures, vocalizations, and arousal-affect were unlearned, (according to Mason), their effectiveness in social interaction was dependent upon experience. This rule applied to the receiving as well as to the sending of social signals. Messages sent can be effective only if the receiver responds to them as signals or knows their meaning. The signals reared under socially restricted situations were neither good senders nor good receivers and were incapable of response integration in an appropriate context. As noted later, there has been some recent direct support for this idea.

D. THREE-MONTH ISOLATES

Following the Mason studies at Wisconsin, research on the effects of early social isolation on later behavior increased rapidly. Several studies were immediately undertaken at the University of Wisconsin Primate Laboratory under the direction of Dr. H. F. Harlow. Rowland, Griffin, Boelkins, Arling, Mitchell, Sackett, Clark, Cross, Pratt, and Senko all conducted studies on various aspects of the Wisconsin-reared social isolates (*M. mulatta*).

Rhesus monkeys reared in enclosed isolation chambers for the first 3 months of life were compared with 3-month wire-cage-reared social isolates on social and learning behaviors (Griffin and Harlow, 1966). The 3-month enclosed isolates showed extreme withdrawal when they were first removed from their isolation chambers. This withdrawal was so severe that one monkey died of starvation because it refused to eat, and another had to be force fed. The enclosed isolates exhibited a decrease in oral and manual exploration of the cage upon emergence and an increase in self-directed orality. They also showed difficulty in adapting to new situations. No long-term differences in social or learning behaviors were found between 3-month enclosed isolates and 3-month wire-cage isolates, however. Both groups showed a drop in disturbance behaviors during 3 months of peer interaction immediately following their isolation. Boelkins (1963) presented data which were in agreement with the Griffin and Harlow study. While the enclosed isolates were a little slow to develop play, they progressively improved in play throughout the first year of life. An autistic posture of self-clutch of the head and body by the arms and legs while remaining in a prone position developed in both groups of 3-month isolates and persisted throughout the second 3 months of peer interaction. The animals were definitely not behaviorally normal. In addition to the effects of isolation itself, Harlow *et al.* (1964) believed that a drastic environmental shift when the monkey was between 50 and 100 days of age may have produced a condition of maximal withdrawal. Normal heterosexual posturing did not appear in these animals at 1 year of age (Harlow *et al.*, 1964 and Griffin, 1965).

E. SIX- AND TWELVE-MONTH ISOLATES

Rowland (1964) reared two groups of four rhesus monkeys in enclosed isolation chambers from a few hours after birth until either 6 or 12 months of age. Another group of four was kept in bare wire cages for the first 6 months and then placed in the enclosed isolation chamber for the next 6 months. The behaviors of these three groups of enclosed isolates were compared to the behaviors of rhesus monkeys raised in bare wire cages for the first year. Although an emergence phenomenon occurred at 6 months, the emergence effect involving extreme emotional anorexia that appeared in 3-month isolates was not observed in the Rowland isolates. Apparently, removal from isolation at 3 months of age is particularly disturbing even though the effects are to some degree reversible.

F. THE MITCHELL FOLLOW-UP STUDIES

1. The Appearance of Aggression

Rowland (1964) reported that enclosed social isolation of rhesus monkeys for 6 or 12 months after birth had a severe debilitating effect on subsequent social behavior. He found that his isolate monkeys were fearful, disturbed, and sexually abnormal when observed 12–20 months after birth. A follow-up study on these same isolates at 28–44 months of age was conducted by Mitchell *et al.* (1966a). Eight of Rowland's enclosed isolates were compared to eight socially sophisticated controls in brief cross-sectional pairings with 12 stimulus strangers: 4 adults, 4 age-mates, and 4 juveniles. The isolates were characterized by infantile disturbance, low environmental orality, high fear, high aggression, low sexual behavior, low play, and by idiosyncratic bizarre movements. The 12-month enclosed isolates were primarily fearful and nonagressive at this age, yet they threatened many attacks. The 6-month enclosed isolates were both fearful and physically aggressive. Thus 6 months of enclosed isolation during the first year had negative effects on social behavior at least up to puberty; abnormal aggression appeared in the 3-year-old 6-month isolates; and, 12 months of enclosed isolation seemed to suppress or at least delay the appearance of abnormal physical aggression (Mitchell *et al.*, 1966a).

The aggression of the 6-month enclosed isolates were often suicidal attacks against huge adult male stimulus animals or brutal beatings of juveniles. Such behaviors were never seen in normal females and were essentially nonexistent in normal 3-year-old males. The number of coo vocalizations emitted by the isolates was only one-third as great as the number emitted by the controls, and the coo appeared to occur out of normal context. Many of the differences between enclosed isolates and controls in the follow-up studies were found in pairings with adult and juvenile stimulus animals but *not* with age-mates (Mitchell *et al.*, 1966a).

2. Rowland's Isolates after Puberty

Mitchell (1968a) continued to follow the Rowland enclosed isolates after they had reached puberty. The females were cycling and the males' testes had descended, but the canines of the males had not completely grown out. This long-term Mitchell study indicated that the enclosed isolates were still socially inactive, fearful, and disturbed at 4½ years of age. Other studies on wire-cage isolates have found social exploration, gregariousness, (Mason, 1963a), social play (Harlow *et al.*, 1966), and sexual behavior (Senko, 1966) to be markedly depressed and qualitatively abnormal in adult and near-adult isolate rhesus monkeys.

Behavioral aberrations apparent immediately following the period of enclosed isolation gradually changed as Rowland's animals matured. For example, during the second year of life monkeys, that had been enclosed isolates during months 1–6 were fearful and subordinate, showing few signs of hostility. These same monkeys continued to demonstrate patterns of fear and submission during the third year; however, at this time the fear became associated with abnormal hostility and hyper-aggression. Abnormal hostility appeared soon after the first year of life (Rowland, 1964), but the hostility was suppressed or delayed by fear when the duration or severity of isolation was extreme (Mitchell *et al.*, 1966a). When the increased aggression became obvious, both the quality and quantity were abnormal. The fights of restricted monkeys have been described as being more frequent, longer in duration, and of greater severity than fights occurring in feral animals (Mason, 1963a). Attacks against huge adult males or brutal beatings of infants were not uncommon in the Rowland isolates' interactions with strangers (Mitchell, 1968a).

The dominance orders in groups of enclosed isolates were also abnormal. While aggressiveness and dominance are in many ways quite different types of behavior, the normal expression of dominance often depends on an appropriate restraint of aggression (Etkin, 1964). It has been shown that monkeys reared in individual wire cages have unstable dominance orders (Mason, 1963a), and Mitchell (1968a) demonstrated that Rowland's 4½-year-old enclosed isolate males were inferior to controls in dominance. In direct contests of dominance between isolates and controls, the control males were significantly more dominant than the isolate males, but the isolate and control females did not differ. Not one of eight enclosed isolate males won more than one-half of its pairings with eight control males; in fact, the isolate males were slightly less dominant than the isolate females (Mitchell, 1968a). One recent study on the relation between early social restriction and later dominance status has failed to replicate the original Mason (1961b) finding that early isolation is related to unstable dominance. However, these researchers (Angermeier et al., 1967) did not isolate their monkeys until 2–4 months of age, thus the failure to find similar results is confounded by this factor.

3. Coprophagy

According to Hill (1966), coprophagy or the eating of excrement, is one of the most difficult problems to deal with in a zoo primate collection. In follow-up studies by Mitchell et al. (1966a) of the Rowland (1964) enclosed isolate rhesus monkeys, it was found that coprophagy often occurred in the isolation-raised animals whereas it seldom if ever occurred in the controls (see also Senko, 1966). It was interesting that the monkeys that had been in the enclosed isolation chamber for the entire first year exhibited the most coprophagy, those isolated for the first 6 months exhibited the second highest amount, and those isolated between 6 and 12 months the third highest amount. It apeared that the longer, the earlier, and the more severe the captive situation in monkeys, the more frequently coprophagy occurred.

Although coprophagy is a common problem in gorillas in captivity, Schaller (1963), in his study of the gorilla in the wild, did not see evidence of coprophagy in 466 hours of direct observation. It must be assumed that coprophagy is a condition occurring only in captivity. The impulse may be related to boredom in the case of captive adults (Hill, 1966), or related to excessive early experience with excrement in the case of isolation-reared monkeys.

4. Abnormal Stereotyped Movements in Rowland's Isolates

Several other behavioral abnormalities appeared in 4½-year-old monkeys reared in enclosed chambers for 6 or 12 months during the first year. The amount of stereotyped pacing, flipping, or jumping was enormously high in these animals, as was self-clasping and self-biting. Even rocking and crouching, infantile behaviors that usually wane with age in wire-cage isolates, still occurred frequently in some of the enclosed cage isolates when they were nearly adults.

Bizarre movements occurred frequently in the enclosed isolates but in situations different from the rocking, pacing, or flipping. While the latter stereotyped movements varied directly with increasing fear, bizzare movements varied inversely with fear (Mitchell, 1968a) and appeared to be stimulated by "events" occurring within the animal. Bizzare movements have been labeled "habit residuals," "catatonic contractures," "ritualistic movements," "abnormal limb posturings," "floating-limb phenomena," and "nonrepetitive stereotyped movements" (cf. Foley, 1934; Cross and Harlow, 1965; Mitchell et al., 1966a; Berkson et al., 1963). Such movements are rigidly performed out of context of other ongoing behaviors of the monkey.

One of the isolates observed by Mitchell et al. (1966a) provides a good example of what is meant by a bizzare movement. One 6-month enclosed isolate male slowly moved his right arm toward his head while in a rigid seated pose and, upon seeing his own approaching hand, suddenly appeared startled by it. His eyes slowly widened and he would at times fear grimace toward, threaten, or even bite the hand "sneaking up on him." If he did not look directly at the hand or did not bite it, "it" would continue to move toward him as he looked at "it" out of the corner of his eyes. As "it" approached him, his eyes became wider and wider until the hand was entirely clasping his face. There he would sit for a second or two, with saucer-sized eyes staring in terror between clutching fingers (Mitchell et al., 1966a). No matter how objective the observer, or how much the observer let Morgan's canon control his comments, a diagnosis of "mental illness" was inevitable.

G. OTHER STUDIES OF ABNORMAL MOVEMENTS

1. Berkson's Studies

Berkson (1967) pointed out the similarity between the stereotyped acts isolation-reared monkeys and apes perform and the stereotyped acts of human beings with certain abnormal conditions. In his excellent review of abnormal stereotyped motor acts, he emphasized the intriguing nature of the source of such movements: "The effects of general arousal and of competing behaviors are significant but tend to be ephemeral since they do not involve the fundamental organization of the behaviors . . . . Most stereotyped acts are organized without important reference to the environment. Thus the study of stereotyped behaviors presents the special problem of attempting to gain control over internal stimuli" (Berkson, 1967, p. 92). Berkson also states that the "deprivation" stereotype represents a level of behavioral complexity somewhere between postural mechanisms and patterns of ideation. He noted that the number of different stereotyped behaviors that have been observed increases from monkeys to apes to humans. In lower monkeys such as marmosets, only "cage stereotypes" occur (Berkson et al., 1966) in isolation-reared animals. Berkson (1967) differentiates between "cage stereotypes" exhibited by wild animals brought into the laboratory and "deprivation stereotypes" displayed by isolation-reared animals. Cage stereotypes develop in many species of animals kept in small cages (e.g., pacing in lions; pacing, back-flipping, and jumping repetitively on all fours in young macaque monkeys). Although both deprivation stereotypes and cage stereotypes are repetitive, only cage stereotypes involve locomotion, and only deprivation stereotypes occur *despite* freedom of movement. In addition, deprivation stereotypes develop only in isolation-reared animals, while cage stereotypes have been seen in both isolation-reared and mother-reared animals (Berkson, 1967). Among the many abnormal stereotyped movements listed by Berkson are body rocking, body twirling, head rolling, head banging, self-clasping, weird limb and body posturing, digit sucking, eye poking and gouging, self-biting, and complex hand movements.

Although infant chimpanzees and monkeys deprived of the mother very early in life (the first year), develop odd and persistent stereotyped movements, infant chimpanzees and monkeys isolated during the second year of life for 6 months or more do not develop abnormal deprivation stereotypes (Davenport et al., 1966; Joslyn, 1967).

2. Mason Rocking Study

Quite recently, Mason (1968) pointed out that: ". . . under natural conditions the mother ordinarily provides a great deal of passive movement stimulation to her infant in the course of her routine activities. . . . We might conjecture that the infant monkey or ape that is deprived of such passive movement stimulation may supply it for himself through self-rocking or similar repetitive activities" (Mason, 1968, p. 83).

Mason compared two groups of rhesus monkeys, both separated from their mothers at birth. One group was reared on a moving cloth-covered dummy and the other on a stationary dummy. All of those reared on the immovable dummy developed persistent stereotyped rocking, while none of the monkeys reared on moving surrogates rocked. In addition, the monkeys reared on moving dummies were less fearful and more active than those reared with stationary surrogates: "Robot-reared monkeys spend less time in contact with the social surrogate; they more often move about the cage and they are quicker to approach and to interact with people. . . . We can be sure that the differences were produced by adding movement to one of the dummies but beyond that it is difficult to be more precise" (Mason, 1968, p. 87).

Mason (1968) described what he called the "primate deprivation syndrome," the symptoms of which are: (1) abnormal postures and movements, for example, rocking, (2) motivational disturbances, for example, excessive fearfulness or arousal, (3) poor integration of motor patterns, for example, inadequate sexual behavior, and (4) deficiencies in social communication, for example, threat by an aggressive animal does not produce withdrawal by a subordinate animal. He noted that the addition of movement in early life did away with at least the first two symptoms of the syndrome. But what about integration of motor patterns and deficiencies in communication?

H. Comments Concerning Communication and Response Integration

Following Mitchell's (1968a) second follow-up of the Rowland enclosed isolates, three of Rowland's 12-month enclosed isolates were sent to Dr. Robert E. Miller and Dr. I. A. Mirsky at the Laboratory of Clinical Science, School of Medicine, University of Pittsburgh. These three animals and three feral monkeys of the same age (5 years) were trained to perform an instrumental avoidance response to a visual stimulus. The isolates and controls performed equally well in acquiring the response. The animals were then paired in all possible combinations for communication of affect tests which utilized a cooperative avoidance technique devised by Miller and his associates. The face of a monkey that received the visual conditioned stimulus to avoid was visible (via closed circuit television) to another monkey that had access to the response bar. The isolate monkeys were not capable of utilizing facial expressions of other monkeys to perform appropriate avoidance responses and, in addition, were also defective in exhibiting facial expressions (Miller *et al.*, 1967). These results demonstrate that the fourth symptom of Mason's primate deprivation syndrome is a real one and they support Mason's hypothesis that isolate monkeys behave abnormally in part at least because they lack the opportunity to acquire communicative skills (Mason, 1963a).

I. Emergence from Isolation and Arousal

1. Dog Studies

The phenomena associated with emergence from an enclosed isolation chamber after a period of isolation early in the life of dogs prompted Fuller and L. D. Clark (1966) to adapt dogs to their future test environment prior to emergence. Their view, more explicitly stated by Fuller (1967), was that abnormalities following isolation result from stress at emergence. The isolation-reared animal allegedly does not habituate in ontogenetic sequence to more and more complex environmental stimuli and, therefore, is overwhelmed by a sudden bombardment by novel stimulation at emergence. Fuller (1964) found that brief periods of adaptation to the future test environment prevented the appearance of increased disturbance and emotional behavior in dogs. There was no evidence for a postisolation depression as was seen in Rowland's (1964) isolate monkeys.

2. Monkey Studies

The above studies of the emergence phenomenon in dogs prompted D. L. Clark at the University of Wisconsin to attempt the Fuller and Clark adaptation procedure with isolate monkeys (*M. mulatta*). Clark (1968) divided 16 infant monkeys into four sex-balanced groups of 4 animals each. A control group was provided extensive mother and peer experience for the first 15 months of life. The three isolate groups were isolated in enclosed chambers for the first 6 months of life (early), between 3 and 9 months of age (intermediate), and for the last 6 months of the first year (late) respectively. Social experience was provided to each isolate group when they were not in isolation and, during isolation periods, all groups of isolates were continuously adapted to the testing environment.

The nonsocial adaptation to the testing situation failed to prevent the development of crouching and rocking behaviors in the 6-month early isolates, but Clark's adapted early isolates recovered from their postisolation depression much sooner than had Rowland's (1964). The 3- to 9-month intermediate isolates did not develop crouching and rocking patterns, yet they were abnormally assertive upon emerging from the chamber. The 6-month late Clark isolates, which had been exposed to social experience for the first half-year of their lives before being sentenced to 6 months in the enclosed chamber, exhibited neither abnormal rocking and crouching nor hyperassertiveness. Follow-up testing (Mitchell and Clark, 1968) of all the Clark animals revealed abnormal

emotional and assertive behaviors in all three groups of isolates as compared to the extensively socialized controls. Clark (1968) concluded that:

> Isolation resulted both in maladaptive and disrupted social stimulus-
> response acquisition and in disrupted and abnormal emotional mecha-
> nisms. . . . Physical social interaction experience was a crucial factor in
> normal social development in the rhesus monkey and . . . early social
> experience prevented the development of maladaptive behavioral ten-
> dencies. Thus the effect of social experience prior to 3-months of age was
> the inhibition of the development of abnormal crouching and rocking be-
> havior, while the effect of social experience prior to 6-months of age was
> the inhibition of both crouching-rocking and abnormal or increased
> emotionality and assertiveness. (Clark, 1968, p. 120.)

Berkson (1968) recently reported that crouching and rocking develop
in isolated crabeating macaques as well as in rhesus monkeys. His data
regarding the development of abnormal stereotyped movements in
general support the findings of D. L. Clark at Wisconsin regarding the
critical ages for the development of rocking and hyperassertiveness:

> Groups of five carabeating macaques were separated from their mothers
> at 0, 1, 2, 4, or 6 months of age and observed during their first year of
> life. Abnormal stereotyped behaviors developed in all groups, but the
> frequencies of different patterns were modified by age at isolation. Self-
> sucking, crouching, and self-grasping developed almost immediately after
> birth, while body rocking and repetitive locomotion began later. Animals
> separated at a later age were more active and aggressive in a novel en-
> vironment than were those separated early. (Berkson, 1968, p. 118.)

3. Follow-up Monkey Studies

A report resulting from follow-up tests of some of Clark's isolates
(Mitchell and Clark, 1968) indicated that something more than simple
adaptation to stimuli of increasing complexity or a decrease in emotional
arousal is required to preclude the development and persistence of all
abnormal behaviors in isolates. Even though continuous adaptation may
help control emotional or arousal levels and decrease postisolation depression and disturbance, aberrant social behaviors such as hyperassertiveness still appear following social isolation. These later effects result
from trouble with response integration or communication, and it is
doubtful that the communicative deficiencies are simply secondary to
excess arousal.

J. Comparative Data

Isolation of rhesus monkeys and crabeating macaques early in life evidently interferes with the development of responsiveness to social cues. This is true for other macaques as well. C. S. Evans (1967) compared laboratory-reared and socially deprived pigtail macaques to a control group which had been reared in a freely interacting colony whose social conditions were thought to simulate those of natural troops. The socially deprived infants were taken from their mothers on the day of birth and reared without macaques maternal care under nursery conditions. From day 2 until 22 the deprived infants were kept in a standard hospital incubator and thereafter isolated in wire-mesh cages of 10 ft³ with solely visual access to other infants. From day 22 the restricted infants were also given daily physical access to one another for 60–90 minutes per day until day 296. Play and communicative behaviors occurred significantly less frequently in the group of socially deprived infant pigtails than they did in the free-colony controls (which were brought into a playroom and tested under similar conditions). Only in the deprived group did fear-grimacing, aggression, digit sucking, and self-clasping appear. Sexual dimorphisms were evident only in the more socially reared group. The brief daily play periods provided for these deprived pigtails did not facilitate their social development to the control level when tested at less than 1 year of age. Social deprivation appears to affect *M. nemestrina* as it does *M. mulatta* and *M. fascicularis*.

K. Age Differences

It has already been noted that behavioral abnormalities change with age. Several studies have shown that the behaviors of enclosed isolate monkeys change as the animal matures (Mitchell *et al.*, 1966a); Mitchell, 1968a; Mitchell and Clark, 1968). Cross and Harlow (1965), in an extensive cross-sectional study of 84 rhesus monkeys, demonstrated that behavioral abnormalities change with age in wire-cage-reared isolates as well as in enclosed cage isolates. Eighty-four laboratory-born rhesus monkeys ranging in age from 1 to 7 years were observed in their home cages. The results were analyzed under broad categories of orality, disturbances, and aggression. In infant (1-year-old) wire-cage-reared monkeys, digit sucking occurred very frequently but this behavior steadily declined with age until it almost never appeared in 7- and 8-year-olds. The curves for vocalizing, grimacing, self-clasping, crouching, rocking, scratching, and convulsive jerking were essentially the same. All occurred more frequently in infants than in adults. Although sucking orality steadily declined with age, self-chewing and self-biting orality increased. Aggression was a late-maturing variable. External threats and self-directed threats increased steadily with age after 3 years. Pacing and yawning also increased with the age of the animals being observed.

Cross and Harlow (1965) compared these age changes in wire-cage isolates to changes in mother-peer-reared animals. There was an almost total absence of digit sucking in mother-raised monkeys. Environmentally directed chewing increased with age in both isolates and controls but was consistently higher in controls. Self-clasping was essentially non-existent in the mother-raised macaque. Grimacing appeared more frequently in the isolates than in the mother-raised monkeys. With regard to differences in aggression, the mother-raised animals showed many more externally directed threat responses than the wire-cage isolates, but self-directed threats and self-biting were essentially nonexistent in the mother-raised animals. It was in this study that Cross and Harlow reported the occurrence of what they called "catatonic contracture": "While the monkey sat in a quiescent state, an arm slowly and gradually floated upward with concurrent flexion of the wrist and fingers—a movement made as if the limb were not an integral part of the monkey's own body (Cross and Harlow, 1965, p. 47). Such behaviors were found almost exclusively in the wire-cage isolates and not in the mother-peer-reared animals. Wire-cage isolates also engaged in autoeroticism more frequently than the controls, while controls exhibited more lipsmacking, cage shaking, pacing, and jumping (Cross and Harlow, 1965). Thus the wire-cage-reared isolate gradually changed as he matured from a rocking, digit-sucking, grimacing, self-clutching recluse to a pacing, self-threatening, masturbating, self-mutilating menace who often made bizarre movements.

L. VISUAL ISOLATES

1. The Use of Slides during Rearing

We have already discussed the effects of visual access versus nonvisual access to other monkeys early in life. There appear to be differences between enclosed isolates and wire-cage isolates (Rowland, 1964), but both kinds of early social isolation produce marked abnormalities. Pratt (1969) reared rhesus monkeys in enclosed isolation chambers from birth to 9 months but presented the enclosed isolates with colored slides of other monkeys during their sentence in isolation. After emergence their nonsocial exploration was essentially equivalent to that of wire-cage isolates. Pratt found, however, that such isolates were more disturbed, more fearful, and less active in social situations than were wire-cage isolates. They were quite clearly nonaggressive at the end of the second year of life and remained crouched in one position during much of a given social test situation. The behaviors of these so-called "visual isolates" were perhaps a little more positive yet very much like the behaviors of Rowland's (1964) 2-year old animals that had been housed in enclosed chambers for the entire first year. Pictures of monkeys alone do not help to prevent the appearance of excessive arousal and aberrant disturbance in social situations.

2. Factors Related to the Appearance of Fear

The response of the visual isolates in the isolation chamber varied with age and according to which pictures were shown to them (Sackett, 1966b). Exploration, play, vocalization, and disturbance occurred most frequently with slides of a monkey threatening and with pictures of infants. Between 2½ and 4 months, threat pictures yielded a high frequency of disturbance. Lever-touching to turn on threat pictures was very low during this age period. Sackett (1966b) concluded that pictures of infants and of threat appear to have prepotent general activating properties and that between 2½ and 4 months of age slides depicting threats appear to release a developmentally determined, inborn fear response. This may be related to the appearance of intense disturbance upon emergence from isolation at 3 months of age (Griffin and Harlow, 1966).

M. SEXUAL BEHAVIOR

Wire-cage-reared isolate monkeys (*M. mulatta*), visual isolates, and enclosed-chamber-reared isolate monkeys do not develop normal sexual behavior (Harlow, 1962). Their social-sexual behavior throughout life is abnormally low frequency and obviously abnormal in form (Senko, 1966; Mitchell, 1968a). While female isolates may at times be impregnated by tolerate and patient feral males, isolation-reared males "are completely expendable" (Harlow, 1962). Attempted mountings by isolation-reared males, which occur rarely, generally are not properly oriented with regard to the feasibility of intromission. The target is often the side of the female's body. As a consequence none succeed in impregnating a partner. Many of the isolation-reared females try to avoid males, threaten them, or collapse under the weight of the mount, making it extremely difficult to complete the act (Harlow and Harlow, 1962b).

1. The Senko Studies

Senko (1966) compared wire-cage adult isolate female monkeys (*M. mulatta*) to feral adult females in pairings with adult males. The deprived adult females displayed more self-clasps, more self-bites, threatened the male more often, and aggressed the adult male more often than did the feral females. In addition, the deprived females were significantly deficient relative to feral females in social proximity, social grooming, sexual presenting, hind-quarter contact, support for male, and incidence of insemination. Senko compared deprived males to feral males and found the ferals to be significantly superior to the deprived males in social proximity to females, social grooming, response to female present, and in incidence of insemination. The deprived males displayed significantly more aggression toward females, self-biting, self-clasping, threatening of females, coprophagy, sexually abnormal responses, and sexually evoked aggression than did the ferals. Number and duration of erections (Senko, 1966) and the practice of masturbation (Mitchell *et al.*, 1966a) indicated that the deprived males had a normal level of sexual motivation.

2. The Puerto Rico Studies

In 1964 Meier presented surprising data which, contrary to the Wisconsin data, indicated that wire-cage isolation-reared rhesus monkeys did not differ from ferals in adult sexual behavior. Meier's (1965) data indicated that "the significance of tactual stimulation, peer contact and play suggested by the Wisconsin investigators had been greatly overemphasized." Missakian examined the data utilized by Meier, however, and concluded that

> . . . There is evidence that the records referred to by Meier are unreliable indicants of reproductive capacity. Instances of misidentification of vaginal lavage with regards to the female and multiple matings of a female during one menstrual cycle have produced errors in records of positive sperm outcome and conception rate. In light of this new evidence, the reproductive behavior of cage-reared males in the Puerto Rico colony was re-examined. . . . Results from this study supported Mason and Harlow's findings. The extent and severity of the deficit produced by social deprivation was reflected in the failure of any cage-reared male to execute a normal mount. (Missakian, 1968, p. 23.)

Missakian (1968) concluded her paper with the statement that periodic attempted matings of wire-cage-reared males with sexually receptive females is not an effective means of revising the effects of social deprivation. A wide variety of social experiences must be provided to such males in order to determine those experiences that are most critical.

Clearly, early social isolation of rhesus monkey males interferes profoundly with adult sexual behavior.

3. Sexual Behavior in Isolate Apes

Kollar *et al.* (1968) have published evidence in which they claim to show that early social isolation is not necessary to interfere with normal sexual behavior in chimpanzee males. Although many of their chimpanzee males were sexually inept, they were feral-born and thus had experienced a normal mother-infant relationship for at least a few months. According to Kollar *et al.* (1968), the majority of the animals were African-born, separated early from their mothers, and raised in isolation or semiisolation. There were no records regarding the age of separation. They believe, however, that conditions of early deprivation existed that interfered with the development of normal patterns of behavior. If chimpanzees reached maturity without an opportunity to copulate, they did not learn to mate properly. The authors suggested that there was apparently a "critical period for learning mating behavior between puberty and the attainment of full anatomical maturity" in the chimpanzee. It is worth noting in comparison that in the rhesus monkey the *first* year of life *seems* to be more important than the period after puberty. Second-year isolation in the monkey in one study had only transient effects on behavior, including sexual behavior (Joslyn, 1967).

At the Yerkes Regional Primate Research Center, Rogers and Davenport (1969) took five male and seven female chimpanzees from their mothers within 12 hours after birth and reared them in closed boxes in isolation from humans and other chimpanzees. At age 3 they were placed in pairs or in larger groups including ferals. Social isolation had a drastic negative effect on chimpanzee sexual behavior; yet some of the isolation-reared chimpanzees recovered and were able to copulate. Rogers and Davenport attributed this recovery to a less rigid or stereotyped pattern in the chimpanzee in contrast to the rhesus monkey. The chimpanzee, according to them, was more amenable to modification by experience.

An interesting serendipitous finding in the Rogers and Davenport study was that the sexual behavior of the chimpanzee reared in enclosed isolation was *less* drastically affected than that of the chimpanzee for whom human maternal care was substituted. Thus the important factor in primate socialization was again *not* simply the amount of stimulation or the amount of social experience: "Qualitative appropriateness of the social companion during early life for the development of later species—

specific behavioral organization is extremely important" (Rogers and Davenport, 1969, p. 203).

It follows therefore that inappropriate early stimulation may in some cases be *worse* than none at all. Early attachment to unnatural objects at least in this case apparently resulted in a preference for the unnatural object over the natural object. Evidently, generalization from human to chimpanzee was much more difficult than from no object of attachment to chimpanzee. This hypothesis is also in keeping with the preference data obtained by Sackett *et al.* (1965) utilizing rhesus monkeys. Rhesus monkeys reared by humans moved toward the human and hence *away* from the monkey in a preference apparatus. Monkeys reared in isolation had no real clear-cut preference but, if anything, chose the *monkey* rather than the human.

IX. MATERNAL PUNISHMENT OR MATERNAL INDIFFERENCE

Sexually mature female rhesus monkeys raised in social isolation exhibit inappropriate and infrequent adult sexual behavior (Mitchell, 1968a); yet such isolation-reared females are often impregnated, either by very sophisticated wild-reared males or through the use of a restraining apparatus which permits copulation (Allen *et al.*, 1967).

A. ORIGINAL "MOTHERLESS MOTHER" STUDY

Seay *et al.* (1964) first studied abnormal maternal behavior in four socially deprived female rhesus monkeys. All four "motherless mothers" were inadequate and none of their infants would have survived without intervention by the laboratory staff. Two of the mothers were brutal toward their infants and two were indifferent. One mother initially abandoned and retreated from her infant. The infant was then hand fed (Blomquist and Harlow, 1961) until day 3 when the mother in question finally passively accepted the infant to her breast. Further attempts to remove the infant resulted in violent attacks directed by the mother against her infant including jumping up and down with her full weight on the infant. Yet this infant and all the infants of brutal and indifferent mothers continued to approach their mothers for physical contact.

When the infants of some other motherless mothers were permitted access to a chamber that housed a standard time recorder to an inanimate cloth surrogate mother, the infants still showed a marked preference for the real mother, even when the real mother was indifferent or abusive (Arling, personal communication). Apparently, a small amount of physical contact with a real monkey mother, however inadequate, is far more reinforcing than unlimited contact with a cloth surrogate.

Amount of abusiveness and hostility is not necessarily inversely related to the amount of nursing and cradling displayed by a mother monkey (Mitchell, 1968b). The most abusive mother in the Seay *et al.* (1964) study exhibited more nursing and more cradling than the other three socially deprived females in the study. In addition, the most brutal mother gradually came to restrain her infant from leaving her and going into the play area. Yet she would violently abuse the infant following nursing or suddenly strike it without provocation, although the severity of the beatings decreased as the infant grew older. The responses of the infant to this abuse were marked disturbance and attempted approaches to the mother.

Each motherless mother and her own idiosyncratic and arbitrary type of inadequacy; but, as a group, motherless mothers cradled and nursed their infants less than did the controls (feral-reared females). In the first 90 days of the infants' lives, the motherless mothers rejected and punished their infants significantly more frequently than did the control mothers, although the difference disappeared in the second 3 months. Normal mothers also punished their infants quite frequently but their punishment was always gentle. Male infants of both motherless mothers and adequate mothers were punished more frequently than females (Mitchell, 1968c).

During the first 2 months of life, the infants of motherless mothers cooed and self-mouthed significantly more frequently than the infants of control mothers. Oral social exploration was heightened in infants of motherless mothers when they interacted with each other. This may have been a compensation for the restriction of mother mouthing and nursing by their motherless mothers but, if so, it was only socially directed because there were no differences between control and motherless-mothered infants in the frequency of mouthing inanimate objects. Control infants engaged in significantly more self-play, but otherwise the infants of motherless mothers were at control levels in all play behaviors.

B. Hyperaggressiveness in the Infants

A subsequent study of still more motherless mothers (Arling and Harlow, 1967) showed that the infants of motherless mothers exhibited a higher frequency of clasps, pulls, and bites in infant-infant interactions than did control infants. As can be seen in the results of the long-term studies that follow, this high level of clasp-pull-bite appearing in infancy was the first indication that motherless-mother rearing facilitated aggression.

In an assessment of the long-term effects of motherless mothering (Møller *et al.*, 1968), it was found that some dominance gestures (yawn, crook tail) that usually serve to redirect hostility away from another animal were deficient in 26-month-old male infants of motherless mothers. Physical aggression was never seen in normal mother-peer-reared control males in the test situation in question, but it was observed frequently in the motherless-mothered males. The near-adolescent motherless-mothered males attacked small and large stimulus animals without warning and without provocation. One of the same males often walked past an infant stimulus animal while frequently striking, biting, or nonchalantly pulling out handfuls of fur and, in another test situation the same male bit a finger off an infant (Chamove *et al.*, 1967). Another motherless-mothered male bit four fingers off a motherless-mothered female cage mate and killed a second motherless-mothered female cage mate. Both of these events occurred prior to adolescence (Mitchell *et al.*, 1967a). Early maternal punishment was positively correlated with later aggression and this appeared to be true within the normal range of maternal rejection and punishment as well as within motherless-mother punishment (Mitchell *et al.*, 1967a).

C. SUBSEQUENT INFANTS OF BRUTAL MOTHERS

As noted in a previous section, normal primiparous rhesus mothers are only slightly different from normal multiparous rhesus mothers (Seay, 1966). The adequately socialized primiparous female is a little more nervous and tends to stroke and protect her infant more frequently than the multiparous female (Mitchell and Stevens, 1970). Multiparous females, however, reject and punish their infants sooner and more frequently than primiparous females do (Seay, 1966; Mitchell, 1969a). Since the motherless mothers discussed above were all primiparous mothers and yet were already very rejecting, it would be expected that if normal rules applied such motherless-mother females should become extremely violent toward their second infants. Yet this is not the case because normal primiparous-multiparous rules do not apply to motherless mothers. Subsequent infants of brutal or indifferent motherless mothers are often treated adequately and are usually treated less abusively than first-born infants (Harlow *et al.*, 1966). Why is this true? Does the experience with the first infant have some therapeutic value?

This question has not been adequately answered but there are at least two studies that have considered the problem. Mitchell (1968a) noted that although adolescent *nulliparous* isolate-reared females were afraid of and aggressive toward infant and juvenile stimulus strangers, immature strangers elicited significantly more positive social responsiveness from the isolates than did age-mate and adult strangers. This suggested than an infant may indeed provide the type of stimulation that would make good therapy for abnormal maternal or paternalistic behavior (Mitchell, 1969b) feasible. Mitchell (1968a) also noted, however, that infant-directed aggression spontaneously decreased between 2½ and 4½ years of age in nulliparous isolation-reared females. Perhaps the decreased abusiveness in the motherless mothers was more a function of age than of the rehabilitative value of the first infant.

Arling *et al.* (1969) designed an experiment to determine if there were any improvements in the infant-directed behaviors of 8-year-old isolation-reared nulliparous rhesus females in the course of four individual 5-hour sessions with a 6-month-old rhesus infant. They also wanted to determine whether or not isolate females still displayed unprovoked physical aggression toward infants when these females had reached 8 years of age. The results were as follows. There were no significant improvements in the infant-directed behaviors of these 8-year old females during the therapy sessions but, unlike 4½-year-old nulliparous isolate females, 8-year-old nulliparous isolate females rarely exhibited unprovoked aggression toward infants.

A second experiment (Arling *et al.*, 1969) showed that the absence of aggression in 8-year-old isolate females was specific to infants; there was no doubt that these isolate females were hyperaggressive toward age-mates. As the nulliparous isolate female matures, she apparently becomes less aggressive toward infants and, if anything, more aggressive toward age-mates. Although these data did not completely preclude the proposal that the first-born provides therapeutic experience for motherless mothers, they did suggest that the changes in the infant-directed behavior of the isolate rhesus females was at least related to her age.

In checking the ages of primiparous motherless mothers that had been studied in the past, it was found that the degree of brutality was inversely related to the mother's age (Arling *et al.*, 1969). Since Cross and Harlow (1965) reported an almost linear increase in externally directed aggressions between 1 and 8 years of age in isolation-reared females, infant-directed aggression appears to be a special case and a striking exception to the rule. Our explanation for this difference starts with the assumption that the amount of aggression in a dyad decreases with increasing dominance stability. Dominance relations between adult animals and infants are in the normal case determined immediately and remain stable, whereas dominance relations between two age-mates are not immediately settled. Isolate rearing is not the normal case, however. The female isolate's behavior upon emerging from isolation, and for some time after emergence, is infantile in form. Since the isolate is herself a kind of infant, it is really not surprising to find that the dominance relations between a real infant and an oversized imitation of an infant are not immediately settled. In playroom tests of isolate females, we have often observed an infant threatening a 4-year old isolate female while the female fear-grimaced in response (Mitchell, 1968a). Young infantilistic female isolates (4 years old) evidently respond to real infants as though they were a threat, while old (8 years old) isolates do not. Since there is a decrease in fear and aggression toward infants with age, apparently some of the devastating effects of early isolation are not completely irreversible in the rhesus monkey (Mitchell, 1968a; Arling et al., 1969). There is evidently even some spontaneous recovery. Findings such as these leave comparative psychopathologists optimistic that the isolation-raised monkeys' persistent pathology may not necessarily have to remain permanent.

X. SUMMARY

Abnormal behavior is a relative term. It must be defined by using some reference control group, and the definition should include statements concerning the ecological validity and specific effects of the test situations employed. The persistence of the behavior is also of some importance to its definition.

It has been noted that abnormalities meeting these defining conditions often appear as a function of unfortunate, yet often uncontrollable, conditions in captivity. Low levels of stimulation, lack of space, presence of human observers, and absence of friendly ties between animals are included in such conditions.

Many behaviors appearing to be abnormal are often normal behavioral states related to the age or the sex of the animal. Even in abnormal animals the abnormality changes with age and is different in males and females.

Abnormalities related to birth involve causative factors such as asphyxia neonatorum, age and parity of the mother, mode of birth, length of labor, and difficulty of labor.

Subtle but sometimes severe behavioral abnormalities arise during and following manipulation of social experiences early in life. Peer deprivation, maternal deprivation, and temporary maternal separations produce pathologies that are usually quite persistent in primates.

Rearing in social isolation produces severe behavioral pathology in all primates. The most important source of stimulation that is absent in such a rearing condition is physical contact from another animal involving a complex combination of skin or fur contact, clinging, movement, oral contact, and warmth. Isolation-reared primates change as they mature from rocking, digit-sucking, grimacing, self-clutching recluses to pacing, socially aggressive, self-threatening, masturbating, self-mutilating menaces who often make bizarre movements. When the isolate animal reaches sexual maturity, its sexual behavior is abnormal. The maternal behavior of the isolation-reared female is indifferent or brutal, yet appears to improve with time. The infants of brutal mothers also show some signs of behavior pathology so that experimentally produced pathology can be "passed-on" to the next generation. The fact that maternal brutality in isolation-reared mothers slowly wanes with age and/or experience leaves comparative psychopathologists reasonably confident that experimentally produced behavior pathology can be alleviated.

ACKNOWLEDGMENT

Dr. G. D. Jensen made valuable suggestions after a critical reading of the manuscript.

REFERENCES

Abrams, P. S. (1969). Age and the effects of separation on mother and infant rhesus monkeys (*Macaca mulatta*). *Annu. Meet. Amer. Ass. Phys. Anthropologists, Mexico City.*

Alexander, B. K. (1966). The Effects of Early Peer-Deprivation on Juvenile Behavior of Rhesus Monkeys. Doctoral dissertation, Univ. of Wisconsin, Madison, Wisconsin.

Alexander, B. K., and Bowers, J. M. (1968). The social structure of the Oregon troop of Japanese macaques. *Primates* 8, 333–340.

Allen, J. R., Schiltz, K. A., Ripp, C., Eisele, S. C., and Johnson, L. C. (1967). Laboratory Procedures. Regional Primate Research Center and Laboratory, University of Wisconsin, Madison, Wisconsin.

Altmann, S. A. (1968). Sociobiology of rhesus monkeys. IV. Testing Mason's hypothesis of sex differences in affective behavior. *Behaviour* 32, 49–69.

Angermeier, W. F., Phelps, J. B., Murray, S., and Reynolds, H. H. (1967). Dominance in monkeys: Early rearing and home environment. *Psychonomic Sci.* 9(7B), 433–434.

Angermeier, W. F., Phelps, J. B., Murray, S., and Howastine, J. (1968). Dominance in monkeys: Sex differences. *Psychonomic Sci.* 12, 344.

Arling, G. L., and Harlow, H. F. (1967). Effects of social deprivation of maternal behavior of rhesus monkeys. *J. Comp. Physiol. Psychol.* 64, 371–378.

Arling, G. L., Ruppenthal, G. C., and Mitchell, G. D. (1969). Aggressive behavior of the eight-year old nulliparous isolate female monkey. *Anim. Behav.* **17**, 109–113.

Autrum, H., and von Holst, D. (1968). Sozialer, Stress bei Tupajas (*Tupaia glis*) und sein Wirkunfanf Wachotum, Kiërpergeurcht und Fortpflanzung. *Z. Vergl. Physiol.* **58**, 347–355.

Berkowitz, L. (1964). Aggressive cues in aggressive behavior and hostility catharsis. *Psychol. Rev.* **61**, 104–122.

Berkson, G. (1967). Abnormal stereotyped motor acts. *In* "Comparative Psychopathology" (J. Zubin and H. F. Hunt, eds.), pp. 76–94. Grune & Stratton, New York.

Berkson, G. (1968). Development of abnormal stereotyped behaviors. *Develop. Psychobiol.* **1**, 118–132.

Berkson, G., Mason, W. A., and Saxon, S. V. (1963). Situation and stimulus effects on stereotyped behaviors of chimpanzees. *J. Comp. Physiol. Psychol.* **56**, 786–792.

Berkson, G., Goodrich, J., and Kraft, I. (1966). Abnormal stereotyped movements of marmosets. *Perceptual Motor Skills* **23**, 491–498.

Bernstein, I. S. (1967). A field study of the pigtail monkey (*M. nemestrina*). *Primates* **8**, 217–228.

Bernstein, I. S., and Mason, W. A. (1962). The effects of age and stimulus conditions on the emotional responses of rhesus monkeys: Responses to complex stimuli. *J. Genet. Psychol.* **101**, 279–298.

Blomquist, A. J., and Harlow, H. F. (1961). The infant rhesus monkey program at the University of Wisconsin Regional Primate Research Center and Laboratory. *Proc. Anim. Care Panel II* No. 2 (April), pp. 57–64.

Boelkins, R. C. (1963). The Development of Social Behavior in the Infant Rhesus Monkey Following a Period of Social Isolation. MS. Thesis, Univ. of Wisconsin, Madison, Wisconsin.

Bowlby, J. (1961). Separation anxiety: A critical review of the literature. *J. Child Psychol. Psychiat.* **1**, 251–269.

Chamove, A. S. (1966). The Effects of Varying Infant Peer Experience on Social Behavior in the Rhesus Monkey. M.A. Thesis, Univ. of Wisconsin, Madison, Wisconsin.

Chamove, A. S., Harlow, H. F., and Mitchell, G. (1967). Sex difference in the infant-directed behavior of preadolescent rhesus monkeys. *Child Develop.* **38**, 329–335.

Clark, D. L. (1968). Immediate and Delayed Effects of Early, Intermediate, and Late Social Isolation in the Rhesus Monkey. Doctoral dissertation, Univ. of Wisconsin, Madison, Wisconsin.

Cross, H. A., and Harlow, H. F. (1965). Prolonged and progressive effects of partial isolation on the behavior of macaque monkeys. *J. Exp. Res. Pers.* **1**, 39–49.

Davenport, R. K., Menzel, E. W., Jr., and Rogers, C. M. (1966). Effects of severe isolation on "normal" juvenile chimpanzees. *Arch. Gen. Psychiat.* **14**, 134–138.

DeVore, I. (1963). Mother-infant relations in free-ranging baboons. *In* "Maternal Behavior in Mammals" (H. L. Rheingold, ed.), pp. 305–335. Wiley, New York.

Draper, W. A. (1965). Sensory stimulation and rhesus monkey activity. *Perceptual Motor Skills* **21**, 319–322.

Draper, W. A. (1966). Free-ranging rhesus monkeys: Age and sex differences in individual activity patterns. *Science* **151**, 476–478.

Etkin, W. (1964). Cooperation and competition in social behavior. *In* "Social

Behavior and Organization Among Vertebrates" (W. Etkin, ed.), pp. 35–52. Univ. of Chicago Press, Chicago, Illinois.

Evans, C. S. (1967). Methods of rearing and social interaction in *Macaca nemestrina*. *Anim. Behav.* **15**, 263–266.

Fiennes, R. N. T.-W. (1968). Ecological concepts of stress in relation to medical conditions in captive wild animals. *Proc. Roy. Soc. Med.* **61**(2), 161–162.

Foley, J. P., Jr. (1934). First year development of a rhesus monkey (*M. mulatta*) reared in isolation. *J. Genet. Psychol.* **45**, 39–105.

Foley, J. P., Jr. (1935). Second year development of a rhesus monkey reared in isolation. *J. Genet. Psychol.* **47**, 39–105.

Fuller, J. L. (1964). The K-puppies. *Discovery* (Feb.).

Fuller, J. L. (1967). Experimental deprivation and later behavior. *Science* **158**, 1645–1652.

Fuller, J. L., and Clark, L. D. (1966). Genetic and treatment factors modifying the postisolation syndrome in dogs. *J. Comp. Physiol. Psychol.* **61**, 251–257.

Gantt, W. H., Newton, J. E. O., Royer, F. L., and Stephens, J. H. (1966). Effect of person. *Conditioned Reflex* **1**, 18–35.

Gartlan, J. S. (1968). Structure and function in primate society. *Folia Primatol.* **8**, 89–120.

Green, P. C. (1965). Influence of early experience and age on expression of affect in monkeys. *J. Genet. Psychol.* **106**, 157–171.

Griffin, G. A. (1965). Effects of Three Months of Total Social Deprivation on Social Adjustment and Learning in the Rhesus Monkey. M.S. Thesis, Univ. of Wisconsin, Madison, Wisconsin.

Griffin, G. A. (1966). The Effects of Multiple Mothering on the Infant-Mother and Infant-Infant Affectional Systems. Doctoral dissertation, Univ. of Wisconsin, Madison, Wisconsin.

Griffin, G. A., and Harlow, H. F. (1966). Effects of three months of total deprivation on social adjustment and learning in the rhesus monkey. *Child Develop.* **37**, 533–547.

Hall, K. R. L., and Mayer, B. (1967). Social interactions in a group of captive patas monkeys (*Erythrocebus patas*). *Folia Primatol.* **5**, 213–236.

Hansen, E. W. (1966). The development of maternal and infant behavior in the rhesus monkey. *Behavior* **27**, 107–149.

Hansen, E. W., Harlow, H. F., and Dodsworth, R. O. (1966). Reactions of rhesus monkeys to familiar and unfamiliar peers. *J. Comp. Physiol. Psychol.* **61**, 274–279.

Harlow, H. F. (1958). The nature of love. *Amer. Psychologist* **13**, 673–685.

Harlow, H. F. (1962). The heterosexual affectional system in monkeys. *Amer. Psychologist* **17**, 1–9.

Harlow, H. F., and Harlow, M. K. (1962a). Social deprivation in monkeys. *Sci. Amer.* **207**, 136–146.

Harlow, H. F., and Harlow, M. K. (1962b). The effects of rearing conditions on behavior. *Bull. Menninger Clin.* **26**(5), 213–224.

Harlow, H. F., and Harlow, M. K. (1969). Effects of various mother-infant relationships on rhesus monkey behaviors. *In* "Determinants of Infant Behaviors" (B. M. Foss, ed.), Vol. 4, pp. 15–36. Methuen, London.

Harlow, H. F., and Zimmerman, R. R. (1959). Affectional responses in the infant monkey. *Science* **130**, 421–432.

Harlow, H. F., Rowland, G. L., and Griffin, G. A. (1964). The effect of total social deprivation on the development of monkey behavior. *Psychiat. Res. Rep. Amer. Psychiat. Ass.* **19,** 116–135.

Harlow, H. F., Harlow, M. K., Dodsworth, R. O., and Arling, G. L. (1966). Maternal behavior of rhesus monkeys deprived of mothering and peer associations in infancy. *Proc. Amer. Phil. Soc.* **110,** 58–66.

Held, R., and Bauer, J. A. (1967). Visually guided reaching in infant monkeys after restricted rearing. *Science* **155,** 718–720.

Hill, C. A. (1966). Coprophagy in apes. *Int. Zoo Yearb.* **6,** 251–257.

Hill, C. W., Greer, W. E., and Felsenfeld, O. (1967). Psychological stress, early response to foreign protein, and blood cortisol in vervets. *Psychosom. Med.* **29,** 279–283.

Hinde, R. A., Spencer-Booth, Y., and Bruce, M. (1966). Effects of 6-day maternal deprivation on rhesus monkey infants. *Nature* (*London*) **210,** 1021–1023.

Jensen, G. D. (1968). Reaction of monkeys mothers to long-term separation from their infants. *Psychonomic Sci.* **11**(5), 171–172.

Jensen, G. D., and Bobbitt, R. A. (1968). Monkeying with the mother myth. *Psychol. Today* **1,** 44.

Jensen, G. D., and Tolman, C. W. (1962). Mother-infant relationship in the monkey, *Macaca nemestrina.* The effect of brief separation and mother-infant specificity. *J. Comp. Physiol. Psychol.* **55,** 131.

Jensen, G. D., Bobbitt, R. A., and Gordon, B. N. (1967). Sex differences in social interaction between infant monkeys and their mothers. *Recent Advan. Biol. Psychiat.* **21,** 283–292.

Jensen, G. D., Bobbitt, R. A., and Gordon, B. N. (1968). Effects of environment on the relationship between mother and infant pigtailed monkeys (*Macaca nemestrina*). *J. Comp. Physiol. Psychol.* **66,** 259–263.

Joslyn, W. D. (1967). Behavior of socially experienced juvenile rhesus monkeys after eight months of late social isolation and maternal-offspring relations and maternal separation in juvenile rhesus monkeys. Unpublished Ph.D. dissertation. Univ. of Wisconsin, Madison, Wisconsin.

Joslyn, W. D. (1968). Social Adjustment in Two Rhesus Monkeys After Prolonged Peer Privation. Oregon Regional Primate Research Center, Univ. of Oregon, Eugene, Oregon.

Kaufman, I. C., and Rosenblum, L. A. (1966). A behavioral taxonomy for *M. nemestrina* and *M. radiata*: Based on longitudinal observations of family groups in the laboratory. *Primates* **7,** 205–258.

Kaufman, I. C., and Rosenblum, L. A. (1967). The reaction of separation in infant monkeys: Anaclitic depression and conservation-withdrawal. *Psychosom. Med.* **29,** 648–576.

King, J. A. (1958). Parameters relevant to determining the effects of early experience upon the adult behavior of animals. *Psychol. Bull.* **55,** 46–58.

Kollar, E. J., Edgerton, R. B., and Beckwith, W. C. (1968). An evaluation of the ARL colony of chimpanzees. *Arch. Gen. Psychiat.* **19**(5), 580–595.

Kummer, H., and Kurt, F. (1965). A comparison of social behavior in captive and wild hamadryas baboons. *In* "The Baboon in Medical Research" (H. Vagtborg, ed.), pp. 65–80. Univ. of Texas Press, Austin, Texas.

Lashley, K. S., and Watson, J. B. (1913). Notes on the development of a young monkey. *J. Anim. Behav.* **3,** 114–139.

Lindburg, D. G. (1970). Grooming behavior as a regulator of social interactions in rhesus monkeys *In* "Social Regulatory Mechanisms in Primates" (C. R. Carpenter, ed.), Univ. of Pennsylvania Press Philadelphia, Pennsylvania.

McCulloch, T. L. (1939). The role of clasping activity in adaptive behavior in the infant chimpanzee: I. Delayed response. *J. Psychol.* **7,** 283–292.

McCulloch, T. L. (1939b). The role of clasping activity in adaptive behavior of the infant chimpanzee: IV. The mechanics of reinforcement. *J. Psychol.* **7**, 305–316.

McCulloch, T. L., and Haslerud, G. M. (1939). Affective responses of an infant chimpanzee reared in isolation from its kind. *J. Comp. Psychol.* **28**, 437–445.

Mason, W. A. (1960a). The effects of social restriction on the behavior or rhesus monkeys: I. Free social behavior. *J. Comp. Physiol. Psychol.* **53**, 583–589.

Mason, W. A. (1960b). Socially mediated reduction in emotional responses of young rhesus monkeys. *J. Abnorm. Soc. Psychol.* **60**, 100–104.

Mason, W. A. (1961a). The effects of social restriction on the behavior of rhesus monkeys: II. Tests of gregariousness. *J. Comp. Physiol. Psychol.* **54**, 287–290.

Mason, W. A. (1961b). The effects of social restriction on the behavior of rhesus monkeys: III. Dominance tests. *J. Comp. Physiol. Psychol.* **54**, 694–699.

Mason, W. A. (1961c). Effects of age and stimulus characteristics on manipulatory responsiveness of monkeys raised in a restricted environment. *J. Genet. Psychol.* **99**, 302–308.

Mason, W. A. (1963a). The effects of environmental restriction on the social development of rhesus monkeys. *In* "Primate Social Behavior" (C. H. Southwick, ed.), pp. 161–173. Van Nostrand, Princeton, New Jersey.

Mason, W. A. (1963b). Social development of rhesus monkeys with restricted social experience. *Perceptual Motor Skills* **16**, 263–270.

Mason, W. A. (1968). Early social deprivation in the nonhuman primates: Implications for human behavior *In* "Environmental Influences" (D. C. Glass, ed.), pp. 70–100. Rockefeller Univ. and Russell Sage Found., New York.

Mason, W. A., and Green, P. C. (1962). The effects of social restriction on the behavior of rhesus monkeys: IV. Responses to a novel environment and to an alien species. *J. Comp. Physiol. Psychol.* **55**, 363–368.

Mason, W. A., and Sponholz, R. R. (1963). Behavior of rhesus monkeys raised in isolation. *J. Psychiat. Res.* **1**, 299–306.

Mason, W. A., Green, P. C., and Posepanko, C. J. (1960). Sex differences in affective social responses of rhesus monkeys. *Behaviour* **16**, 74–83.

Masserman, J. H., Wechkin, S., and Woolf, M. (1968). Alliances and aggressions among rhesus monkeys. *Sci. Psychoanal.* **7**, 95–100.

Meier, G. W. (1965). Other data on the effects of social isolation during rearing upon reproductive behavior in the rhesus monkey (*M. mulatta*). *Anim. Behav.* **13**, 228–231.

Meier, G. W., and Garcia-Rodriguez, C. (1966). Continuing behavioral differences in infant monkeys as related to mode of delivery. *Psychol. Rep.* **19**, 1219–1225.

Menzel, E. W., Jr. (1968). Primate naturalistic research and problems of early experience. *Develop. Psychobiol.* **1**, 75–184.

Miller, R. E., Caul, W. F., and Mirsky, I. A. (1967). Communication of affects between feral and socially isolated monkeys. *J. Pers. Soc. Psychol.* **7**, 231–240.

Missakian, E. A. (1968). Reproductive behavior of socially deprived male rhesus monkeys. *Abstr. 2nd Int. Congr. Primatol. Atlanta, Ga., July.*

Mitchell, G. (1968a). Persistent behavior pathology in rhesus monkeys following early social isolation. *Folia Primatol.* **8**, 132–147.

Mitchell, G. (1968b). Intercorrelations of maternal and infant behaviors in *Macaca mulatta*. *Primates* **9**, 85–92.

Mitchell, G. (1968c). Attachment differences in male and female infant monkeys. *Child Develop.* **39**, 611–620.

Mitchell, G. (1969c). Responses of experienced and inexperienced monkey mothers to social and nonsocial stimuli. *West. Psychol. Ass. Ann. Meet., Vancouver, B. C., Can., June.*

Mitchell, G. (1969b). Paternalistic behavior in primates. *Psychol. Bull.* **71**, 399–417.

Mitchell, G., and Clark, D. L. (1968). Long term effects of social isolation in nonsocially adapted rhesus monkeys. *J. Genet. Psychol.* **113**, 117–128.

Mitchell, G., and Stevens, C. W. (1969). Primiparous and multiparous monkey mothers in a mildly stressful social situation: I. First three months. *Develop. Psychobiol.* **1**(4), 280–286.

Mitchell, G., Raymond, E. J., Ruppenthal, G. C., and Harlow, H. F. (1966a). Long term effects of total social isolation upon behavior of rhesus monkeys. *Psychol. Rep.* **18**, 567–580.

Mitchell, G., Ruppenthal, G. C., Raymond, E. J., and Harlow, H. F. (1966b). Long term effects of multiparous and primiparous monkey mother rearing. *Child Develop.* **37**, 781–791.

Mitchell, G., Arling, G. L., and Møller, G. W. (1967a). Long term effects of maternal punishment on the behavior of monkeys. *Psychonomic Sci.* **8**, 209–210.

Mitchell, G., Harlow, H. F., Griffin, G. A., and Møller, G. W. (1967b). Repeated maternal separation in the monkey. *Psychonomic Sci.* **8**, 197–198.

Møller, G. W., Harlow, H. F., and Mitchell, G. (1968). Factors affecting agonistic communication in rhesus monkeys (*Macaca mulatta*). *Behaviour* **31**, 339–357.

Nissen, H. W., Chow, K. L., and Semmes, J. (1951). Effects of restricted opportunity for tactual, kinesthetic, and manipulative experience on the behavior of chimpanzee. *Amer. J. Psychol.* **64**, 485–507.

Pratt, C. L. (1969). Social Behavior of Rhesus Monkeys Reared with Varying Degrees of Early Peer Experience. M.A. Thesis, Univ. of Wisconsin, Madison, Wisconsin.

Pratt, C. L., and Sackett, G. P. (1967). Selection of social partners as a function of peer contact during rearing. *Science* **155**, 1133–1135.

Rogers, C. M., and Davenport, R. K. (1969). Effects of restricted rearing on sexual behavior of chimpanzees. *Develop. Psychol.* **1**, 200–204.

Rosenblum, L. A. (1961). The Development of Social Behavior in the Rhesus Monkey. Doctoral dissertation, Univ. of Wisconsin, Madison, Wisconsin.

Rosenblum, L. A., and Kaufman, I. C. (1968). Variations in infant development and response to maternal loss in monkeys. *Amer. J. Orthopsychiat.* **38**(3), 418–426.

Rosenblum, L. A., Kaufman, I. C., and Stynes, A. J. (1964). Individual distance in two species of macaque. *Anim. Behav.* **12**, 338–342.

Rowell, T. E. (1966). Forest living baboons in Uganda. *J. Zool.* **149**, 344–364.

Rowell, T. E. (1967). A quantitative comparison of the behavior of a wild and caged baboon group. *Anim. Behav.* **15**, 499–509.

Rowell, T. E., and Hinde, R. A. (1963). Responses of rhesus monkeys to mildly stressful situations. *Anim. Behav.* **11**, 235–243.

Rowland, G. L. (1964). The Effects of Total Social Isolation upon Learning and Social Behavior in Rhesus Monkeys. Doctoral dissertation, Univ. of Wisconsin, Madison, Wisconsin.

Sackett, G. P. (1965). Effects of rearing conditions upon monkeys (*M. mulatta*). *Child Develop.* **36**, 855–868.

Sackett, G. P. (1966a). Development of preference for differentially complex patterns by infant monkeys. *Psychonomic Sci.* **6**, 441–442.

Sackett, G. P. (1966b). Monkeys reared in isolation with pictures as visual input: Evidence for an innate releasing mechanism. *Science* **154**, 1468–1472.

Sackett, G. P. (1968). Abnormal behavior in laboratory reared rhesus monkeys. *In* "Abnormal Behavior in Animals" (M. W. Fox, ed.), pp. 293–331. Saunders, Philadelphia, Pennsylvania.

Sackett, G. P., Porter, M., and Holmes, H. (1965). Choice behavior in rhesus monkeys: Effect of stimulation during the first month of life. *Science* 147, 304–306.

Sackett, G. P., Griffin, G. A., Pratt, C. L., and Ruppenthal, G. C. (1966). Mother-infant and adult female choice behavior in rhesus monkeys after various rearing experiences. *Midwest. Psychol. Ass. Meet., Chicago.*

Saxon, S. V. (1961). Effects of asphyxia neonatorum on behavior in the rhesus monkey. *J. Genet. Psychol.* 99, 277–282.

Schaller, G. B. (1963). "The Mountain Gorilla, Ecology and Behavior." Univ. of Chicago Press, Chicago, Illinois.

Schlottman, R. S., and Seay, B. M. (1968). Mother-infant separation in *Macaca irus. Southeast. Psychol. Ass. Meet., Roanoke, Va.*

Seay, B. M. (1966). Maternal behavior in primiparous and multiparous rhesus monkeys. *Folia Primatol.* 4, 146–168.

Seay, B. M., and Harlow, H. F. (1965). Maternal separation in the rhesus monkey. *J. Nerv. Ment. Dis.* 140, 434–441.

Seay, B. M., Hansen, E. W., and Harlow, H. F. (1962). Mother-infant separation in monkeys. *J. Child Psychol. Psychiat.* 3, 123–132.

Seay, B. M., Alexander, B. K., and Harlow, H. F. (1964). Maternal behavior of socially deprived rhesus monkeys. *J. Abnorm. Soc. Psychol.* 69, 345–354.

Senko, M. G. (1966). The Effects of Early, Intermediate, and Late Experience Upon Adult Macaque Sexual Behavior. M.S. Thesis, Univ. of Wisconsin, Madison, Wisconsin.

Simons, R. C., Bobbitt, R. A., and Jensen, G. D. (1968). Mother monkeys (*Macaca nemestrina*) responses to infant vocalizations. *Perceptual Motor Skills* 27, 3–10.

Singh, S. D. (1966). The effects of human environment upon the reactions to novel situations in the rhesus. *Behaviour* 26, 243–250.

Singh, S. D. (1968). Effect of urban environment on visual curiosity behavior in rhesus monkeys. *Psychonomic Sci.* 2(3), 83–84.

Sluckin, W. (1965). "Imprinting and Early Learning." Aldine, Chicago, Illinois.

Southwick, C. H. (1966). Experimental studies in intragroup aggression in rhesus monkeys. *Amer. Zoologist* 6(3).

Spencer-Booth, Y., and Hinde, R. A. (1966). The effects of separating rhesus monkey infants from their mothers for six days. *J. Child Psychol. Psychiat.* 7, 179–198.

Sugiyama, Y. (1966). An artificial social change in a Hanuman langur troop (*Presbytis entellus.*) *Primates* 7, 41–72.

Thompson, N. S. (1967). Primate infanticide. *Lab. Primate Newsletter* 6(3), 18.

Tinklepaugh, O. L. (1928). The self-mutilation of a male *Macacus rhesus* monkey. *J. Mammal.* 9, 293–300.

Van Wagenen, G. (1966). Studies in reproduction (*Macaca mulatta*). *Proc. Conf. Nonhuman Primate Toxicol., Warrenton, Va.* (C. O. Miller, ed.). U. S. Dep. Health Educ. and Welfare.

Wheaton, J. L. (1959). "Fact and Fancy in Sensory Deprivation Studies." Sch. of Aviat. Med., Brooks Air Force Base, Texas.

Windle, W. F. (1967). Asphyxia at birth, a major factor in mental retardation: Suggestions for prevention based on experiments in monkeys. "Psychopathology of Mental Development." pp. 140–147. Grune & Stratton, New York.

The Nilgiri Langur *(Presbytis johnii)* of South India*

Frank E. Poirier

Department of Anthropology, Ohio State University, Columbus, Ohio

* This research was supported by Public Health Service grant MH 11099-01 attached to Fellowship 2 F1-MH-22, 140-02 (BEH).

I. MATERIALS AND METHODS

A field study of the Nilgiri langur (*Presbytis johnii*) was conducted primarily in the Nilgiri district of Madras State, South India, from September 1965 through August 1966. The results of a total of 1250 hours of direct observation were compiled. This figure does not represent time spent in surveys, and so on. Observations were made in various ecological niches and during all weather conditions.

Once individual troops were recognized, the first task was to establish an observational pattern yielding the most reliable information. During the first month of study, a 7-day observational week was maintained. Eventually, however, a 5-day week averaging 9 or 10 hours of observation per day was followed. Time of entrance into the field varied, but observations typically began at approximately 0730 or 0800 and terminated at 1730 or 1800. During the course of the study, however, observations were made from 0530, before sunrise, until dark. Nighttime observations were of variable effectiveness because of the dark, black hair coloration and arboreal habitat. Even with a full moon the monkeys were almost impossible to spot.

A. CHOICE OF STUDY TROOPS

Thirty troops were positively identified within the major study area. Since most sholas or jungle areas that Nilgiri langurs inhabited precluded observations, the choice of study troops was determined not only by the requirements for representative age/sex compositions but also for suitable study areas. Fortunately, both objectives were satisfied early in the study. Four of the five troops studied most intensely (I, II, III, and V) inhabited sholas (described in Section II,B), and one (IV) an area surrounded by cultivated potato and cauliflower gardens. Three troops (III, IV, and V) were accustomed to people; two (I and II) were not. After a time troops I–IV became accustomed to the observer's presence and were studied from close range. Troops varied in age/sex composition, size, ecological habitat, and adaptation to food plants. Groups I, II, III, IV, and V were observed for approximately 560, 352, 159, 113, and 20 hours, respectively.

B. NOTE-TAKING PROCEDURES

The basic recording technique involved the maintenance of a daily manual of weather and plant conditions. Weather conditions noted included rainfall, temperature, and humidity, plus a check on such ground conditions as dampness and the existence of frost and groundwater. As much as possible of the floral and faunal content of each habitat was classified. Efforts to obtain chemical analyses of food, soil (which was occasionally consumed), and stool samples were abortive.

The recording of social interactions involved noting the time, place, duration, and participants of each social interaction. Daily movement and activity patterns were recorded. Each evening, while the daily activities were still fresh in our minds, notes were typed, indexed, and reviewed. Where useful, sketches were drawn and photographs taken to assist interpretations. The only mechanical aids utilized were a pair of Nikon 7 × 50 binoculars and a Nikon 35-mm camera fitted with 135- and 200-mm telephoto lenses.

C. INDIVIDUAL IDENTIFICATION

Individual identification was extremely arduous. Except for the wounds on some males, Nilgiri langurs were comparatively free of obvious identifying marks. Furthermore, identification was hindered by what appeared, with the help of the arboreal habitat, to be minimal sexual dimorphism. Unless the observer spotted the white thigh patch of a female or a male's penis (which was practically obscured by the black pubic hair and gray-black testicles), sexual identification was often impossible. Tanaka (1965, p. 112) had the same difficulty and wrote: ". . . I was compelled to give up the observation by individual discrimination." Canine and body size also proved to be of only limited value in identifying animals. Thus, following an introductory period which varied from troop to troop, individual behavioral and/or vocal characteristics were the most useful tools for identification. Tonal variations in the adult male whoop display (described in Section VI,F) and individual reactions of animals to each other and to the observer were very helpful. During most of the study, all the animals of two troops were recognized, as well as the adults, subadults, and random individuals in the three remaining troops. Infants were primarily identified as they associated with their mothers. These infant identifications, however, were occasionally confused by the pattern of infant transference.

D. OBSERVATIONAL TECHNIQUES

Once a troop became habituated to the observer, it was possible to observe adult males from as close as 3 or 4 ft. Females, however, remained wary throughout the study and never permitted such close observation. It was advantageous to observe from a supine position from which one could comfortably focus on the troop's behavior in the trees. The principal disadvantage of this method is one's vulnerability to falling branches, urine, and feces.

II. MORPHOLOGY AND TAXONOMY

Nilgiri langur adult males measure approximately 31 inches exclusive of the tail and weigh 30–35 lb. Adult females measure approximately 23 inches exclusive of the tail and weigh 28–30 lbs. The tail is uniformly black and often attains a length of 37 inches (Leigh, 1926a; Pocock, 1928; Prater, 1965).* The hair of the Nilgiri langur is glossy black, coarse, and thick. The long crown and whisker hairs are of a uniform buff-brown tint; there is no contrast between the whiskers and crown. Older adults of both sexes have a triangular gray patch of hair extending

---

* The authors from which the data are derived make little mention of sample sizes.

from the anus to a point above the base of the tail. Females, apparently from birth, have a cream-colored hair patch on the interior of the thigh. The hair of the youngsters is reddish-brown until approximately 10 weeks of age; then it turns the dark, black, adult color. At approximately 14 weeks, infants develop the characteristic hair crown and whiskers (Poirier, 1968d).

Sterndale (1884) first classified the Nilgiri langur as *Presbytis johnii*. Historically, however, the Nilgiri langur has been referred to as *Simia johnii* (1829), *Semnopithecus cacullatus* (1834), *Semnopithecus jubatus* (1840), *Semnopithecus cephaloteru* (1884), and *Semnopithecus johnii* (1888). The form has alternately been referred to the genus *Pithecus* and *Kasi* (Ryley, 1913). Pocock (1928) considered the Nilgiri langur a subspecies of the Ceylon bear monkey. Osman Hill (1934) feels its nearest relative is the Ceylonese colobid *Presbytis vetulus vetulus*.

III. GEOGRAPHICAL DISTRIBUTION AND ECOLOGICAL NICHE

A. DISTRIBUTION

Inhabiting South India, Nilgiri langurs are found in Madras, Mysore, and Kerala states in the Anaimalai, Brahmagiri, Cardamom, Nelliampathy, Nilgiri, Palni, Shevaroy, and Tinnevelly hills. The northern limit of their distribution appears to be the Coorg district in Mysore State; the southern limit, the Cape Comorin district in Madras State. Nilgiri langurs range from approximately 8–13° N and 75–80° E and inhabit elevations from approximately 1500–7700 ft where they prefer the denser forest areas.

The Anaimalai, Cardamom, Nilgiri, Palni, and Shevaroy hills were all raised as great horsts produced by a thermal expansion of the sima, an uplift occurring during post-Jurassic and late Tertiary times. The Anaimalai, Nilgiri, and Palni hills are subdivisions of montane subtropical forests. The lower slopes are "southern subtropical wet hill"; above 5000 feet they are classified as "southern wet temperate" forests. These ranges have a rainfall of 60–250 inches per annum and a monthly mean temperature of 45°–55° F with a maximum of 60°–75° F. Wind is an important environmental factor. The result is a rich savannah or parkland with occasional peat bogs. Vegetation is typically low (50- to 60-ft) forest with much undergrowth and many epiphytes, mosses, and ferns (Spate, 1954).

B. Ecological Niche (see Fig. 1)

Botanically, the Nilgiri hills are divided into four tracts, each with its respective flora. Tract 1 comprises the deciduous forests growing along the slopes. These are basically deciduous forests in the dry months of January through March, but the trees are never entirely bare. Many of the tropical trees characteristic of Madras State grow in tract 1. Tract 2 is comprised of moist evergreen forests on the western slopes at approximately 3000–4000 ft where trees occasionally grow to 200 ft or more. Tract 3, the major ecological niche of the Nilgiri langur, comprises the sholas or woods of the plateau. The grassy downs covering much of the plateau area comprise tract 4; trees and shrubs, such as *Rhododendron nilgirihensis*, are scattered about this tract.

Nilgiri langurs occupy the moist evergreen forests, to a lesser extent than tract 3. Tract 2 is exceedingly moist from the first rains in March until the end of December when leeches abound. Trees are covered with epiphytic orchids, ferns, mosses, and creepers. *Strobilanthus*, the characteristic undergrowth, is represented by 20 species. Two palms, *Caryota urens* and *Arenga wightii*, are conspicuous. Above 4000 ft the forests of tract 2 shrink and gradually yield to the sholas toward the plateau. On the Malabar (western) side of the district, tract 2 forests reach to the plains, as they do in parts of the south Canara, Coorg, and Travancore districts. Elsewhere, they yield at approximately 1000 ft to deciduous forests or tracts of reed bamboo.

Nilgiri langurs in the Nilgiri district primarily inhabit the sholas of tract 3. The shola is a strip of forest surrounded by grassland which is transected by a narrow watercourse. Sholas predominate in the divided valleys dotting the surface of grassy undulating hills typical of the plateau interior. They characterize the sheltered valleys in which the wind is not as strong as that blowing across the open downs. Sholas typically exhibit a three-story vegetational pattern. The shrub story is a closed canopy bound by creepers into a solid mass of foliage. A variety of trees ranging in height from 15 to 30 ft compose the understory. The upper story is characterized by an irregular tree layer from 40 to 70 ft tall. The Nilgiri langur makes various use of the respective canopies, as shown in Table I.

Although superficially similar, sholas vary considerably in height, growth, accessibility, and crop composition, depending on their situation. Trees or shrubs such as *Sympolocus spicata* and *Eugenia* spp. are universal. Others, however, such as *Hydnocarpus alpina*, grow only on the eastern and *Eurya* only on the western plateau. (*Hydnocarpus* is a deciduous tree typical of slopes 6000 ft and below. *Eurya* is a low-

FIG. 1. Ecology of the Nilgiri hills.

growing shrub.) At lower levels sholas exhibit better height, gradually merging with the evergreen forests below. The ratio of sholas to grassland (and thus Nilgiri langur concentration) decreases from the eastern to the western half of the plateau.

Sholas approximate somewhat the moist evergreen forests characteristic of tract 2, however, being situated at higher altitudes the trees are of different genera and species. Shola trees are typically evergreen; Mytraceae, Lauraceae, and Styrcaceae are the families commonly represented. These trees rarely grow above 70 ft. Rubiaceous shrubs and *Strobilanthus* comprise the undergrowth. Ferns and mosses abound; among the former *Alsophilia latebrosa*, a tree fern, predominates. Orchids, *Oberonia*, are poorly represented. There is one species of reed bamboo (*Arundinaria wightiana*) and some shrubby balsams and

TABLE I

CANOPY USAGE

Canopy	Representative flora	Prime hours of usage	Prime patterns of usage
Upper story, 40–70 ft	*Acacia, Eucalyptus*	1700–0800	Sleeping, early morning sunning, feeding
Understory, 15–30 ft	*Cinamomum, Glochidion, Litasae, Loranthus*	0800–1700	Feeding, resting, movement
Shrub story, 5–10 ft	*Berberis, Rubus, Rosae*	0900–1100 1300–1500	Feeding
Forest floor	*Cytis, Ulex, Acacia* seedlings, assorted grasses	Rarely utilized	Feeding, fighting, escape, play

begonias (*Begonia malabarica* or *B. floccifera*). Those sholas into which man has penetrated contain substantial amounts of *Acacia molissima* or *A. melanoxyln*.

C. Climatological Data

The following is a brief summary of average climatic conditions prevailing in the Nilgiri district. The first 3 months of the year are virtually dry and are a procession of bright, clear days during which a dry, northeast wind prevails. Evening hoarfrosts occur in December through February, as a result of which the grassy downs turn brown. The thermometer falls to 20° F when placed in contact with the ground, however, it seldom dips below 38° F when exposed to a breeze a few feet from the surface. The frost kills all the delicate plants, vegetables, and grasses. Northeasterly winds prevail from January through April. Showers are common in April and May and the grasses and trees begin a new cycle. Temperatures reach their highest point in May and the climate is less bracing. Although temperatures in Ootacamund (the base camp) seldom ranged above 70° F in the shade, if exposed to the sun's direct rays, recordings could reach 120° F. The filtering effect of the "Scotch mist", however, prevents serious sun damage to flora and fauna.

Preceding the southwest monsoon, heavy banks of clouds appear over the southwest hills. The southwest monsoon lasts from June through August, blowing strongest on the extreme west of the plateau. On the plateau itself rainy days alternate with days of a heavy mist. Because of the heavy rain and moisture-saturated air, vegetation grows with tropical speed. In August and September breaks occur, the mist dissipates, and the sun appears for several days successively. By the end of August, the monsoon slackens or terminates. The shorter northeast monsoon follows in October and lasts through November, when bright clear days and frosty nights reappear (Francis, 1906; Grigg, 1880).

D. Faunal Complement and Interspecific Conflict

Tract 2 had a full complement of jungle fauna which included elephants, tigers, panthers, snakes, and numerous avian forms. Three species of monkeys were common: *Macaca radiata* (bonnet macaque), *Macaca silenus* (lion-tailed macaque), and *Presbytis entellus* (common langur). In contrast, tract 3, which was the major study habitat, had a limited faunal complement. The restricted faunal complement is attributable in some measure to the altitude and to the pressure of the human population's expansion of agriculture. Tigers and elephants, for example, have been forced into the game reserves of tract 2.

Limited interspecific contact was witnessed in tract 3. Nilgiri langurs did not appear to be in competition with any species over the existent food supply. There was an abundance of leaves, fruits, berries, and buds in the deciduous forest environment despite the fact that 1962–1966 experienced subnormal rainfall. If some unexpected situation were to severely limit the food supply, however, Nilgiri langurs could find themselves in competition with the giant squirrel, sambar, barking deer, hare, and domestic cow and buffalo, for all regularly ate many of the same plants as the langurs. The arboreal langur could, however, readily minimize or escape competition by feeding higher.

At the border of tracts 2 and 3 Nilgiri langurs overlapped with two species of Indian macaques, *M. radiata* and *M. silenus*. There was also occasional contact with the common South Indian gray langur *P. entellus*. Most contact occurred in the forests surrounding the town of Gudalur ca. 4000 ft. Where overlap occurred it seemed, although information is insufficient, that each species inhabited a restricted ecological niche, seemingly vertically restricted, reducing interspecific contact. Preliminary data suggest that interspecific contact is characterized primarily by peaceful avoidance.

There was occasional overlap with the bonnet macaque, *M. radiata*, in tract 3 on the northern and eastern slopes at approximately 6000 ft. The overlap occurred primarily in forests ringing tea plantations. The nature of the interaction between the bonnet macaque and the Nilgiri langur varied. On the one hand, Hutton (1951, p. 681) noted that when the two meet the macaques ". . . are speeded on their way by the Nilgiri langurs." On the other hand, Simonds (personal communication) saw both species feeding peacefully in the same bushes, and we have observed bonnets and Nilgiri langurs feeding peacefully within 20 yd of each other. (It can be noted, however, that where bonnets and common langurs overlap, primarily in the Mudamulai Game Reserve, langurs give way to the macaques.) Hutton also noted that when Nilgiri langurs and lion-tailed macaques meet ". . . some terrific battles take place in the trees, and there are casualties on both sides. . . ." Hutton's observations could not be checked during the present study. Sugiyama (1968) notes, however, that lion-tailed macaques are not aggressive to solitary Nilgiri langurs who wander into their range.

E. RELATIONS WITH NONPRIMATES

The relationship between Nilgiri langurs and the majority of the species inhabiting tract 3 was that of mutual tolerance. The langur's daily range often brought it into proximity with both ungulates and ruminants. This was occasionally mutually beneficial, for each acted positively to the sudden flight of the other. The following is illustrative:

> A huge sambar feeds across the stream from troop I which is feeding low in the *Rubus* shrubs. A hunter quietly stalks the sambar; neither the langurs nor the sambar are immediately cognizant of his presence. Suddenly, the sambar breaks the bushes and charges from the area. The monkeys immediately move into the tree tops where the male emits the alarm call (November 2, 1965).

It is very unlikely that healthy langurs fell victim to predators on more than infrequent occasions. Sometimes dogs (wild and domestic) and an occasional panther seem to have succeeded in capturing a monkey, however, for Nilgiri langur hair was occasionally found in carnivore droppings. Once a domestic dog caught a subadult male. Rather than try to escape, the langur lay helplessly on the ground, the dog biting its thighs and neck. Furthermore, neither the captured male nor his troop members vocalized during the interaction. In fact, the troop watched momentarily and quietly moved into the shola. The interaction terminated when the observer chased off the dog. The sequence is reproduced below. In contrast, bonnet macaques were observed to flock above a dog and raise a loud ruckus as the dog jumped for the nearest animal. There is little question, however, that these macaques, which raid cultivated fields, are continually harassed and are potentially easier prey than the langurs studied.

> A subadult male of troop VII runs from a dog across the open area separating the home ranges of troops III and V. The dog catches the subadult and immediately begins biting it about the neck. The bites seem to do little harm, however, as the thick mat of hair about the neck protects him. Rather than attempt to fight or escape, the langur lies helplessly on the ground. The remainder of his troop escapes silently; none looks back or makes an attempt to aid the subadult. After approximately 5 minutes, the observer chases the dog away. The subadult is submissive as it is examined and photographed. Physical injury is minimal, there being only a few superficial bites on the neck and thigh. First attempts to place the male in a tree fail, for he immediately falls either from lack of strength, fright, or both. I replace him in the tree where he sits. Not once during the entire sequence does he vocalize. When I return an hour later he is gone (June 18, 1966).

F. Man as a Predator

Life in association with man seems disastrous for many Nilgiri langur populations which are unable to withstand civilized settlements within their midsts. Those managing to survive have to a large extent retreated before man into the deepest recesses of the sholas. Man proves to be the langur's major predator. The langur's flesh and glands are highly prized as food and for preparing medicines in the ayurvedic medicinal system. The medicinal preparation, known in local vernacular as *karum kurangu rasayanam* or "black monkey medicine," is prescribed for whooping cough and other lung ailments, as well as for general debility. Some jungle tribes are said to drink the fresh blood for its supposed rejuvenatory powers. (After drinking the blood one runs a mile to prevent coagulation.) The langur's skin is also used for fashioning drum heads, and its beautiful, glossy hair is highly regarded for decorative purposes.

Table II summarizes the interspecific relationships in tract 3. Table III summarizes the major aspects of the tract-2 and -3 ecological niches in the Nilgiris.

IV. TROOP DYNAMICS

A. Troop Size

Thirty-three Nilgiri langur troops containing 212 animals were counted in the study area. An average bisexual troop contained 8.9 animals, a figure tabulated from 14 bisexual troops on which accurate counts of all age and sex categories were obtained. The largest bisexual troop con-

TABLE II

INTERSPECIFIC RELATIONS OF NILGIRI LANGURS IN TRACT 3

Species	Relation	Competition for food
Homo sapiens	Predator	Very slight[a]
Felis tigris (tiger)	Unknown	None
Felis spp? (panther)	Predator	None
Canis aureus (jackal)	Predator	None
Canis domesticus (domestic dog)	Predator	None
Cuon dukhumensis (wild dog)	Predator	None
Ratufa indica (giant Indian squirrel)	Peaceful	Slight
Lepus spp– (hare)	Peaceful	Slight
Cervus unicolor (sambar)	Peaceful and complementary	Slight
Muntiacus muntjak (barking deer)	Peaceful and complementary	Slight
Sus scrofa (wild bear)	Peaceful	Slight
Gallus ferrugineus (jungle fowl)	Peaceful	None
Macaca radiata (bonnet macaque)	Peaceful?	Slight
Domestic cow and buffalo	Peaceful	Slight

[a] In a sense man is, however, a very real competitor for he destroys many langur food trees to prepare his garden plots or to plant cash trees.

TABLE III
COMPARISON OF ECOLOGICAL NICHES IN THE NILGIRIS

	Tract 2 (evergreen forests)	Tract 3 (shola)
Altitude	3000–4000 ft	5000–7500 ft
Rainfall, average of 1962–1965	81.44 inches	59.57 inches
Temperature range	50°–98° F	30°–70° F
Main monsoon	Southwest	Southwest
Humidity	40–91%	41–93%
Wind	No information	Inhibited
Vegetation	3-Story	3-Story
Human population	Sparse	Sparse, but increasing
Overlap with non-human primates	Bonnet and lion-tailed macaque, common langur	Bonnet macaque, but limited
Faunal content	Full complement of jungle fauna	Limited
Possible predators	Man and some carnivores	Mainly man
Comments on ecology	Malarious jungle, some areas border plantations	Some areas border plantations; habitat being rapidly destroyed in some areas

tained 25 animals and the smallest 3; however, solitary animals and unisexual pairs and trios were also observed (Table IV). Because Nilgiri
langurs typically moved single file through the trees, with each individual
passing over the same branches as the preceding individuals, fairly accurate troop counts were obtained.

B. Age and Sex Ratio

Adults constituted the bulk of the membership in Nilgiri langur troops.
Of the 14 troop samples on which the most accurate data were obtained,
adults constituted 60% of the 108 animals positively identified according
to age category (Table V). Adult males comprised 46% of the adult
population; thus the sex ratio was 1.2 adult females to 1 adult male. The
ratio of young (infants and juveniles) to the adult/subadult (reproductive or potentially reproductive) population was approximately 26 to
74%, or 2.9 adults or subadults to 1 infant or juvenile. The greatest variation occurred in the subadult class; only 3 of the 14 troops had subadults.

C. One-Male Troops

Most Nilgiri langur troops had only one adult male, differentiating
the Nilgiri langur troop from the multiple-male troops characteristic

TABLE IV
Troop Counts[a,b]

Troop	AM	AF	A	SM	SF	S	JM	JF	J	IM	IF	I	Total
I*	1	4	—	—	—	—	—	—	—	1	3	—	9
II*	1	3	—	—	—	—	1	—	—	1	—	1	7
II (after change)	4	5	—	—	—	—	—	—	—	1	—	—	10
III*	3	4	—	3	2	4+	1	2	—	1	—	2	25
IV*	3	1	—	4	1	—	1	1	—	1	—	—	12
V*	1	2	—	—	—	X	—	—	X	—	—	X	7–9
VI*	1	2	—	—	—	—	—	—	—	—	—	1	4
VII*	2	—	—	—	—	—	—	—	—	—	—	—	2
VIII*	1	2	—	—	—	—	—	—	—	—	—	2	5
IX*	1	1	2	—	—	—	—	—	2	1	—	2	9
X*	3	—	—	—	—	—	—	—	—	—	—	—	3
XI*	1	1	2	—	—	—	—	—	1	—	—	1	6–7
XII*	1	—	2	—	—	—	—	—	2	—	—	—	5
XIII*	2	1	?	—	—	X	—	—	X	—	—	X	16
XIV*	1	4	—	—	—	—	—	—	—	—	1	—	6
XV	—	—	4–5	—	—	—	—	—	X	—	—	X	10+
XVI	—	—	X	—	—	X	—	—	X	—	—	X	10–15
XVII	—	—	1	2	—	—	—	—	—	—	—	—	3?
XVIII	—	—	1	—	—	—	—	—	—	—	—	—	1
XIX	1	1	—	—	—	—	—	—	—	—	—	1	3
XX	—	—	4–5	—	—	—	—	—	—	—	—	—	4–5
XXI	—	—	4	—	—	—	—	—	—	—	—	—	4
XXII	—	—	5	—	—	—	1?	—	—	—	—	—	6
XXIII	—	—	X	—	—	X	—	—	X	—	—	X	10+
XXIV	—	—	2	—	—	—	—	—	—	—	—	—	?
XXV	1	—	—	—	—	—	—	—	—	—	—	—	21
XXVI	—	—	X	—	—	X	—	—	X	—	—	X	15
XXVII	—	—	5	—	—	—	—	—	—	—	—	—	5
XXVIII	—	—	4–5	—	—	—	—	—	—	—	—	—	4–5
XXIX	—	—	X	—	—	—	—	—	—	—	—	—	?
XXX	—	—	3	—	—	—	—	—	—	—	—	—	3?
XXXI				Heard only whooping									
XXXII				Heard only whooping. Found skull of langur									
XXXIII				Heard only whooping									
XXIV				Found langur stool									
XXXV				Found langur stool									

[a] The above counts were obtained from the tract-2 and -3 niche in the Nilgiri district.

[b] An asterisk indicates the groups on which the most accurate counts were obtained; X indicates that the age class was present but their number was not determined; A, adult; S, subadult; J, juvenile; and I, infant; M, male; F, female.

of most species. One-male troops might result from several factors, none of which were positively confirmed. A differential birth and mortality rate favoring adult females or a severe antagonism between males are the most obvious possibilities.

TABLE V
CRITERIA FOR AGE CLASSIFICATION

Age group	Physical criteria
Infant 1 (birth to 4 months)	The hair is reddish-brown until 10 weeks old at which time it turns dark black. The pink skin darkens, and the hair tuft develops at 14 weeks. Female infants have a white patch on their thighs by 10 days old
Infant 2 (4 Months to 12–14 months)	The color change is completed during this time
Juvenile (14 Months to 4 years)	This is a period of rapid growth. The canines are not developed
Subadult (4 Years to completion of maturity at 6–7 years for males and 4–5 years for females)	This period is characterized by the near completion of growth. In the final stages the males possess the adult the adult canine teeth
Adult (6–7 years for the males and the time of first birth for the females)	Fully developed body stature. The males may weight 30–35 lb the females average 25–28 lb (Leigh, 1926; Pocock, 1928; Prater, 1965). Graying is evident up the base of the tail in both sexes

Age group	Behavioral criteria
Infant 1	The infant-1 remains closely attached to the mother who nurses it and carries it in the ventral position. There is little peer orientation
Infant 2	At 10 weeks the infant begins to break the ties with the mother. By 4 or 5 months infants spend a good deal of time playing with their peers. In case of danger they still seek the mother for protection. Weaning is one of the most important patterns in the infant-2 stage
Juvenile	Weaning is completed if an infant is born in the succeeding year. Most of the day is spent within the peer group. Juveniles flee from danger on their own accord
Subadult	Subadults spend most of the day in the adult or subadult subgroup. Younger subadults are still involved in play behavior, especially the subadult males
Adult	An adult female spends much of her time caring for and raising infants. Adult males and adult females enter the breeding cycle

Some evidence on the role of male/male antagonism is available. Several trends became apparent in comparisons of the social relationships of adult males and adult females, both in one- and multiple-male troops. Dominance interactions were more frequent and severe in the adult male than the adult female hierarchy. Considering a multiple-male, multiple-female troop, such as troop II, adult male–adult male encounters totaled 84% of the adult-adult dominance sequences, or one interaction per 7.7

hours of observation. Adult female–adult female encounters totaled 16% of the adult encounters, or one sequence per 39 hours of observation. As one might expect, in multiple-male troops subordinate adult males directed a high percentage of their dominance interactions outside the male hierarchy at animals, such as infants and juveniles, that were not disposed to retaliation. Such aggression could prove injurious to the animal attacked, especially since the alpha male made no effort to aid animals involved in intratroop aggression. Over time, the presence of more than one adult male in a troop might prove detrimental to younger troop members. Considering multiple-male troops of three or four adult males, the gamma or delta males directed 88% of their aggression outside the male hierarchy. The delta male of troop II, for example, directed 57% of his dominance interactions at an infant 2 (Poirier, 1969a,b). Table VI compares the frequencies with which troop-II males directed their dominance interactions within and without the male hierarchy.

Another trend appeared in grooming behavior. Grooming, which suggested that positive social bonds existed between the participants, was more frequent between adult females than adult males. Duration was a major differentiating feature in grooming; female-female bouts lasted longer than either female-male or male-male sequences. These data may also reflect the relative inability of adult males to establish and maintain contact with each other when compared to adult females.

Whereas it is not possible to state emphatically that adult male Nilgiri langurs leave a troop because of male-male antagonism, that possibility certainly exists. The apparent lack of male-male hostility in all-male groups, however, remains to be explained; there may be a positive attraction among males which breaks down in the presence of females. Sugiyama's data (1967) on South Indian langurs and Kummer's (1967b) data on hamadryas baboons suggest a similar interpretation.

Further factors lessening the selective value of multiple-male Nilgiri langur troops are the lack of predation and the capability for rapid arboreal escape. The adult male Nilgiri langur played a minor role in

TABLE VI
FLOW OF DOMINANCE INTERACTIONS IN TROOP II

Male rank	Intersubgroup interactions (%)	Intrasubgroup interactions (%)
Alpha	15	85
Beta	32	68
Gamma	55	45
Delta	88	12

protecting the troop against danger. Each animals' prime defense was its own ability to flee. In contrast to the protective role assumed by adult males in some terrestrial species (i.e., macaques and baboons), the Nilgiri langur male was often the first individual to escape prospective danger.

The one-male troop generally characteristic of Nilgiri langurs is similar to that reported for South Indian langurs (Sugiyama, 1965a,b, 1967), but differs from gibbon (Carpenter, 1964) and hamadryas baboon (Kummer and Kurt, 1963; Kummer, 1968) one-male organizations. The gibbon group is characterized by an association consisting of one adult male, one adult female, and their offspring. In contrast to the langur, many one-male hamadryas baboon troops come together at night to sleep. Although gibbon and hamadryas baboon social orders superficially resemble that of the langur, they are therefore qualitatively quite different.

Sugiyama's (1965a,b) analysis of one-male South Indian langur troops suggests that the exclusion of male langurs, including juveniles, from a bisexual troop was attributable to a severe attack by a male living in an all-male group. The attack deprived the resident male of his status and drove all the males from the troop. Thus ". . . repeated attacks of the male group help the troop to maintain the one-male troop pattern" (Sugiyama, 1965b, p. 403). The one recorded instance in which a Nilgiri langur male group joined a bisexual troop was nonaggressive, suggesting that Sugiyama's analysis may not be applicable to the Nilgiri langur situation (Poirier, 1968b, 1969a).

Whatever the mechanisms favoring one-adult-male Nilgiri langur troops, a number of adult males lived outside bisexual troops. Eight, or 36%, of the 22 males counted in the study area lived either as solitaries or in male pairs or trios.

D. Unisexual Groups or Solitary Animals

Unisexually grouped or solitary animals lived apart from bisexual troops for varying lengths of time. All-male and all-female groups were encountered, the latter being an unusual occurrence among nonhuman primates. Most individuals outside bisexual troops joined as pairs or trios; in all cases these were composed of adults. Three adult males were recorded as being solitary. The accompanying tabulation indicates the occurrence of nonbisexual troop individuals.

The daily routine of solitary animals, or of unisexually grouped individuals, mimicked that of animals residing in bisexual troops with the exception that unisexual troop residents fed slightly more often. This

Mode of existence	Occurrence	Remarks
Solitary male	3	—
Male pair	1	—
Male trio	2	One trio joined troop II
Female pair	1	Joined troop II

may be correlated with a lack of opportunity for social interaction with others; for example, the time bisexual troop members spent in social interactions such as grooming or dominance was spent instead feeding. No agonistic or grooming sequences were observed in unisexual groups.

The relative facility with which a male trio joined troop II suggests that for some a solitary or unisexual group existence was a temporary condition (this is discussed in detail in Poirier, 1969a). Because some animals left or were driven from their parental troop, lived for various periods of time apart from bisexual troops, and eventually joined different troops there existed a behavioral mechanism for interbreeding among semiclosed or isolated troops.

E. SUBGROUPS

Individuals living with bisexual Nilgiri langur troops were organized into fairly consistent patterns of subgroups. Regular assemblages of animals within a troop having a recognizable affinity from day to day were designated as subgroups (Southwick et al., 1965). The subgroup was a social aggregate of individuals of similar age and/or sex. There were subgroups of adult males and older subadult males; adult females, adult females with infants and juveniles, and subadult females; and subadults and older juveniles.

Most interanimal relationships occurred within rather than between subgroups; this was especially true of grooming and dominance sequences. Table VII gives the percentages of interactions between ani-

TABLE VII
SUBGROUP INTERACTION

Interaction	Homosexual (%)	Heterosexual (%)	Like age (%)	Different age (%)
Grooming	48	27	71	29
Dominance	40	27	85	15
Play	20	12	42	58

mals within and without subgroups. Excluding play behavior occurring among animals of unlike age (an artifact of troop structure), most social interactions took place within respective subgroup assemblages. Percentile discrepancies (i.e., the failure in all cases to total 100%) result from the fact that many animals involved in interactions could not be accurately sexed. The high percentage of play behavior between members of different age classes is attributable to the fact that there were often few youngsters (those of playing age) of similar age class within a troop.

Most social interaction occurred between animals of like age and sex, resulting in a weakening of overall troop integration. Adult males and adult females rarely interacted with one another, especially in one-male troops in which the adult male normally remained socially and physically on the troop's periphery. Infants and juveniles had little opportunity to interact with adult males. In contrast with macaque, baboon, and common langur troops, in which most troop members had some opportunity to associate with other members, Nilgiri langur individuals seldom interacted with animals outside their subgroup.

Such social interactions as grooming, which functions in most non-human primate societies to strengthen troop cohesion by creating positive social bonds between the participants, occurred principally between subgroup members. This seemed to bind subgroup members more closely together while it excluded those outside its borders. In our opinion, Nilgiri langur subgroups created a situation that insulated their members from contact with others. This resulted in a number of autonomous units acting more in reference to animals within it than to the troop as a whole. This point is developed further in Section XIV.

V. DIET AND FEEDING HABITS

Nilgiri langurs are vegetarians whose main staples were the leaves, flowers, buds, seeds, and the bark or stem of various plants. They rarely consumed nonvegetable matter. Feeding was the most conspicuous time-consuming behavior witnessed; feeding began almost immediately upon rising and occurred intermittently throughout the day. Overall, a total of 7 or 8 hours daily might be spent feeding.

A complete listing of the Nilgiri langur diet appears in Poirier (1968b). A total of 52 floral specimens were consumed. The largest portion of the diet was derived from shrubs, with leaves representing the most preferred portion of the plants consumed.

In an attempt to determine whether or not the plants eaten by the

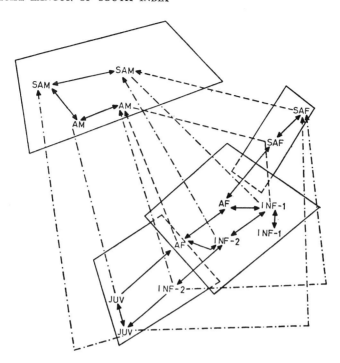

Fig. 2. Flow of social interactions. The continuous lines indicate the most frequent social interaction, broken lines occasional social interaction, and the dash-dot lines minimal interactions. Arrows indicate the direction in which most social interactions commonly flow. Boxed-in areas designate subgroup assemblages. AF, adult female; AM, adult male; SAF, subadult female; SAM, subadult male; JUV, juvenile; INF, infant.

langurs were likewise palatable to man, we sampled 25 of the 52 specimens consumed. Most plant specimens had a distinctly sour or bitter taste, a point made by Schaller (1963) concerning gorilla foods. This, of course, is not to say that the plants tasted the same to the langurs.

A. Specifics of Feeding Behavior

Most feeding occurred in the trees; seldom did the animals eat directly on the ground. If an individual fed on the ground, it was constantly alert and obviously nervous. It continually looked around and any sudden noise or movement sent it immediately into the nearest tree. An animal wishing a food plant growing near the ground, such as leaves of a young *Acacia* tree or the stem of a *Rubus* vine, descended to the ground, broke off a desired portion, and carried it back into the tree. The food was carried in either the hands or mouth. In its arboreal refuge, the langur fed leisurely.

Most food was conveyed by the hand to the mouth. Branches were bent toward the animal and held in one hand or foot, the free hand picking off the desired portion. Food was seldom taken directly into the mouth; however, one consistent exception occurred when the flowers of the *A. molissima* tree were eaten. Attempts to strip the *A. molissima* flowers from the stem and convey them to the mouth were wasteful, for the flowers immediately fell into many small pieces that slipped through the fingers. *Acacia* flowers were most easily eaten directly from the stem by drawing the stem through the mouth between the canines and premolars. *Cytis* flowers were similarly consumed.

Nilgiri langurs were discriminating in their choice of food sources. Shrubs such as *Mahonia, Rubus,* and *Litasae* received special attention. For example, only the soft, succulent, red leaves of *Mahonia* were eaten. The untouched green leaves were very tough and dry and were covered with sharp spines which easily pierced the skin. When a *Mahonia* leaf was desired, the langur carefully picked its way past the green leaves by brushing them aside with the back of the hand which is well protected by a thick matting of hair. When a *Mahonia* leaf was in an area requiring entrance into a shrub, it was often reached by "riding" a young sapling over the shrub. This eliminated climbing into the shrub and becoming stuck with the spines.

The seasonal fruiting and flowering of many food sources had a marked affect on troop movement, that is, there were definite seasonal patterns of home range utilization. Particularly during the early days of the fruiting of *Acacia*, many troops modified their daily movement pattern in order to take advantage of this food source. It was not uncommon for a troop to spend as long as 1 week feeding in one stand of *Acacia* trees.

B. LOCAL FOOD PREFERENCES—DIETARY ADAPTATIONS

Local troop food preferences were common. For example, the *Acacia* tree common to the home ranges of troops I, II, III, and IV was highly regarded by the members of troops I, II, and III but was merely a supplementary food source for troop IV. This selection from the available foods suggests that "cultural" differences in food habits existed. Viewed in broad terms, the most important aspect of this diversity in food preference is the ability of the langur populations to adapt to various habitats, choosing from whatever food the specific area offered.

Much was learned about the Nilgiri langur's ability to adapt to changing ecological conditions by noting their acceptance of new food items.

Within the past 85–90 years, many natural sholas in the Nilgiris were destroyed and numerous tracts replanted with *Eucalyptus globulus* trees (Francis, 1906; Grigg, 1880; Krishnamurthi, 1958). Langurs seldom ate (to this author, the medicinal-smelling foul-tasting) *Eucalyptus* leaves. Occasionally, adults in troops I and IV consumed the leaf petiole, but the leaf itself was seldom ingested. In both troops, however, infants sometimes consumed the entire leaf. The present evidence suggests that Nilgiri langurs are quite capable, particularly during infancy, of accepting new foods. [Japanese investigators, such as Itani (1959), Tsumori (1967), and Yamada (1957), discuss the acceptance of new dietary items, or as they phrase it, "subcultural propagation." The acquisition of new feeding habits often commences either with the infant or adult male age group (Kawamura, 1963). The rapidity of acceptance of new food items depends upon the animal(s) initiating its acceptance.]

Of considerable import in discussing the acquisition of new dietary items is the habituation of troop IV to the domestic potato and cauliflower. Both crops, introduced to the Nilgiris not more than 80–100 years ago, are now the main agricultural stays. Troop IV's home range encompassed large cultivated tracts of potato and cauliflower. Local knowledgeable informants related that the farms in this area were at most 25–35 years old. Thus, considering also that these are the sole farms in the area, this local langur population has only recently accepted this new dietary item. In addition, troop IV also ate cultivated cabbage and the ornamental garden poppy. In all cases the roots, fruits, leaves, and stems were consumed.

A further illustration of willingness to accept new dietary items occurred when troop II vacated its home range because of human destruction of the habitat. Troop II not only had to accommodate itself within its new home range, which primarily meant recognizing the home range boundaries of the four adjacent troops (Poirier, 1968c), but also had to adjust to new and different food plants. The food staples in the original home range were *Acacia* leaves, flowers, and fruits, which in the new area were superseded by *Litasae* and *Loranthus*. Furthermore, the variety of plants in the new home range was wider and offered several new species for consumption (see Table VIII).

Further information on the acceptance of new dietary items is found in Poirier (1968b, 1969b).

C. Acquisition of Food Habits

From birth until 5 weeks, the infant's sole nourishment was its mother's milk. The first instance of an infant observed with solid food occurred at approximately 5 weeks; a leaf was stripped from a branch and

TABLE VIII
Troop II Dietary Adaptations

Major food plants in original home range	Major food plants in new home range
Acacia melanoxyln	Acacia melanoxyln
Acacia molissima	Acacia molissima
Glochidion neilgherense	Cinnamomum wightii
Litasae ligustrina	Eunoymus crenulatus[a]
Litasae wightiana	Evodia lunar-ankenda[a]
Mahonia leschenautti	Litasae ligustrina
Rubus rugosae	Michelia nilagirica[a]
	Microtopis ovalifolia[a]
	Photinia lindleyana[a]
	Psychotria spp.[a]

[a] Food source not present in original home range.

carried about in the mouth. It was not masticated and soon dropped. Such behavior occurred rather frequently from the fifth through tenth weeks of life. Much of this early familiarization with edibles without actual ingestion appeared to occur during play behavior.

Acquisition of food habits did not involve food sharing between a mother and her infant. A mother never offered her infant food. On five occasions, however, an infant 1 attempted to take food from its mother; only one attempt succeeded. A mother usually turned her back on an infant when it reached for something she was eating.

The earliest period in the acquisition of feeding habits was basically one of trial and error. Occasionally, infants placed matter into their mouths which adults never ate. Twice an infant put moss in its mouth which was chewed briefly but then ejected.

Nine-week-old infants transported food from one place to another. Infants often left their mothers to obtain a desired food substance which was carried back, usually in the mouth, to feed close to her. Infants readily supplemented their milk diet with solid foods by 3 or 4 months of age.

D. Nonvegetable Foods

Little information was gathered on nonvegetable food sources, and it is likely that they constituted a minute portion of the diet. Nilgiri langurs were not observed to seek out deliberately and eat insects. It is quite likely, however, that they inadvertently ingested (unidentified) insect eggs and larvae which were abundant on the undersides of *Rubus, Toddalia,* and *Compositae* leaves. Stool depositions were constantly examined for evidence of feeding on insects. Only one sample in the hundreds examined showed evidence of nonvegetable food. This specimen revealed four unidentified beetles which were eliminated without having been digested. Numerous times langurs passed by bird nests containing eggs, through cobwebs in which flies and other insects were trapped, and past beehives without attempting to disturb them.

E. Geophagy

Five times during the study animals in three different troops ate dirt. [Many herbivorous mammals, especially ungulates, have a special preference for salt and other minerals contained in soil (Bouliere, 1962).] In all cases the individual was a male, twice it was a subadult and three times an adult. The soil consumed was reddish in color and came from a depth of not more than 3 inches. Twice an animal scraped the dirt with its incisors, rolled it in its hands into a ball, and then carried it into a tree to eat. It is interesting to note the manner in which members of troop IV obtained soil that they ate. Monkeys in troop IV scraped the soil with their hands, rather than their incisors, and then rolled it into a ball to eat. Either the habit of digging potatoes and similar forms from the ground has been carried over by members of troop IV to digging dirt, or the dirt-digging habit has been carried over to digging food. In either instance, digging behavior was characteristic only of troop IV.

Not enough instances of dirt eating were recorded to determine if certain soil-eating locales were preferred (as seems to be the case, for example, with the Cayo Santiago rhesus). Investigation of soil-eating sites indicated, however, that they had been previously visited. There was no evidence at these sites that suggested that the langurs were searching for insects. An attempt to obtain a chemical analysis of the soil was abortive; however, a member of the Geological Survey of India reported that the soil was of clay-humic origin with a high aluminum (alkaline) content.

Morris (1927) noted that South Indian elephants ate red clay soil which upon analysis was found to contain a high percentage of aluminum. (Morris's report came from the Cardamon hills which are of similar geological origin as the Nilgiris. Since a chemical analysis of the soil eaten by the Nilgiri langurs was not obtained, Morris's analysis or soil eaten by elephants is presented in hopes that a similar origin might indicate similar chemical content: Loss on ignition, 12.5%; alumina, 21.4%; iron as Fe-23, 14.1%; silica, 9.3%; sand and other siliceous matter, 43.1%.) Morris observed that elephants often suffered from "pica," an unusual appetite. Pica resulted from overacidity which produced a desire for an alkaline substance, such as those contained in the soil.

It is possible that Nilgiri langurs also suffered from overacidity. Numerous instances of belching and flatulation, outside of social inter-action, were recorded, indicating that the large quantities of leaves and vegetable matter consumed daily occasionally created gastric distress. Such signs of gastric distress were especially noticeable following a hurried feeding bout, for example, before retiring or moving. We observed similar instances of geophagy among South Indian langurs and bonnet macaques (who were also observed to eat charcoal). McCann (1933) reported that the golden langur sometimes consumed salty-tasting earth [Schaller (1963) reports geophagy for gorillas].

F. DRINKING BEHAVIOR

Nilgiri langurs seemed to obtain most of their water indirectly from the leaves they ate. Most food plants that we sampled contained some moisture. Drinking behavior was only recorded five times, and licking water from leaves or tree trunks after a rain four times. Twice an animal licked the water adhering to its own hair. When individuals drank directly from a water source, it was from a stream or a pool which had collected in a ground depression, such as that left by a footprint in the mud. Langurs drank by sucking water through their lips; they were not observed dipping their hands into water and then licking the water from the hair.

Water requirements seem to vary among langur species, a phenomenon apparently related to regional differences in rainfall and abundance of succulent vegetation. Ceylon gray langurs (Ripley, 1967), South Indian langurs (Sugiyama, 1965a), and the Nilgiri langur pass several months without visiting water sources. On the contrary, langur groups in Central India tend to be spaced in order to take advantage of artificial water sources used in crop irrigation. In the water scarce areas of North India, common langur troops ". . . that otherwise might not have any contact with each other come together at wells and reservoirs" (Jay, 1965a, p. 206).

VI. THE COMMUNICATION MATRIX

A. INTRODUCTION

Communication is the mechanism through which organisms interact; it is a system of message transmission integrating and controlling social behavior. The communication matrix is the system maintaining social order (Emlen, 1960). Patterns of primate communication ". . . initiate, channel, and direct the behavioral expression of previously established 'sets,' predispositions or motives" (Carpenter, 1963, p. 49).

Nilgiri langur communicative behavior manifests certain properties which are to a great degree common to all nonhuman primate communication. The communicative repertoire of the species is part of the behavioral repertoire of all troop members, at least at some stage in their ontogenetic development. There are restraints upon the recombination of communicative acts; certain sequences (communication chains) are far more evident than others. The sequence in which communicative units are strung together is often or perhaps always learned; recombination depends for each animal, upon previous experience. In the face of a dominant animal's threat, subordinates recombine a series of learned subordinate responses.

The relative success or failure of any social animal relates to its ability to associate properly with its role: to communicate with the appropriate signals at the appropriate moment. Each monkey assumes various roles during the day; each role entails a set of communicative actions. The totality of these communicative roles may be designated an individual's "communication matrix" (Gumperz, 1962; Morris 1946). Individuals alternate between what may be termed "primary" and "secondary" responses. For example, a beta animal's primary responses are associated with its beta dominance rank. Its secondary responses are those associated with its subordinate position relative to the alpha animal. Every animal must switch from primary to secondary responses and back again with interference. This may be designated "code switching" (Haugen, 1953).

Messages in the nonhuman primate signaling system may be directed to one particular animal or to the group en masse. The former is designated "one-to-one" signaling, the latter "to-whom-it-may-concern" signaling (Altmann, 1962a). The Nilgiri langur warning bark and whoop vocalization, which are sounded throughout territorial battles, are examples of the latter. Most messages, however, are directed. Additionally, certain messages may be directed to no one particular animal, yet still be directed. Actions appearing to occur *in vacuo,* such as the nervous limb twitching that Nilgiri langur infants often exhibited during weaning, fall into this category.

B. SEGREGATION OF COMMUNICATION ELEMENTS

Defining the parameters of a communicative unit is one of the most difficult problems the investigator encounters in his attempt to analyze nonhuman primate communication. At what temporal point in a sequence does the communication chain begin and end? At best, the investigator makes a subjective judgment based on his observations, but there is bound to be variance between different researchers. As a case in point, Altmann (1962b) divides rhesus behavior into finer units than most investigators; he considers, for example, rhesus mounting behavior as five separate behavioral patterns. Where possible, Nilgiri langur communication chains were segregated into meaningful (again, a subjective judgment) segments. Continual reduction of each communicative chain into its smallest components, however, could easily end in confusion and misinterpretation rather than clarification.

A second problem is the dichotomy of communicative versus noncommunicative behavior. Struhsaker (1967, p. 5) notes that ". . . patterns of feeding and drinking undoubtedly convey some information to other monkeys about the behavioral state of the individual concerned, but they are not of obvious and immediate social significance. The separation of behavior patterns into communicative and non-communicative patterns thus involves a rather subjective attempt to distinguish those patterns that have an obvious and immediate social consequence from those that do not." Nilgiri langur patterns were considered to be communicative when they met this same criterion.

A fully detailed description of the Nilgiri langur communication matrix deserves a broader treatment than is the focus of this chapter. In lieu of a detailed analysis, an overall picture of the communicative process is included. Details are included in various chapter subsections, for example, patterns involved in dominance, play, and grooming are discussed under these headings. A complete listing of the Nilgiri langur communication matrix is given in Table X.

Most of the Nilgiri langur communication repertoire was witnessed in intratroop situations. The system was primarily composed of gestures, movements, and body postures; auditory and tactile cues were less important. Intratroop signals functioned primarily to mediate agonistic situations, during movement, to inform the troop of impending danger, in peaceful situations occurring during the daily routine (e.g., grooming, play, and feeding), and between a mother and infant. Discounting agonistic situations, however, intratroop communication was limited.

C. Displacement Activities

Some situations arouse conflicting tendencies during which animals sometimes manifest behaviors not obviously pertinent to the situation at hand. Such behaviors are termed "displacement activities" (Tinbergen, 1951). Nilgiri langurs exhibited two displacement activities, scratching and sham or symbolic feeding. Sham or symbolic feeding was recorded among adult males during moments of tension, especially during territorial encounters or in response to extragroup disturbances. Males frequently interrupted their territorial whoop displays to stuff two or three handfuls of leaves into their mouths. Occasionally, leaves were not placed in the mouth but just held in the hands, or the leaves were placed between the lips and subsequently dropped (see Fig. 3b). The following is a typical example.

> The troop-II delta male sits in an *Acacia* tree over the potato farm vocalizing loudly at the presence of a dog. He is extremely nervous and urinates and defecates frequently. He continues to emit the alarm call for almost 10 minutes. He pulls some leaves from a branch beneath the one on which he sits and places them in his mouth. He summarily spits them out without mastication. He again begins to vocalize . . . (July 20, 1966).

Scratching occurred in similar conflict situations. Animals occasionally interrupted threat sequences to scratch themselves. Nursing mothers sometimes scratched in a cursory manner when the investigator watched them.

D. Internal Expression Phenomena

This category includes all reactions, *sensu lato*, of the digestive tract to disturbances. Frequent urination and defecation were highly consistent signs of tension. Elimination was practically universal during or following such stressful situations as dominance sequences or territorial encounters. Fright situations, such as the sudden appearance of dogs or the unexpected appearance of the observer, produced involuntary releases of the sphincter muscles. Quite possibly olfactory communication was involved during stressful situations causing defecation. Defecation

FIG. 3a. Bark threat of adult male.

FIG. 3b. Stare threat of adult male. Note leaves in mouth.

was especially noticeable during territorial encounters when the over-powering fecal odor nauseated the observer.

Regurgitation in stress conditions was not recorded.

E. Displays

The most spectacular display in the Nilgiri langur repertoire is the jump display component of territorial battles. Characteristics of the display are described in Section VIII, A.

F. Intertroop Spacing Mechanisms

The prime intertroop spacing mechanism is the whoop vocalization. The vocalization is a loud, sonorous, prolonged, violent expiration of air emitted through rounded lips. The vocalization is a relatively pure drawn-out sound approaching a musical tone. The onset of the vocalization is characterized by a violent rasping inspiration of air. The abdomen pulsates under the pressure of the expulsion of air. Although the call is usually emitted from standing or a moving position, it is occasionally emitted from a sitting position.

The whoop vocalization was differentiated into various components, each in their own right a communicative element. Depending upon the context, the act may terminate with any of its components.

(1) The onset of the whoop vocalization is signaled by a rapid inspiration of air reminiscent of a snore.

(2) The initial phrases of the actual vocalization are discernibly shorter than succeeding phrases. Individual tonal differences exist which help the observer identify the vocalizing animals.

(3) Immediately following the first frames of the vocalization, the call is interrupted, the tail hangs over the back (the tip pointed toward the head), and calm prevails. The interruption seems to be a brief interlude during which others focus their attention on the vocalizing male. The length of the break differs; there seem to be consistent individual variations in the duration of the interlude.

(5) A most impressive component of the behavioral pattern is the jump display described in Section VIII, A.

(6) Upon termination of the jump display, the vocalizing animal sits and emits the hiccup vocalization for as long as 5 minutes. This vocalization seems to be a vehicle for releasing energy or tension. Hiccuping is frequently interrupted by the ho-ho vocalization described below. Following a whoop display, the adult males involved often sat as long as

10 minutes, sometimes for as long as an hour, hiccuping intermittently. The vocalizing animal was very nervous and any slight provocation was likely to stimulate frenetic jumping through the trees. Consequently, other troop members avoided it.

Marler (1969) describes a pattern of colobus tongue clicking in situations analagous to those in which Nilgiri langurs hiccup. The sound appears to be made in a similar fashion. It is possible that the Nilgiri langur hiccup was tongue clicking, however, we are unable to confirm this.

The Nilgiri langur whoop vocalization/jump display was evoked by extratroop or occasionally intratroop stimuli. Males almost religiously responded to the whooping of males in surrounding troops. The sight of adjacent troops, the falling of a tree, the beeping of a claxon horn or occasionally an intratroop dominance sequence all elicited the call. In the following example the displaying animal utilizes all the elements commonly accompanying the whoop vocalization/jump display.

> The troop-II alpha male begins to whoop in response to a jump display by the male in troop I. He rises and begins the violent, rasping inhalation of air. He then makes a leap of 20 ft to an adjacent tree, grunting harshly as he lands. He emits a short whoop from a static standing posture, stops, stands tautly with his tail arched high above his back, and freezes for approximately 10 seconds. The sudden lull is impressive. He begins to move rapidly about the tree, his weight breaking off dry, dead branches which crash thunderously to the ground. He moves to the treetop and branch ends where he sits momentarily. He then continues to whoop. He jumps to an adjacent tree and moves rapidly through the branches, never breaking stride. This portion of the act is silent except for the noise of cracking branches and the contact of the male's feet with the heavy limbs. Approximately 3 minutes into the display the male desists, sits, and begins to hiccup. He interrupts the hiccup with a ho-ho vocalization and harsh grunt. He hiccups for another 30 seconds, then stops. During the sequence other troop members avoid his careening, frenetic movements. The male sits and begins to eat *Acacia* leaves. He urinates and defecates. The sequence ends after 5½ minutes (January 26, 1966).

The ho-ho vocalization was emitted in situations similar to those evoking whooping, however, it indicated a lower degree of excitement. The vocalization itself is less resonant than the whoop; the frames are shorter and more staccato. In contrast to the whoop vocalization which was performed only by adult males, both adult males and females emitted the ho-ho vocalization. Adult males emitted the call more frequently, however.

The final auditory cue recorded during intertroop agonism is canine grinding. By slowly rubbing the canine teeth together during exaggerated opening and closing of the mouth, adult male Nilgiri langurs created a grating sound. The noise carried over rather long distances. Only adult males, who possessed the largest canines, were recorded canine grinding. Agonistic males faced each other, or occasionally away from each other, and ground their canines for as long as 20 or 30 minutes. During one territorial battle, the opposing males ground their canines for 90 minutes, intermittently punctuating the act with ho-ho vocalizations

All colobids studied to date emit vocalizations similar either in form or function to the Nilgiri langur whoop vocalization. The South Indian common langur vocalization is lower-pitched, hollower sounding, and emitted in slower phrases than the Nilgiri langur call (Poirier, 1968b). This observer had absolutely no difficulty distinguishing the vocalization of these respective species. Furthermore, the frenetic movement characteristic of Nilgiri langur jump displays was less apparent in South Indian langur displays. Ceylon gray langurs (Ripley, 1967) emitted similar vocalizations. Among *P. entellus* and *P. johnii*, only males vocalized; in contrast, Ceylon langur females as well as the males commenced the display. The displays of the Ceylon and Nilgiri langurs were elicited by both intratroop dominance interactions and extratroop stimuli, the latter being most common. North Indian langurs, however, displayed only in response to extratroop disturbances. A similar intertroop male vocalization is reported for *Presbytis cristatus* (Bernstein, 1968).

African colobids emit analogous vocalizations. Recently, Marler (1969) published an account of *Colobus guereza* territorial behavior. An integral part of *C. guereza* territorial displays includes loud vocalizations which Marler terms "roars." Marler has discussed the advantage of loud sounds in a forest environment. To be accurately identified by adjacent troops in the presence of wind and other forest animals, the sound must have a distinctive structure.

> Such sounds usually are among the loudest that the species utters, softer sounds usually being used to increase distance. They tend to be more structured than other sounds in the repertoire in the sense that temporal organization is often elaborate. They sometimes have a purer tone than the grunts and barks used in closer range signaling. Thus they are clearly identifiable at a distance, even with a considerable level of background noise. (Marler, 1968, p. 429.)

Lion-tailed macaques (*M. silenus*) emit a vocalization very similar to the Nilgiri langur whoop vocalization (Sugiyama, 1968). That the lion-tailed macaque developed this vocal signal is interesting, for it is rare among other macaques. The lion-tailed macaque is the most arboreal macaque, however, spending its life in the thick jungle treetops of South India (see Section III,D). Lion-tailed macaques and Nilgiri langurs are physically very similar. Local inhabitants, in fact, refer to both as the "black monkey" or *karum karungu*. Both species overlap in the same environment, and Sugiyama poses the question whether or not the physical resemblance and certain behavioral traits result from an adaptation to a similar environment.

G. MOVEMENT-COORDINATING SIGNALS

Only one Nilgiri langur vocalization, labeled the "grunt," appeared to contribute to troop coordination during movement. As the name implies, the grunt is a deep guttural call of short duration. The call is strictly an adult vocalization. The initial call elicited similar responses from nearby animals following which the initiator and recipient(s) often continued to vocalize. The exchange might last as long as 10 minutes. The call indicated temporal proximity and thus facilitated troop cohesion, especially during movement through thick foliage and when the troop was spread out with troop members out of sight of each other. Adult males frequently grunted when returning to the troop after territorial conflicts. In this case the grunt seemed to signal their return.

The lack of coordinating vocalizations is most probably related to the fact that Nilgiri langurs seldom moved at a pace requiring vocalizations which help maintain cohesion. In contrast, a terrestrial species such as the Japanese macaque has 10 distinct vocalizations which are emitted during wandering (Itani, 1963).

H. AGE, SEX, AND STATUS DIFFERENTIATION

Differences existed in the utilization of Nilgiri langur communicative patterns according to age, sex, and status. Differentiation was especially obvious in the adult age category. For example, only adult males emitted the whoop vocalization and only adult females presented patterns contingent upon their status as mothers. There were also differences in response according to the age, sex, or status of the animal issuing a cue. For example, adult male alarm calls evoked stronger responses than adult female alarm calls. There was a noticeable prolongation of tension when males alarm-called.

I. Specialization of Vocalizations

While many communication patterns are specialized in the sense that they convey different meanings, the dichotomy of the gruff bark-alarm call in predaceous situations is interesting. Adults emitted the gruff bark in response to extratroop dangers such as a dog, the observer or strange person, or the unexpected appearance of a strange langur. The alarm call functioned similarly to the gruff bark and was also emitted in tension situations. The call located the vocalizing animal by attracting the attention of other troop members. All activity momentarily ceased until the troop discerned the excitement-producing stimulus. In contrast to the wide range of stimuli evoking the gruff bark, the alarm call was emitted only in the presence of village dogs. Although Nilgiri langurs are potentially threatened by a number of predators, man and his dogs are potentially the most dangerous enemies.

Struhsaker (1967) notes that vervets respond differently to male and female predator vocalizations. Female vervet vocalizations elicit stronger responses than male vocalizations in the presence of mammal predators. Vervets also emit different vocalizations when the predator is a mammal, bird, human, or a snake.

J. Possible Dialect Differences

In keeping with the isolation of various Nilgiri langur troops, certain components of the communication matrix appeared to be emphasized in some troops and not in others (Poirier, 1968b, 1969b). Whether or not this is an artifact of sampling is not known, but if the repertoire is composed of learned patterns, it would not be surprising to find that certain communicative elements were less emphasized or were almost absent in some troops. As the isolation process continues (through human destruction of the niche) it is not unlikely that some Nilgiri langur populations will modify existing behavioral patterns, emphasizing whatever communication differences currently exist.

K. Size of the Repertoire

Compared to macaques and baboons, the Nilgiri langur repertoire of facial gestures is less varied. Facial expressions, namely, those characteristic of the ears and scalp, such as ear flattening and scalp wrinkling, are missing. The absence of such gestures may be attributable to a lack of facial musculature specialization. Furuya (1961–1962) noted a similar dearth in variety of facial expressions among Malaysian silvered leaf monkeys.

The present study distinguished 19 distinct audible patterns (including canine grinding), however, Tanaka (1965, p. 116) states "the vocalizations of Nilgiri langurs is very poor in variety . . ." Tanaka recognized four groups or classes of vocalizations: (1) Demonstrative sounds; included in this category are sounds, such as the whoop vocalization, which are emitted in intertroop conflicts. (2) Threatening sounds at an attack; this category subsumes vocalizations labeled "barks" by the present author. (3) Defensive sounds similar to a scream; these are submissive vocalizations which are emitted by females and youngsters. Adult males never emitted such cues. (4) Sounds emitted quietly; females and infants emitted these sounds while feeding or moving, or when they were slightly surprised or frightened. Tanaka's classes 3 and 4 include vocalizations that this study designated squeals, screeches, grunts, and the subordinate segmented vocalization. Tanaka distinguishes canine grinding as an aggressive behavior common in intertroop encounters.

The communication matrix of *P. entellus* (Jay, 1965a) and *P. johnii* langurs is broadly analogous. A comparison of the respective systems is presented in Table IX. A complete listing of the Nilgiri langur repertoire is given in Table X.

VII. CHARACTERISTICS OF THE HOME RANGE

A home range ". . . is that area traversed by the individual in its normal activities of food gathering, mating and caring for the young" (Burt, 1943, p. 347). The home range is usually constant for a given period of time and does not include areas crossed on occasional long forays of possibly exploratory nature. A familiar setting, the home range is an area of knowledge with which its members identify through daily usage (Washburn and Hamburg, 1965). Through continuous association the area becomes familiar to its inhabitants, and through continuous association the inhabitants become familiar with each other. A shared common knowledge of the home range influences social integration because many individuals are adapted to and continually interact in the same environment.

A. EXTENT

Nilgiri langurs restricted their daily activities to definite areas varying from approximately ¼ to 1 sq mi. The extent of a troop's home range

TABLE IX

COMPARISON OF *Presbytis entellus* AND *Presbtyis johnii* COMMUNICATION SYSTEMS[a]

Patterns	*Presbytis johnii* D	S	E	*Presbytis entellus* D	S	E	Comments
Anal inspection	X	X	X	—	—	—	—
Biting air	X	—	X	X	—	X	—
Crouch, then stand suddenly	—	—	—	—	X	X	—
Displacement	X	—	X	X	—	—	—
Dominance pause	—	—	—	X	—	X	—
Embrace	X	X	X	X	X	X	—
Embrace/groom	X	—	X	—	—	—	—
Grin	X	—	—	X	—	X	*Presbytis entellus* grimace?
Head bob	X	—	—	X	X	X	—
Looking away	—	X	X	—	X	—	—
Lunge in place	X	—	X	X	—	X	—
Mount	X	X	X	—	—	—	—
Observe dominance distance	—	X	X	—	X	—	—
Open mouth	X	—	X	—	—	—	—
Present	X	X	X	—	X	X	—
Stare threat	X	—	X	X	—	X	—
Shaking head	—	X	X	—	—	—	Only estrous *P. entellus* female utilizes
Slap at	X	X	X	X	X	X	—
Slap ground	—	—	—	X	—	X	—
Sniffing mouth	X	—	X	—	—	—	—
Touch	X	X	X	X	—	—	—
Tongue in and out	—	—	—	—	X	X	—
Turn back on	X	X	X	X	X	X	—
Vocal cues							
Alarm call	X	X	X	X	X	X	—
Canine grind	X	—	X	X	X	X	—
Cough	—	—	—	X	X	X	—
Chuckle	—	—	—	—	—	—	*Presbytis johnii* mothers use it
Growl	X	—	—	—	—	—	—
Gruff bark	X	X	X	X	X	X	*Presbytis entellus* male and females both utilize it
Grunt	X	X	X	X	X	X	—
Hiccup	X	X	X	X	X	X	*Presbytis entellus* belch?
Hoho	X	X	X	—	—	—	—
Hollow subordinate sound	—	X	X	—	—	—	—
Pant	X	—	X	—	—	—	—
Roar	X	—	—	—	—	—	—

TABLE IX (*Continued*)

Patterns	*Presbytis johnii*			*Presbytis entellus*			Comments
	D	S	E	D	S	E	
Scream	—	X	X	—	X	X	Only *P. johnii* infants use it
Screech	—	X	X	—	—	—	—
Squeal	—	—	X	X	X	X	Only *P. johnii* females use it
Subordinate segmented sound	X	X	X	—	—	—	—
Whooping	X	X	X	X	X	X	—
Warble	—	—	—	—	—	—	Only *P. johnii* mothers use it

^a D, dominant; S, subordinate; E, equal.

was determined by noting daily locations on a hand-drawn map. Troop home range sizes differed markedly for various reasons; those that were determined are herewith outlined.

(1) The size and composition of a troop was related to the size of its home range. Larger troops containing 8 to 10 adults occupied larger areas, presumably to meet increased nutritional requirements.

(2) The size and proximity of neighboring troops influenced home range size. Troops inhabiting high-population-density areas occupied smaller home ranges than troops in low-density areas. In high-population-density areas, a troop's range probably represented a combination of behavioral-ecological requirements. Smaller home ranges in densely populated areas reduced the incidence of intertroop encounters (because of the reduction or elimination of overlapping home ranges, which is discussed in Section VIII, A) which were typically aggressive.

(3) The concentration and type of food plant in the home range played a major role in determining its size. Troops inhabiting sholas lacking a major food source, such as the *A. melanoxyln* or *A. molissima* tree, had larger home ranges than troops occupying areas in which *Acacia* was abundant. One hearty *Acacia* stand containing four or five trees readily supported 10 to 15 animals for at least 4 months per year. This reduced the necessity of traveling over a wide area in order to fulfill dietary requirements. In contrast, troops relying mainly on shrubs to fulfill their dietary needs covered a wider (yearly) range. A male group's home range appeared to be larger than that of a bisexual troop; this probably served to reduce contact with males living in bisexual troops.

TABLE X

Summary of Stimulus Situations Evoking Communicative Act, Age, and Sex of Sender, Message, and Most Frequent Responses

Stimulus	Act	Age and sex of sender	Message	Most frequent response
Intra- or intertroop agonism	Attack	A through I 2	High state of tension	Return attack or run away
	Bite	A through J	High state of tension	Attack or run away
	Biting air	AM, AF, SA	Threat	Return threat, present, run away
	Chase	A through I 2	Threat	Run away
	Displacement	A, SA, J (rare)	Mild dominance assertion	Move away, present
	Embrace	A through I 2	Mild state of tension	Embrace and groom
	Face sniffing	A through I 2	Uncertain	Just sit, move away
	Form of progression	A through I 2	Status	Variable
	Freezing	A through I 2	Subordination	Ignore
	Genital inspection	AM, AF, SA	Mild dominance assertion	Allow inspection, move away
	Grin	AM, AF	Threat	Present, move away
	Head shaking	AM, AF	Subordination	Mild threat
	Jump display	AM	Agitation, intimidation, location	Variable—usually a similar response
	Look away	A through I	Mild subordination	Ignore
	Look threat	A through J	Mild dominance	Subordinate oneself
	Mount	A through I 2	Usually dominance	Subordinate oneself
	Observe dominance distance	A through I	Mild subordination	Avoid proximity, present
	Open mouth	A, SA, J	Threat	Present, move away
	Present	A through I 2	Subordination	Terminate threat, move away
	Run away	A through I	Subordination	Chase, terminate threat
	Slap at	A through I 2	Aggressive threat	Vocalize, run away, slap back
	Stare	A, SA	Threat	Look away, present
	Touch	A through I 2	Mild assertion or comforting	Present, calm down, move away
	Turn back to	A, SA, J	Subordination	Terminate threat
Play	Chase	SA, J, I	Indicates playfulness	Play
	Gamboling	SA, J, I	Indicates playfulness	Play

Context	Behavior	Meaning	Age/Sex[a]	Response
Grooming	Open mouth	Invitation to play	SA, J, I	Play
	Wrestle	Rough play	A through I	Play
	Cursory grooming	Invitation to groom	A through I 2	Groom
	Present	Invitation to groom	A through I 2	Groom
Mother-infant relationship	Approach toward	Signal to cling	AF	Clinging
	Outstretched arm	Signal to cling	AF	Cling
	Pat on back	Signal of impending movement	AF	Cling tightly
	Standing over	Signal to cling	AF	Cling
		Vocalizations		
Intra- and intertroop agonism	Canine grind	Extreme tension	AM	Return threat
	Grunt	Call attention to	AM, AF	Return call
	Hiccup	Tension	AM, AF	Move away
	Hoho	Call attention to	AM, AF	Return call
	Hollow subordinate sound	Subordination	A, SA	Terminate threat
	Pant threat	Aggressiveness	AM, AF	Subordinate itself
	Roar	Unclear	AM	Attack, terminate attack
	Screech	Extreme submissiveness	A through J	Embrace
	Subordinate segmented sound	Appeasement	AM, AF, SA, J	Return call
	Whoop	Location, intimidation	AM	Return call
Progression	Grunt	Location	AM, AF	Flight and search for stimuli
Presence of predator	Alarm call	Warning	AM, AF	Flight and search for predator
	Gruff bark	Tension, warning	AM	Move toward, punish
Mother-infant relationship	Scream	Intense agitation	I	Move toward, no response
	Squeak	Call for help	I	Move toward
	Squeal	Call to infant (?)	AF	Move toward
	Warble	Unclear	AF	Move toward
Undetermined	Chuckle	—	—	—
	Growl	—	—	—
	Roar	—	—	—

[a] A, Adult; SA, subadult; J, juvenile; I, infant (1 or 2); M, male; F, female.

B. Extent of Movement

The extent of movement within the home range was greater for some animals than others. Adult females, especially mothers, remained in the most familiar areas of the home range, which were often synonymous with core areas. Adult males, however, often passed from one region to another, especially during intertroop altercations when they frequently left their home range to penetrate that of an opposing troop to challenge the resident adult male(s). The greater range of movement of the males was also apparent when a troop moved adjacent to a portion of the home range overlapping that of another troop. Adult males were often the first and sometimes the only animals to enter these areas; adult females and youngsters frequently remained behind.

Elsewhere (Poirier, 1968b, 1969a) it has been suggested that the reluctance on the part of females to leave the most familiar, secure areas of the home range is adaptive. Presumably, a female provides her infant the best possible protection within an area with which she is most familiar. (This is an especially cogent argument in regard to Nilgiri langurs since adult males avoided physical contact with infants and assumed little if any direct reponsibility for their safety and, apparently, socialization.) If taken by surprise in a familiar area, a mother rapidly can find a refuge site or escape route. Additionally, within familiar surroundings, a mother best transfers troop traditions to her infant such as the whereabouts of arboreal pathways and the location of feeding and sleeping sites.

C. Core Areas

Nilgiri langur troops utilized certain portions of their home range more than others. Areas of frequent use containing sleeping trees, resting sites, and preferred food sources were designated "core areas." A troop spent as much as 70% of its time within the core area. Areas of heavy temporary use, which were frequented while a desired food source was available, but not before or after, were not included. Within the core area travel routes formed a dense, tangled web of tracings. Outside the core area activity was less conspicuously concentrated (Kaufmann, 1962). The core area always contained a few tall trees, providing a good view of the surrounding area and from which the alpha male could perform the morning whoop display. Although considerable portions of a home range often overlapped the home ranges of neighboring troops, core areas never overlapped.

D. OVERLAPPING OF HOME RANGES

There was considerable overlap of neighboring home ranges, especially in more densely populated areas (illustrated in Poirier, 1968a). Neighboring troops seldom occupied overlapping areas simultaneously. Areas of overlap were often foci of territorial battles, especially if they contained remnant stands of fruiting *Acacia* trees. A troop often avoided an area of home range overlap if it had been previously occupied by another troop.

E. HOME RANGE SHIFTS

One example of a home range shift was noted during the course of this study (see Fig. 4). The evacuation of the home range was necessitated by human destruction of the habitat. The particular locale that this troop inhabited was undergoing rapid and marked change; the area was constantly being defoliated and replanted in stands of *E. globulus* trees. Consequently, further population dislocations are likely to continue into the near future. Since the specifics of the shift are discussed elsewhere (Poirier, 1968c), only the highlights are presented here. At the time of the shift, troop II contained four adult males, two adult females, and an infant 2.

Many interesting patterns characterized the home range shift, one being the very evident reluctance to vacate the original home range until the last tree was cut, suggesting a strong attachment to a familiar area, surely a selective trait (Poirier, 1968d). The first exit from the home range was temporary. Troop II returned as soon as conditions were more favorable, that is, the termination of work and the departure of the forest crews. Boulière (1962) notes that a prime advantage of the home range is the fact that it assures individuals frequenting it not only daily food, but also increased predator protection.

Immediately prior (2 weeks) to the home shift, troop II was joined by a male trio. Twice, the first animal vacating the home range was a member of that trio, suggesting that the newest arrivals to the area were the least reluctant to leave. This again emphasizes the strong troop ties to a familiar area. It should also be noted that the two adult females of the troop were more reluctant to leave the home range for a new living area than were the males. Females appeared more tightly bound to familiar surroundings than males, the selective advantage of which is discussed in Section XIV,E. The female's hesitancy to leave a familiar

surrounding is one of many possible explanations for the greater proportion of nonbisexual troop males.

VIII. TERRITORIAL BEHAVIOR

Using Burt's (1943, p. 351) definition of territory as protection or
defense of ". . . some part of the home range, by fighting or aggressive
gestures, from others of their kind, during some phase of their lives . . . ,"
it can be shown that Nilgiri langurs manifest territorial behavior. The
terms "protection" or "defense" have a dual interpretation. Although
both terms usually refer specifically to fighting, their meaning may be
broadened to include any behavior resulting in exclusion of conspecifics
(Carpenter, 1964). Excepting the temple-dwelling Achal tank rhesus
macaques (Southwick, 1962), defensive fighting is rare among nonhuman
primates. There is, however, a definite tendency for intertroop avoidance,
facilitated by variable daily movement patterns, use of specific areas
(often designated core areas), and visual and auditory location.

A. DYNAMICS OF TROOP SEPARATION

Nilgiri langur troop separation was maintained through variable movement patterns, the male whoop display, and male vigilance behavior. The
daily routine usually maintained troop separation; seldom did daily or
seasonal movement patterns of adjacent troops overlap. The existence
of alternate core areas further reduced intertroop contact. When a troop
occupied a core area near one edge of its home range, a neighboring
troop often moved to a portion of its range removed from that vicinity.

The morning whoop, a sonorous, booming male vocalization audible
over long distances, located troops in their respective home ranges. Adult
males vocalized shortly after rising; the first call elicited similar responses
from surrounding adult males and the forest soon echoed with their
booming vocalizations. Although the vocal pattern was essentially in-
distinguishable, the morning whoop differed motivationally from the
whoop produced when adult males of different troops vocally challenged
one another. Morning whoops were unaccompanied by the frenetic movements characteristic of the display whoop. A male emitting the morning
whoop simply moved to the treetops and vocalized, and other nearby
males responded. In contrast, an integral part of the whoop display,
emitted by males during territorial altercations, was the display jump.

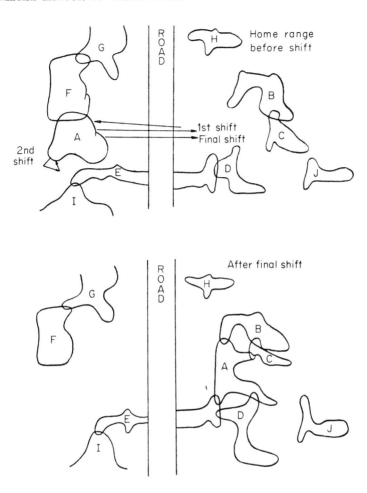

FIG. 4. Home range boundaries before and after shift: The arrows in the first drawing indicate the direction of the shift. Note that the first shift was temporary, the troop returning to the original home range. The letters within the home ranges serve to distinguish them. The home ranges are drawn to approximate scale.

Similar locating vocalizations are described for gibbons (Carpenter, 1940), howler monkeys (Carpenter, 1934), *Callicebus* (Mason, 1966, 1968), North and South Indian langurs (Jay, 1965a; Sugiyama, 1967; Poirier, 1968b), the Ceylon gray langur (Ripley, 1967), lion-tailed macaques (Sugiyama, 1968), and black and white colobus (Marler, 1969).

The pattern of male vigilance behavior described by Ripley (1967) for Ceylon gray langurs and by Hall (1960, 1965), who calls it "dominance vigilance" for chacma baboons and patas monkeys, agrees with the behavior manifested by Nilgiri langur males. [Preliminary examination of data recently collected suggests a similar pattern for the St. Kitts green monkey (Poirier, 1970b).] Adult Nilgiri langur males were alert to any and all disturbances; they often occupied the treetops and scanned the surrounding environs. In contrast to females, which spent much of the nonfeeding period resting, males maintained a constant alert. Nilgiri langur males seemed to be on the alert for other troops, rather than for danger from other sources. Male vigilance was not a means of troop avoidance; rather, it often resulted in intertroop vocal battles. A Nilgiri langur male spotting a male of another troop fixed his gaze in its direction and emitted a low grunting vocalization, which was frequently supplanted by the whoop display.

B. Intertroop Relations

Nilgiri langur intertroop relations assumed various forms, including peaceful feeding in proximity, peaceful withdrawal of one troop, or most often exchange by the males of visual and/or vocal displays and occasionally chasing. Eighty percent of the 109 communicative acts (e.g., any gestural, vocal or postural exchange, or combination of these, involving two or more animals) occurring in 84 aggressive encounters were an exchange of vocal and gestural patterns, 18% involved chasing, and in 2% physical contact occurred. Four times one troop peacefully withdrew from another and 10 times two troops fed in proximity.

It was not positively ascertained what influenced the mode of interaction between adjacent troops or, more specifically, between their adult males. Some males constantly displayed against and challenged one another; others were rarely involved in exchanges and ignored each other's presence. For example, the males of troop II maintained an aggressive attitude toward the males of three of the four troops in the immediate vicinity; however, they often fed peacefully within 15 ft of the male of the fourth troop. [Troop II was unusual because it contained four adult males in a total population of 10 animals (Poirier, 1969a).] It is significant in this regard that the majority of the territorial interactions involved alpha male versus alpha male. Thus even though a male belonged to a larger multiple-male troop, this had little influence on his effectiveness in territorial battles.

Occasionally, a resident male tolerated minor intrusions in his territory, or he delayed his response until another troop had been in his territory awhile. In the latter instance, it was as if the continuous or minor stimuli caused by the intruder's presence finally triggered a reac-

tion in the resident males. This is somewhat analogous to "nervous sum-
mation" (Koford, 1957). Ellefson (1968) suggests in his discussion of
white-handed gibbons that an excitation threshold must be reached
before a chase or other aggressive manifestations can occur.

Although territorial encounters occurred throughout the day, the first
being observed at 0605 and the last at 1800, most altercations were witnessed between 0800 and 1600. Periods of peak excitement occurred
between 0900 and 1200 and 1400 and 1600. Excluding the interval between 1000 and 1200, all were periods of major movement and/or
feeding. Monthly heights of antagonism occurred in January, March, and
April. There is a suggestion that the clustering of encounters was related
to the birth cycle. January immediately followed the winter birth peak;
March and April immediately preceded the spring peak.

Elasped encounter time varied from about 1 minute to over 4 hours;
the latter occurred once. Most interactions lasted less than 1–15 minutes.

Territorial encounters rarely involved physical contact; in fact, contact
was established just twice. Most interactions were simply male exchanges
of visual and/or vocal displays. Approximately 91% of all the vocalizations recorded in 84 intertroop encounters were attention-calling vocalizations or calls indicative of aggression. The major vocalization was
the whoop display in which opposing males moved to the treetops and
limb ends where they were clearly visible and vocalized. Upon completion of the vocalization, the challenged male(s) responded likewise; occasionally, opposing males displayed simultaneously. As much as 500 yd
often separated the displaying males. The male display was differentiated into two components, the whoop vocalization and the bounding
about the trees [which Ripley (1967) labels the "display jump" for
Ceylon langurs]. Displaying males jumped between branches and trees
in a taut, heavy manner, producing a noisy clamor. Branches often broke
under the weight of their impact and came crashing groundward. The
objective appeared to be to create as much noise as possible, discernible
over considerable distances. The rapid bounding through the trees also
produced an unavoidably striking visual contrast between the buff-brown
hair tuft and black body hair. After completing the display the male
returned to his troop, vocalizing loudly. The display whoop manifested
more tension than the morning whoop; upon completion of the former,
the male often sat hiccuping and grunting (both vocalizations indicated
tension) for several minutes. Urination and defecation frequently accompanied the display whoop.

Chasing was rare during territorial encounters, but when it occurred,
it was often on the ground. When opposing males chased each other, it
appeared to be ritualistic. During the chase the males frequently main-

tained a separation of about 3 ft. At times physical contact appeared imminent, but it occurred only twice. The males frequently interrupted the chase and sat a few feet apart exchanging threats. A pursuing male often traveled far from its home range, but upon approaching the opposing troop's core area there was a reversal of the pursuer-pursued roles. Penetration by the pursuing male often included a series of displacements in which he occupied the branch or approximate spot on the ground vacated by the pursued male. Displacements continued until the pursued male turned and attacked the intruder. There was repeated alternation until both males stopped, threatened each other, and eventually returned, noisily, to their respective troops. Such interactions could last, over an hour. The following example taken from our field notes illustrates "ritual chasing."

> The alpha male of troop II comes to the ground and faces the adult male of troop V; they exchange grunts and bite air at each other. The alpha male rises and begins to chase the troop-V male. They bound across the ground with their tails arched over their backs. The alpha male approaches to within 3 ft of the troop-V male and maintains this distance for 20 yd before sitting. The troop-V male stops running and turns to face him; they again bite air at each other. The alpha male turns and runs back to his own home range; the troop V male follows. At one point they are little more than an arm's reach apart but, again, they never touch. Once within his home range the alpha male stops running and sits facing the onrushing troop-V male, who stops about 8 ft from him. They face one another and grunt loudly. The alpha made rises and begins to chase the troop-V male; they penetrate troop-V's home range. At one point during the chase they are within touching distance. Once in his home range, the troop V male stops running and they face one another for the fourth time. They exchange grunts and bite air. Following 20 minutes of jockeying about, the alpha male returns to his troop (March 31, 1966).

Excitement was high during chasing bouts; defecation and urination were frequent, not only by the males involved, but occasionally by members of their respective troops. At this time the fecal odor was overpowering, raising the possibility that olfaction was a component of the interaction. If chasing occurred in the midst of one of the opposing male's troops, his troop members abandoned all caution and the observer was pressed to dodge them as they scurried and jumped about him, raining dead branches, moss, and lichens to the ground. In fact, at one time an animal ran into my legs, almost knocking me to the ground. Such close approaches to the observer were not seen at other times.

An example of a prolonged encounter between males of opposing troops is presented in Poirier (1968e).

Eighty percent of all intertroop encounters involved solely adult males. In one multiple-male troop, the alpha male assumed the aggressive role

45% of the time, nine times more frequently than the delta male. The frequency of male involvement in territorial battles in multiple-male troop II is given in Table XI. Although the situation obtaining in troop II may not be exampler of all Nilgiri langur troops, unfortunately, troop II is the only multiple-male troop on which enough information was gathered.

In multiple-male troops, more than one adult male was rarely involved in any single encounter and unless an encounter occurred in proximity to them, other adult males avoided it. Troop members not involved ignored the commotion. Females, especially, ignored the male encounters and continued to feed unconcernedly at the height of the activity. Just once was a female involved in an intertroop encounter. She directed a short sequence of facial threats at a female of an opposing troop. A male penetrating the interior of an opposing male's troop occasionally chased a female or youngster briefly, but he never physically attacked them.

C. DISCUSSION

The large amount of antagonism between Nilgiri males of different troops was totally unexpected. As arboreal animals occupying upper story vegetation providing an unobstructed view of the surroundings, they could easily avoid encounters. It is therefore of considerable importance that adult male Nilgiri langurs regularly sought conditions for displaying against adult males of other troops. An adult male frequently deserted his troop to challenge another adult male. This is striking because Nilgiri langurs possess means of signaling troop locations and avoidance which could substitute for adult male aggression in the spacing of troops. Furthermore, within the troop, Nilgiri langurs were lethargic and seldom aggressively involved. Although encounters between males of different troops were frequent, participant cost was minimal because physical contact and injury were rare.

Nilgiri langur intertroop encounters frequently occurred well within a home range, did not result from accidental meetings, and were not

TABLE XI

MALE INVOLVEMENT IN TERRITORIAL BATTLES: TROOP II

Status	Percentage of encounters
Alpha male	45
Beta male	30
Gamma male	20
Delta male	5

directly related to crowding or to competition for food or water. There-fore, the precise functions of Nilgiri langur territories are unclear. Nilgiri langur territorial behavior does protect core areas against entry by neighboring troops and thus indirectly prevents or minimizes over-feeding and overcrowding. Furthermore, territorial displays of Nilgiri langur males might provide them with social stimuli helpful in reproduc-tive success (Darling, 1952). Nilgiri langur territorial behavior may also be a factor in the distribution of the adult males and the exclusion of some from the genetic pool; the troops between which there was con-tinual antagonism were mostly one-male troops. Nilgiri langur male en-counters also suggest defense of the social integrity of the troop (Ripley, 1967).

D. Comparative Data

The Nilgiri langur spacing mechanisms of varied movement patterns and intertroop vocalizations are shared wholly or partially by other monkeys and apes. The predominant use of vocal patterns is primarily an arboreal characteristic, however. In macaques and baboons, visual communication often substitutes for loud vocalizations in the spacing of troops.

Southwick (1962) attempted to classify patterns of intertroop relations in some nonhuman primates by distinguishing passive and active be-havioral types. The African red-tailed monkey, gorilla, and common langur were listed as examples of the former; the temple-dwelling rhesus of North India, gibbons, and howlers were presented as exemplary of the latter. In view of Sugiyama's (1967) and Ripley's (1967) material, the inclusion of *P. entellus* in the passive category is not entirely justified.

There is a strong resemblance to the relationship characteristic of Nilgiri langur males of adjacent troops (also see Tanaka, 1965) and those described for the Ceylon gray langur (Ripley, 1967), *P. cristatus* (Bernstein, 1968), and the South Indian langur (Sugiyama, 1965a,b, 1967). Ceylon langur encounters are more aggressive than Nilgiri langur encounters, however, and Ceylon females are often involved. The peace-ful avoidance noted for North Indian langurs contrasts sharply; when two North Indian langur troops were in proximity, they did not threaten each other, and fighting between troops was never witnessed. "Fighting between two troops was never observed; if they both happen to be nearby, the larger group usually takes precedence and the smaller re-mains at a distance until the larger moves away" (Jay, 1965a, p. 212).

In contrast to the chief role assumed by the alpha male during inter-troop encounters in multiple-male Nilgiri langur troops is the role played by other males among the Ceylon langur and rhesus macaques. The gamma and, to a lesser extent, the beta male accounted for most of the aggressive intertroop encounters among Ceylon gray langurs. "The alpha male became involved in chases and at times, was an instigator, but he frequently took second place in the attacks" (Ripley, 1967, p. 247). Among rhesus on Cayo Santiago, the beta or gamma male more often assumed the belligerent position. "It is a remarkable fact that the male of highest prestige . . . frequently does not engage immediately in inter-group fights" (Carpenter, 1964, p. 381).

Although Nilgiri langurs are territorial, evidence is insufficient to relate adult male antagonism directly to either breeding and rearing or to the food and shelter type of mammalian territorial behavior. Rather, a prime function of Nilgiri langur territories seems to be the spacing of males and the indirect prevention of overfeeding and overcrowding. Available information suggests that territoriality characterizes the adaptive behavior of four langur species.

IX. MATERNAL BEHAVIOR AND THE MOTHER-INFANT RELATIONSHIP

A. INTRODUCTION

Jay's (1962, 1963a,b, 1965a) reports on the unusual aspects of common langur mother-infant dyads, have focused much attention on the langur mother-infant relationship. While field and laboratory studies confirm that the mother-infant bond is one of the strongest and most persistent, important variations occur (Jay, 1965b; Tinklepaugh and Hartman, 1932; Tomilin and Yerkes, 1935). The quality of maternal care depends in part upon the experience of the mother, the nature of the social situation, and the developmental status of the youngster.

Mother-offspring interactions help prepare the infant for subsequent social development. The maternal relationship is the first affectional bond for the youngster (Harlow, 1962, 1963) and is perhaps the prototype of all later such bonds (Jensen et al., 1967). The species-characteristic social milieux in which the dyad develops exerts a prominent behavioral force:

> The endless process of social adjustment begins with the mother, whose behavior is, in many ways, representative of the larger group. From her, the infant learns to perceive the meaning of a gesture or glance, discovers

what food may be taken with impunity from a larger animal, and finds
that bites and slaps will be returned in kind. (Mason, 1965, p. 531.)

The following observations of the mother-infant dyad are mainly from
four troops, that is, Groups I, II, III, and IV, whose composition are
indicated in Table IV; however, troop I supplied the majority of the
information. The mother-infant relationship was traced from 10 days
after birth until 8 months of age.

B. REPRODUCTIVE SUCCESS

The reproductive success of the observed troops, expressed by the
ratio of births to adult females (Koford, 1963), was fairly high. Twenty-
one of 30 adult females that were counted were associated with infants
under 1 year, a reproductive success of 70%. Six females gave birth during
the study period, none of whom were previously associated with an
infant. Four of the births occurred in troop I. The time span between
deliveries was estimated at approximately 18–24 months; however, because of the length of the study period this could not be definitely
established.

C. BIRTH SEASONS

A birth season is a "distinct period of the year to which all births are
confined. There must be some months in which no births occur" (Lancaster and Lee, 1965, p. 488). The primary birth season of Nilgiri
langurs, occurring in May and June, corresponded to the southwest
monsoon. A second birth season of lesser intensity, occurring in November, corresponded to the northeast monsoon. One birth was recorded
in February. Upon returning from India, we were notified that two
infants had been born in troop IV early in February, perhaps suggesting
that the second birth season extends into February. No births were witnessed outside these periods. Although an annual reproductive cycle
must include a copulatory season, a gestation period, and a birth season,
only data on birth seasons was gathered. In fact, copulatory activity
was never recorded.

The measurable features possibly influencing the birth seasons of
troops in the Nilgiris are given in Table XII. The incidence of births
seems to have had some correlation with increasing rainfall and daylight hours, a more abundant and varied diet, and decreasing temperatures. The effect of environmental factors is probably quite complex and
not always direct, however.

TABLE XII
BIRTH SEASONS AND CLIMATIC CORRELATIONS[a]

Month	Amount of daylight (hours)	Humidity (%)	Rainfall	Level pressure (mm)
November	12–13	70	Middle to end of northeast monsoon. Average for 1962–1965 was 4.26 inches	780
December	12–13	77	End of northeast monsoon. Average for 1962–1965 was 3.34 inches	780
February	11–12	57	Trace of rain. Average for 1962–1965 was 0.27 inches	780
May	11–12	77	Beginning of southwest monsoon. Average for 1963–1965 was 5.02 inches	779
June	11½–12½	82	Southwest monsoon begins in earnest. Average for 1962–1965 was 6.70 inches	777

Month	Temperature	Food	Wind (mph)
November	Frosts begins at end of month. Average for 1962–1964 was 19.7°C	Beginning of seed season for *Acacia*	2.1
December	Hoarfrosts common. Average for 1962–1964 was 13.3°C	Vegetation slightly thicker because of the northeast monsoon	1.5
February	Hoarfrosts common. Average for 1962–1965 was 13.4°C. Follows the coldest month of the year	Vegetation checked slightly during the first 3 months of the year	1.9
May	Average for 1962–1965 was 16.1°C	Very beginning of budding and flowering season of many of the food plants	1.3
June	Average for 1962–1965 was 14.3°C[b]	Budding and flowering season. Very seasonal growth, abundant and varied diet	4.3

[a] The humdity, rainfall, level pressure, temperature, and wind speed information is from the Government Hydrometric Survey Office, Ootacamund, Madras state.
[b] With the sun vertical, however, the temperature could rise to 18.9°C.

Birth seasons of various Nilgiri langur populations in South India had a broad similarity. This is illustrated in Table XIII. The information for Kodaikanal and Ambasamudram is from Leigh (1926b).

D. Effect of Births upon the Troop

The birth of a Nilgiri langur infant affected troop movement and structure. Prior to and for approximately 1 month after parturition, a mother moved slightly less than the troop. In fact, the troop's wandering as a whole was slightly less and slower in pace, allowing for the mother's slight decrease in movement. This was especially noticeable in one-adult-male troops in which two or three adult females carried infants.

Nilgiri langur mothers fed less than other troop members. This seemed to be related to the fact that the newborn slept a good deal, during which time a mother sat quietly. Mothers, therefore, remaining less active than other troop members, produced a conspicuous mother subgroup which fed, slept, and moved together. In larger troops juveniles and young subadults of both sexes were often drawn to the mother subgroup. The attraction was not the infant but the adult female herself, suggesting in many cases that they were her prior offspring. The adult female–infant and juvenile clusters that were observed may indicate the existence of family groups or subgroups in this species.

E. Physical Appearance of the Nilgiri Langur Infant

A young Nilgiri langur infant was sparsely covered with reddish-brown hair; the skin lacked pigmentation and was a pale pink color. Infant females were readily distinguishable by the presence of a white patch of medial thigh hair, also seen in adult females. The natal coat turned the dark, black color of adults by 10 weeks, and by 14 weeks the buff-brown hair tuft was evident.

Melanization of the skin began slightly later than the natal coat color

TABLE XIII
Correlation of Birth Seasons in South India

Location	Altitude[a]	Birth periodicity
Kodaikanal, 11°N 76°E	7000 ft	End of May and beginning of June. Some in April and December
Ambasamudram, 8.5°N 76.6°E	4000 ft	Mid-December
Ootacamund, 11°37′N 76°E	7200 ft	Major season occurs in May and June; also end of November and beginning of December and February.

[a] Altitude may be a very important factor in understanding the variation of birth seasons among Nilgiri langur populations. Lower altitudes first experience the monsoons and concomitant conditions such as decreasing temperature, daylight, and increasing food supply. If these conditions are indeed correlated with the incidence of births, as preliminary data suggests, the variation in birth patterns between troops living at lower and higher altitudes can be partially understood.

change. The skin acquired the dark adult pigmentation by 2 or 3 months; the palms, soles, and areas about the ischial callousities and eyes darkened last. A langur youngster frequently retained pink rings about its eyes and pink spots on the palms and soles. The areas about the ischial callousities and eyes darkened last, occasionally remaining pink until 4 or 5 months of age. The color-change time table was a good indication of relative age.

F. REACTION TO THE NATAL COAT

The natal coat seems to be an essential element in releasing a female's maternal behavior (Booth, 1962; Jay, 1962). The natal coat is generally present during the first 2 or 3 months of life when an infant most requires its mother's protection and nourishment. It is almost certainly more than an accident that the duration of natal coat color coincides with the period of maximal dependency. Furthermore, a positive correlation existed between the natal coat color of the infant and the protracted concern of adult females. Most infant transferring (discussed in Sections IX,I–L) and the strongest mother-infant bond coincided with the reddish-brown color phase. Coincident with acquisition of the adult skin pigmentation and coat color, the infants became more independent of their mothers, adult females' interest declined, and they sought less contact with the infants. The adults' decline of interest in the infants contrasts sharply with the infants' continuing interest in them.

G. CLINGING POSITIONS

Soon after birth a Nilgiri langur infant clung to its mother so securely that she could leap 10–20 ft without dislodging it. The infant clung by firmly grasping with its hands and feet to the hair on the lateral portion of its mother's chest; the position assumed was a ventral-ventral embrace ["ventral cling" (Rosenblum and Kaufman, 1967)] common to most monkeys. [The African colobus mother's habit of carrying an infant in her mouth is unique among monkeys (Booth, 1957).] An infant never rode "jockey style" (dorsal-ventral) on its mother's back, a highly impractical position for an arboreal animal constantly jumping through trees. (Among arboreal howlers, however, infants do ride on the mothers' backs.)

When a mother rested, the youngster sat at her chest with the mother's nipple in its mouth. The mother hugged the infant with one or both arms and drew her knees to her chest, forming a protective cradle. When it rained, a mother hunched over her infant and placed her back to the rain thus assuming a posture that provided the youngster the greatest warmth and protection. Occasionally, a Nilgiri langur infant lay prone across its mother's lap.

H. Protective Role of the Mother

A Nilgiri langur mother had the prime responsibility for protecting both herself and her infant. While the newborn infant was at its mother's chest or nearby, the mother's most conspicuous role was to provide protection and nourishment. The infant began to leave its mother to explore its immediate surroundings at approximately 3 or 4 weeks of age; however, it was seldom more than 2 ft away. At this point in the infant's early development, the mother's protective role began to diminish.

Although a mother made her presence readily noticeable (i.e., by running toward her infant) if some danger such as a dog or strange human approached, at other times she seemed only passively aware of her 3+-week-old infant. Because the infant's muscular coordination was not fully developed and its movement about the branches clumsy, this lack of maternal attention could prove fatal. Rather often, a month-old infant hung precariously from a branch vocalizing loudly to attract its mother's attention; the mother frequently ignored the distress calls. On a number of occasions an infant plummeted 5 or 10 ft, only to be rescued by grabbing a passing branch to which it clung and vocalized loudly. Once the infant fell, however, the mother almost immediately responded by retrieving it. Other troop members shared a mother's apparent lack of concern for her infant's safety. Rather than aid an infant caught in a precarious position, a nearby animal often left the scene, especially if the infant began to vocalize in distress. The following examples are typical.

> A 3-week-old infant hangs upside down from a branch, vocalizing loudly; the beta female sits approximately 2 inches away ignoring the infant. The delta female with her infant at her chest moves within 6 inches of the infant, looks at it, and then walks away. Following 2 minutes of intense calling, the infant's mother (the gamma female) approaches and pulls the infant to her (December 28, 1965).
>
> The alpha female leaves her 37-day-old infant sitting on a dry, fragile branch and moves to an adjacent tree to eat. A female sitting close by follows, rocking the branch which then cracks. The infant vocalizes loudly, but no one comes to its aid. The branch breaks and the infant falls; it quickly grabs an adjacent branch with its hands to which it hangs, screaming loudly. After a moment its mother approaches and pulls the infant to her chest (January 9, 1966).

Nilgiri langur females frequently left their younsters unattended while they moved to adjacent trees and shrubs to feed. This was first noticed when the infant was approximately 10 days old, and became more apparent with further maturation. One 5- or 6-month-old infant was either abandoned by its mother or abandoned its troop; the former is more likely. The infant was located in a shrub row approximately 1 ft from the ground; its troop was not in the vicinity. When we attempted, unsuccessfully, to capture the infant, no animal came to its rescue. Members of another troop sitting 15 ft away either completely ignored the situation or watched passively. The infant never vocalized all the time we attempted to capture it.

I. INFANT TRANSFERENCE

A distinguishing feature of the Nilgiri langur mother-infant relationship was the transference of one mother's infant to other females. The term "transference" is preferred to that of "passing," which Jay (1962) employed to describe the common langur behavior, because "passing" implies freely giving of an object from one to another. Often, this was not exactly the case with Nilgiri langurs. Transference was first recorded when the infant was approximately 10 days old (remember, however, that the first week of life was not observed). Infant transference was most frequent at 3 weeks of age; it ceased after 7 weeks. Therefore it appeared only for a very limited time during the youngster's early development.

Table XIV shows the relationship between age and infant transference.

J. METHODS OF INFANT TRANSFERENCE

A female obtained another's infant in various ways. Most often a mother simply approached another adult female and deposited her own infant on the branch near her. The female with whom, or by whom, the infant was left need not have indicated a prior intention to care for the youngster. Occasionally, a female obtained an infant from its mother by grooming the youngster. After passively watching the grooming se-

TABLE XIV

FREQUENCY OF INFANT TRANSFERENCE

Age (days)	10	14	21	28	35	42	49
Total infant transference[a]	5	13	18	6	7	5	2
Infant transference prior to feeding[b]	1	6	13	13	5	1	2

[a] Total of 56.
[b] Total of 41.

quence, the mother left her offspring with the grooming female. Some-times an infant left its mother to sit at the chest of a nearby female. The mother rarely attempted to stop the infant. The following illustrates the first mode of infant transference.

> The alpha female of troop I, with her offspring at her chest, sits next to the beta female. The alpha female rises and the youngster drops from her chest; the mother moves away leaving the neonate behind. The infant immediately vocalizes; the beta female approaches, the infant stops vocaliz-ing and goes to her chest. The beta female gave no indication that she wanted the infant prior to the alpha female's departure (January 3, 1966).

Prior to the birth of her infant, the troop-I beta female sometimes moved to another female and simply took the offspring from her. The infant frequently protested loudly, but the mother seldom attempted to interrupt the sequence. The following is typical.

> The beta female approaches the delta female who sits with an infant at her chest. The beta female reaches for the infant and pulls it to her; the infant squeaks loudly and tries to remain with its mother. The mother sits calmly, making no attempt to discourage the beta female. The infant force-fully taken from its mother is carried 3 ft away; it continues calling. The mother leaves her infant with the beta female and moves away to eat (December 9, 1965).

K. Factors Influencing Infant Transference

Infant transference did not produce or reveal differences of social status among the adult females. Any female could take an infant from any other female; there is no evidence that a dominant female could take an infant more readily from a subordinate female than vice versa. In fact, occasionally a dominant female had to forcefully regain her infant from a subordinate.

> The delta female sits about 30 ft from the alpha female with the alpha female's infant at her chest. The infant begins to squeak as his mother approaches. The delta female scoops the infant to her chest and jumps to an adjacent tree; the alpha female follows. The alpha female moves through the tree and sits next to the delta female. She reaches for her infant, but the delta female turns her back. The alpha female desists, then moves in front of the delta female and again reaches for the infant. Grabbing her infant by the arm, she begins to pull, however, the delta female refuses to let go. The delta female finally releases the infant, who goes to its mother. The delta female then moves toward the beta female who carries the former's infant (December 15, 1965).

The patterns that did occur in infant transference seemed to be based more on "friendship" bonds than any other relationship. The degree of intimacy between females appeared to have some relation to the frequency of infant transference between them. In addition nulliparous females seemed to have a strong desire to participate in infant transference sequences. [It has been suggested that this behavior be termed "maternal yearning," that is, a strong desire to cling to an infant, which if frustrated caused minor psychological and physiological distress. Distress was alleviated when a female contacted an infant. This maternal behavior could be similar to the stress-reducing effect of Nilgiri langur embracing behavior common to dominance situations, and in humans under stressful conditions (Poirier, 1966, 1968a).] The beta female of troop I, who did not give birth until approximately 6 weeks after the other females, was often the recipient in infant transference sequences.

Age also influenced the incidence of a female's receipt of an infant. Infant transference occurred only among adult females; subadult or juvenile females never tried to take an infant from its mother, nor were they ever recipients in infant transference sequences. The adult male of troop I was twice the recipient of the alpha female's infant.

L. FUNCTION OF INFANT TRANSFERENCE

The immediate function of infant transference was most obvious when a female wished to feed. When a mother moved lower in a tree or bush to feed, she left her infant with or near another female (see Table XIV). The mother, freed of the infant's encumbrance, was better able to feed in the shrub level, and the infant was relatively free from predation, such as dogs, which might originate from the forest floor. At most other times, for example, during rest and sleeping periods, an infant remained with its mother. Early morning observations confirmed that infants slept only with their mothers.

Infant transference probably influenced the development of the mothering experience. It is reasonable to assume, since most adult females had access to most infants, that most females had some practice at being a mother prior to giving birth. Infant transference may have some relation to the nonaggressive social order characteristic of langurs. Since most adult females had access to most infants, a mother's ranking had little influence upon the infant's future social status.

M. BABY SITTING

It was not uncommon to find an adult Nilgiri langur female associated with two, and occasionally three, infants whose mothers left them, usually to feed. A female need not indicate a desire to mind the infants left in her care; rather, she was often the last individual remaining in a rest or sleeping area. The "baby sitter" role alternated frequently as the original "sitter" left and another female took its place. As many as three females assumed the role in a short period of time. The following example is typical.

> The beta female of troop I sits in the center of three playful infants whose mothers feed in the surrounding shrubbery. After "sitting" for 5 minutes, the beta female joins them; the deserted infants vocalize. The gamma female leaves the shrub and approaches her infant who runs to meet her. Momentarily, the infant drops from her chest and joins the other infants. The gamma female "sits" for 2 minutes and then jumps to an adjacent tree as the delta female moves in behind her. The delta female "sits" for 10 minutes until the other females stop eating and retrieve their infants (December 29, 1965).

The "baby sitter" did not protect a youngster(s) left in her care and the youngster(s) was frequently unattended when she left. Infants often responded to desertion by squeaking and running about the branches; a mother might not retrieve her infant in such a situation. When a female had two infants at her chest, there was often a struggle as to which would nurse. Even when one of the infants was her own offspring, a mother did not help it obtain the nipple. It seemed possible that any lactating female could nurse another's infant.

A female carrying two infants moved through the trees with little difficulty. She often leaped 10 or 15 ft without dislodging them. One female carried three youngsters a short distance; however, her movement was impaired because she supported one infant with her left hand.

N. MOTHER-INFANT SOCIAL RELATIONSHIPS—GROOMING AND PLAY BEHAVIOR

An outstanding characteristic of the Nilgiri langur mother-infant relationship was the lack of interanimal contact in such interactions as grooming and play behavior. Mothers were seldom recorded, or in some groups never recorded, to groom their own infants. Only 7% of the total grooming sequences occurred between an adult female and infant. Female-infant grooming sequences were frequently cursory acts; the longest bout which lasted 5 minutes involved two adult females grooming an infant 2, that is, an animal approximately 14 weeks old which possessed adult skin and hair pigmentation. Adult female–infant 1 grooming was slightly more frequent than adult female–infant 2 grooming. Adult female–juvenile grooming was not recorded, however, adult male–juvenile grooming was. Infants never invited grooming from adult females, in fact they often resisted by moving away. In contrast to common langur mother-infant grooming aggregations, Nilgiri langur mother-infant clusters were rarely characterized by grooming but were instead characterized by a lack of mother-infant interanimal contact, excluding clinging.

A mother rarely engaged her infant in play, despite the fact that she was often the object of the infant's play. Youngsters often climbed and jumped on their mothers, mouthed their arms, swung on their tails, slapped their faces, and pulled their ears or hair. Mothers accepted considerable abuse from exuberant infants until the youngsters were approximately 4 months old. After this time it was not unusual for the mother to rebuke the infant mildly by slapping at it. Except in troop II, an infant never approached an adult male in a playful manner. [Much of the adult male–infant play occurring in troop II coincided with the joining of three adult males to the troop. In this context play behavior appeared to reduce tension and expressed the males' nonaggressiveness (Poirier, 1969b).]

O. Mother-Infant Communication Network

1. Vocal Communication

The vocal patterns of the Nilgiri langur infant were distinct in volume, intensity, and pitch from those of older animals. The infant's vocal repertoire (both males and females) was dominated by squeaking and screaming. Squeaking was designated as follows:

> The squeak vocalization is a high-pitched, grating sound produced with the mouth barely open. The call is composed of long drawn-out phrases repeated continually at approximately the same pitch level. As tension increases, the mouth opens wider and the pitch level rises. No other facial gestures accompany the call. In some instances the youngster runs about the branches in proximity to the mother; when emitted from a seated position, spasmodic limb movements are common. Youngsters look directly at the mother while vocalizing.

The squeak vocalization was emitted whenever an infant was tense or excited. If the desired attention was not forthcoming, the infant often exhibited patterns recognizable as a "temper tantrum," which included violent spasmodic limb movements, running about the branches, and high-pitched screams.

The basic vocalization was divided into two components: the sound produced when separated from the mother and the sound emitted when reunion with her was imminent or accomplished. When the mother and infant were separated, the vocalization was long and drawn-out; when they were reunited the sound was segmented, the tone varied, and the vocalization was lower-pitched. A mother's response to the infant's squeaking depended upon her temperament and the situation provoking the signal (e.g., the presence of another langur troop).

Screaming was primarily a weaning vocalization (see Section IX,R). It was designated as follows:

> The scream is a very shrill, high-pitched, nontonal prolonged call produced with the lips retracted and the mouth half open, resembling the grin mouth. The infant looks in the mother's direction while vocalizing. Spasmodic limb movements are common if the call is emitted from a sitting position, especially during high states of tension. The vocalization is often accompanied by frenetic running about the branches. The overall bearing is one of extreme agitation.

An infant whose attempts to nurse were rebuffed frequently vocalized for as long as 20 or 30 minutes. Mothers generally ignored the vocalizing infant. Other infants occasionally moved to the vocalizing youngster and attempted to engage it in play. The effort was always abortive and resulted in louder calling.

The auditory network coordinating mother-infant behavior was primarily limited to the infant's vocalizations. Only three instances were recorded in which a mother "called" her infant, all in response to the sudden appearance of the observer while the animals fed in low shrubs. Twice an infant reacted to the call of its startled mother by running to her and assuming the clinging position.

2. Gestural Communication

Although the new infant depended upon the mother for its nourishment and transportation, it was not totally passive. It was instead a clinging, squirming, vocal animal, influencing through tactile stimuli its relations with its mother. The infant's movements as it touched its mother, shifted positions, and attempted to nurse stimulated her as she held and adjusted it. If the infant struggled in her arms, the mother placed it in a more comfortable position; when the infant nursed, the mother immediately put one or both arms about it.

The mother's most frequent communicative device was the extension of her arm which signaled the infant to cling. An infant often sat in front of its mother vocalizing loudly, but it did not attempt to force itself to her chest. Only after the mother extended her arm did the infant assume the clinging position. A mother also signaled her infant to cling by standing directly over it. With the infant on her chest, a mother signaled impending movement by lightly touching its back, at which the infant clung tightly. An infant often reacted to the overall emotional state of its mother. When a mother was unduly nervous (which was indicated by head twitching, vocalizations, defecation, and urination), her infant frequently also exhibited tension through vocalizations and/or spasmodic limb movement.

P. WEANING

During weaning ". . . there is a gradual transition of the dyadic relationship in monkeys and apes from an initial period of virtually continuous physical attachment and co-directed attention, through several transitional stages, to an ultimate stage of independent and separate functioning of the offspring and mother. Such independence is obviously necessary for the infants to enter into the adult activities of their species, and for the mother to turn her attention to the next offspring when it comes" (Kaufman and Rosenblum, 1969).

Q. EARLY STAGES OF WEANING

The Nilgiri langur weaning process began early. Beginning from 7 to 9 weeks of age, a mother consistently left her infant for longer and longer time periods. When she returned to the proximity of the infant, the youngster made the final effort to join her.

> The alpha female moves to a branch where her infant sits vocalizing loudly, she looks at the infant who moves toward her. The alpha female moves closer; finally, they sit 1 ft apart. The mother looks at the infant and extends her arm; the infant runs to her (January 29, 1966).

By 9 weeks of age an infant took some solid food, thus facilitating the weaning process as a mother no longer needed to be present to satisfy all the infant's nutritional requirements. Once the infant initiated the break in contact, the mother showed considerable willingness to leave. By 14 weeks a mother continually left her infant who was forced to find ways through the trees to reach her; the mother seldom aided the infant. A mother commonly moved 10–15 ft from her infant and waited for it to reach her. When it did, she moved again, and the frustrated infant followed. This often continued for more than 20 minutes, but the infant persisted. This pattern may have helped teach the youngster the proper routes through the trees, for it ran over the same progression routes as the mother. The following is a typical sequence.

> The gamma female leaves her infant and jumps to an adjacent tree; her infant remains behind vocalizing loudly. After 1 minute of squeaking, the infant moves higher into the tree and looks at the mother who sits 6 ft

away. Moving lower in the tree, the infant jumps toward the mother; she rises and jumps to another tree. The infant sits squeaking for 40 seconds; finally, it again jumps to the mother. It barely clears the jump. Squeaking loudly, the infant runs to its mother; she extends her arm and the infant immediately clings. The mother leaves carrying the clinging infant (March 4, 1966).

In the earliest weaning stages, a mother alternated between accepting and rejecting her infant; advances toward the mother could stimulate positive or negative responses. Where the infant once went immediately to the mother and was assured of her acceptance, it now moved toward her slowly and with hesitation. At this stage the mother loosened her hold upon the infant; the infant decreased its nursing and remained in the embrace position for shorter time periods. The mother's mode of cradling probably had a definite communicative significance for the infant, and the infant's changing from nursing to active behavior had a definite communicative significance for the mother (Jensen *et al.*, 1967).

R. FINAL STAGES OF WEANING

As maternal solicitude waned, maternal behavior appeared which actively encouraged the dissolution of the mother-infant dyad. The two most prevalent forms of separation behavior were nipple withdrawal and punitive deterrence (Kaufman and Rosenblum, 1966). During the final weaning stages, the mother refused with greater frequency and harshness the infant's attempts to nurse; either she pushed the infant's head from the nipple with her hand, or she moved when the infant showed nursing intentions. The mother occasionally allowed the infant momentary access to the nipple which she interrupted by brushing its head aside. At this age the youngster was still somewhat dependent upon the breast for its nourishment and still sought the nipple when distressed; therefore, it didn't easily accommodate to its mother's rejection.

A mother frequently lost patience with a 10- or 11-month-old infant that persisted in its attempts to nurse and dealt harshly with it. Violent muscle spasms, frenetic movement, and loud screaming were generally observed as the infant forcefully attempted to reestablish nipple contact. A mother responded by turning her back, slapping, or biting the infant. The bite was not severe and probably never broke the skin. The infant's outbursts often disturbed nearby animals who threatened the infant or mother. The mother's persistent rejection initially *increased* rather than decreased the infant's dependent behavior and its desire for contact with the mother (Kaufman and Rosenblum, 1969, describe similar behavior for laboratory pig-tailed macaques).

The prolonged period of stressful weaning was completed when the Nilgiri langur was approximately 1 year old. If an infant was not born in the succeeding year, however, the 1-year-old continued close association with its mother, maintaining oral nipple contact while resting and sleeping. Juveniles often retained strong ties with the mother even after a successive infant was born.

In addition to physical rejection, an important correlate of weaning was the mother's emotional rejection of the offspring. The mother, once the major source of comfort, warmth, and food, was now hostile and denying. The severity of rejection varied with the temperament of the females and infants involved. Some mothers rejected their infants more positively than others, and some infants were more persistent in their attempts to resist rejection. Older multiparous females probably weaned their infants with less effort than younger females (Jay, 1962).

Sexual differences between male and female Nilgiri langur infants during weaning are either nonexistent or the sample was too small and attenuated to allow such comparisons. Among laboratory groups of *M. nemestrina*, male infants left the mother more often than female infants, and mothers left male infants more than they left female infants (Jensen *et al.*, 1967). Reports on feral Japanese macaques are consistent with this data (Itani, 1959). Itani noted that infant males left their mothers to form male peer groups at an age when young females remained with their mothers. Perhaps the greater behavioral variation noted for Japanese macaque males and females, as compared to Nilgiri langur males and females, is related to this early developmental period.

S. ADULT MALE–INFANT RELATIONSHIP

The adult male Nilgiri langur showed little interest in newborn or young infants; he seldom touched or approached them. Instead, he maintained a fairly consistent 10–15 ft separation between himself and the mother-infant pair. Young infants likewise avoided adult males, but if one moved too close the male moved away. Adult male–infant avoidance continued until the youngster was 12–15 months old, after which juveniles approached the adult male to sit nearby. There was no standardized adult male–juvenile greeting as reported for North Indian common langurs (Jay, 1965a).

Occasionally, an adult male assumed a protective attitude toward a youngster. Four times the adult male of troop I responded to the continual din of a vocalizing infant by retrieving it and sitting with the infant at his chest. Twice the infant belonged to the alpha female. A male occasionally responded to an infant's continued vocalizations by mildly attacking its mother who reacted by retrieving the infant.

T. COMPARATIVE DATA

1. Care and Protection of the Infant

Among colobids, the social unit for the protection and care of the youngster is the mother subgroup; however, the adult male's role can be placed along a continuum. North and South Indian langur males can be placed at opposite poles; Nilgiri langur and *P. cristatus* males occupy a middle position. The North Indian langur male ". . . plays no active role in the young infant's life other than a very generalized and indirect protective one . . ." (Jay, 1962, p. 469). However, the South Indian langur adult male "joined actively in the protection of a newborn" (Sugiyama, 1967, p. 224). Both Nilgiri langur and *P. cristatus* males played some part in the protection of the infant. "In the case of the [*P. cristatus*] males, however, the immature animals initiated the interaction and the males were passive participants" (Bernstein, 1968). In no instance does the male langur care approach the Japanese macaque adult male–infant relationship (Itani, 1959).

A continuum can also be usefully applied when discussing the protective behavior of the langur mother. The North Indian langur mother occupies one end of the continuum, the Nilgiri and South Indian langur mother occupy the other. The protective behavior of the Nilgiri and South Indian langur mother was not as obvious as that exhibited by the North Indian common langur mother. A North Indian langur mother was constantly aware of her infant "regardless of the mother's location, whether she is within a few feet of her infant or as far as fifty feet, she is always aware of its whereabouts, and at the slightest provocation of danger or harm to the infant, she immediately runs to take it" (Jay, 1962, p. 470). During the time that a South Indian langur infant is being carried, ". . . the mother does not always keep constant and close watch over it, but sometimes the mother starts feeding, resting, or sleeping composedly" (Sugiyama, 1967, p. 233). A female of a nearby South Indian langur troop occasionally gained access to an infant of a different troop, resulting in an intertroop conflict to regain the infant.

2. Mother-Infant Contact

In contrast to the lack of Nilgiri langur mother-infant grooming, the North Indian langur mother "inspects, licks, grooms and manipulates the infant from the hour of its birth" (Jay, 1965a, p. 221). Among South Indian langurs ". . . grooming between mothers and their children is of high frequency, especially grooming done by the mother to her infant" (Sugiyama, 1967, p. 237). Likewise, many *P. cristatus* grooming episodes involve females holding orange infants and the infants themselves (Bernstein, 1968).

3. Infant Transference

Unlike most monkeys, a langur mother allows other females to hold and carry a newborn, a behavior that Nilgiri langurs terminated at 49 days of age. North Indian langur mothers began to pass their infants a few hours after birth, as soon as the infant was dry. "As many as eight females may hold the infant during the first day of life, and it may be carried as far as 50 feet from the mother" (Jay, 1965a, p. 221).

Contrasting with the willingess of North Indian and Nilgiri langur mothers to give up their infants (and the usual nonreluctance of the infants to leave), the South Indian langur infant "resists other females, clinging to the mother's body and squealing loudly and violently" (Sugiyama, 1967, p. 228). A South Indian langur female wishing to hold another's infant had to obtain it forcefully. A South Indian langur mother sometimes rejected the demands, but generally she allowed another female to take her infant. "But she never hands the infant to the other female of her own free will" (Sugiyama, 1967, p. 228). A similar reluctance to allow other females access to one's infant occurred in the South Indian langur troops we observed. In 19 of 49 infant transfers witnessed among *P. cristatus*, the infant vigorously resisted by clinging to the original female and attempting to run from the second female (Bernstein, 1968).

The macaque and baboon mother-infant relationship contrasts with the langur pattern of infant transference. Baboons were attracted to a newborn infant, and soon after birth an animal approached to touch the infant or groom the mother. They did not attempt to take the infant from the mother, however, nor did the mother leave her infant with others (DeVore and Hall, 1965). A bonnet macaque mother did not allow other females to hold her infant ". . . and if a dominant female picks up the infant of a subordinate female, the mother will hold part of her infant even though the dominance distance is very great" (Simonds, 1965, p. 192). A rhesus mother was not the center of attraction of other females or infants, and the infant had very limited contact with the group members other than its mother during the first weeks of life (Southwick et al., 1965).

4. Weaning and an Extended Mother-Infant Relationship

Although the Nilgiri langur weaning process generally began earlier and lasted longer than among common langurs, the mother-offspring bond seemed more persistent. A Nilgiri langur juvenile maintained a close relationship with the mother if she allowed it. In contrast, North Indian langur juveniles were completely independent of their mothers. "When the juvenile is approximately two years old, its mother gives birth to another infant, and all remaining social ties with the last infant are completely severed" (Jay, 1965a, p. 229). The smaller Nilgiri langur troop, lacking large peer groups, may precipitate stronger filial ties when compared to the common langur troop in which a juvenile has ample opportunity for peer interaction.

Nilgiri langur mother-infant-juvenile ties suggest the possible existence of family groupings within the female subgroup. Because individual troop members typically rested in random patterns within their respective subgroups, it was considered highly significant if a juvenile consistently chose to rest with a particular female and her infant. The following observations from troop III were considered as possible indications of family groupings.

> A mother rests with an infant at her chest; a late juvenile lies on the branch immediately behind her (May 27, 1966).
> A mother rests with her infant at her chest; a late juvenile lies next to them. This is the same group as recorded on the 27th (May 31, 1966).
> A mother rests with an infant at her nipple; a juvenile lies against her back (July 8, 1966).
> A mother rests with an infant at her nipple. A juvenile approaches and forces his head next to that of the infant; however, only the infant nurses. The mother makes no attempt to rebuff the juvenile. When the mother shifts positions, the juvenile moves away. This is the same group as observed yesterday (July 9, 1966).

X. DOMINANCE BEHAVIOR

A. Introduction

Dominance has long been recognized as a conspicuous feature of the behavior of many primates. Variation in dominance expression by different primate genera and species does exist, but in every nonhuman primate society now known there is some competition for rights to incentives, such as food and estrous females. "These dominance orders among adult males, among adult females, and among young, markedly affect social integration and group control" (Carpenter, 1964, p. 351). Behavioral patterns maintaining dominance are essential aspects of troop integration.

As an arboreal form manifesting slight sexual dimorphism, it was hypothesized that the Nilgiri langur would exhibit a minimal amount of overt dominance behavior. Indeed, the relaxed tenor of Nilgiri langur life and the minimal amount of physical contact exhibited during dominance encounters does contrast with reports of some terrestrial species (i.e., Altmann, 1962b; DeVore, 1963; Hall and DeVore, 1965).

B. Determination of Ranks

Determination of dominance ranks in various Nilgiri langur troops involved noting individual aggressive and submissive tendencies in various social situations. Observations were made of social interactions within the troop, on external events such as the presence of nonbisexual troop males, and during intertroop territorial altercations.

In many dyadic social relationships, there was a tendency for one participant to use more aggressive signals than the other. All things being equal, the former was designated the "dominant animal," the latter the "subordinate animal." For most animals there were individuals that usually dominated them and those that they usually dominated. There was a third category of neutral relationships involving little or no overt dominance expression. This occurred between a mother and her infant 1 or between young peers. "Viewed from the standpoint of the society as a whole, there are interindividual dominance relations that determine dominance hierarchies" (Altmann, 1962, p. 399).

C. General Characteristics of the Dominance System

Several generalizations may be made concerning the Nilgiri langur dominance structure [as modeled on Carpenter's (1964) presentation].

(1) No single behavioral mechanism, such as fighting, uniformly implements the achievement and maintenance of dominance, rather, various behavioral patterns are employed.

(2) A sexual gradient in dominance expression exists; males are clearly dominant over females. Therefore two dominance hierarchies are present; one for adult males and one for adult females.

(3) Involvement in dominance interactions varies for different age and sex categories. Table XV demonstrates that most dominance interactions in which the age and sex of the participants was determined occurred between animals of like sex and age, reflecting the fact that Nilgiri langur troops are divided into subgroups having restricted social interaction (Poirier, 1969a). The participants' sex could be determined in 70% of the dominance interactions, of which 65% occurred between

TABLE XV

Tabulation of Total Dominance Sequences in Troops I–IV

Dominant	Subordinate[a]							Times dominated another age class
	AM	AF	SA	J	I 2	I	Unidentified	
AM	75	109	12	2	26	0	0	149
AF	0	92	1	1	4	0	0	6
SA	0	0	11	11	1	0	0	12
J	0	0	1	1	0	0	0	1
I 2	0	0	0	0	1	0	0	0
I	0	0	0	0	0	4	0	0
Unidentified	0	0	0	0	0	0	25	—
Total involvement in dominance encounters	224	207	51	15	32	4	25	

[a] For abbreviations see footnote to Table X.

animals of like sex. In all but 25 of the 391 dominance interactions witnessed, ages of the individuals involved were determined. Most interactions occurred between animals of like age. This is summarized as: like-sex dominance encounters, 65%; unlike-sex dominance encounters, 5%; like age dominance encounters, 85%; unlike age dominance encounters, 15%.

(4) Most dominance encounters involve adult males. Adult male/adult male dominance sequences accounted for the majority of the dominance sequences in multiple-male troops. At least one adult was involved in 88% of the total encounters.

(5) Some animals are involved in more dominance encounters than others. This was rather consistently observed, for example, for the gamma male of troop II and the gamma female of troop I. Whether or not this was a result of social structure or individual temperaments was not clearly evident.

(6) The most accurate status measure is not only the number of times one animal dominates another but who else it dominates as well. For example, the alpha and gamma females of troop I dominated each other once. The alpha female dominated the beta female nine times; however, the gamma female dominated her only once.

(7) Dominance interactions entail minimal aggression. Agonistic situations involving physical contact characterize a small part of the dominance sequences.

(8) Expression of dominance behavior is relatively rare. One dominance interaction occurred in each 3.1 hours of observation. This is likely

attributable to four basic factors (also see Schaller, 1963): (a) There was little competition for food or mates, food was abundant, and mating behavior was not witnessed. (b) Animals remained alert to possible encounters and subordinate individuals avoided potentially aggressive situations. (c) Nilgiri langurs were simply not aggressive. (d) Arboreal species manifest less dominance behavior than terrestrial species (Poirier, 1968b).

(9) There were no clear seasonal differences in aggressive tendencies.

D. FEMALE DOMINANCE HIERARCHY—GENERAL CHARACTERISTICS

In contrast to North Indian langurs among which an adult female dominance hierarchy is neither rigid nor well defined (Jay, 1963b), the Nilgiri langur female dominance order was relatively stable and well defined. Even though encounters were rare, adult females could be ranked in a linear order where relationships between females seemed rather clearly defined. Certain females constantly dominated others by assuming priority at desired feeding and resting locations.

Table XVI summarizes adult females dominance interactions in troop I during 561 hours of observation.

E. ALPHA FEMALE—TROOPS I AND II

During the daily routine the alpha female was not readily distinguishable from other troop females. She was not the object of attention for either the males or other females. The alpha female was not necessarily

TABLE XVI
ADULT FEMALE DOMINANCE HIERARCHY OF TROOP I

	Subordinate				Times dominant
Dominant	Alpha	Beta	Gamma	Delta	
Alpha	—	9	1	6	16
Beta	4	—	6	6	16
Gamma	2	1	—	2	5
Delta	1?	0	0	—	1?
Times subordinate	6 + 1?	10	7	14	
Times involved in female-female dominance sequences	22 + 1?	26	12	14 + 1?	
Times dominant (%)	70	61	36	1	
Times subordinate (%)	30	39	64	99	

involved in more dominance interactions than females of descending rank (Table XVI). Similar to the alpha male, however, she could assert her position at will. She took food from where she wished and was avoided by other females when irritated. She occasioned no privileges in infant transference (Poirier, 1968d), however. He infant was taken by all the females in the troop, and she lacked privileged access to infants of subordinate females.

F. Female Dominance Interactions

Adult female dominance interactions were rare and of brief duration. Most female dominance interactions involved behavioral patterns such as displacement, moving away, looking at, or looking away from. Physical contact was restricted mainly to the embrace or embrace-grooming complex (described in the Appendix, Section D). There was little overt aggression and attack patterns were rarely witnessed. Adult female dominance sequences rarely involved wrestling, chasing, or slapping.

G. Alliance Formation

Adult females rarely formed temporary alliances from which two or more females simultaneously chased and/or threatened another. The one temporary alliance observed, involving two adult females chasing a third, lasted approximately 20 seconds. The alliance, which pitted the alpha and beta females against the gamma female, appeared to be spontaneous and voluntary. Members of the alliance acted with little reference to one another, and the alliance terminated as soon as its members directed their attention elsewhere.

H. Male Dominance Hierarchy—General Characteristics

The adult male dominance hierarchy was a relatively rigid and well-defined order which was maintained with minimal aggression. Most male-male dominance interactions were accomplished in a nonviolent manner through visual cues. In a typical sequence the dominant male simply looked at the subordinate animal who responded "properly" (usually by looking away). This promptly terminated the interaction.

When relaxed, adult males in multiple-male troops mingled freely with little aggressive expression. The apparent absence of male-male antagonism in all male groups remain to be explained, however. (Twenty hours were spent observing all-male groups.) Apparently, there is a positive attraction among males which fragments somewhat in the presence of females (Kummer, 1967a; Poirier, 1968b, 1969a). Sugiyama's (1967) data on South Indian langurs suggests a similar conclusion.

Table XVII summarizes the dominance interactions in the troop-II male hierarchy during 340 hours of observation.

I. MEASURE OF MALE STATUS

The principal measure of a male's status was not the number of times he dominated another animal, but rather which males he dominated. In multiple-male troops, a male's position within the male subgroup was the most accurate indication of his status in the group as a whole. For example, the beta male of troop II, who was slightly more dominant than the gamma male within the male hierarchy, appeared subordinate to him when total figures were considered. Because the gamma male was more active than the beta male in dominance interactions outside the male hierarchy, his total dominance frequency rose (Poirier, 1968b, 1969a).

J. CHARACTERISTICS OF THE ALPHA MALE—TROOP II

No immediately obvious morphological characteristics differentiated the alpha, beta, or gamma males in troop II. The alpha male was not the largest animal nor did he possess the largest canines. Observations of multiple-male troops suggested that the alpha male was defined by behavioral rather than physical traits. While the characteristics differentiat-

TABLE XVII
DOMINANCE INTERACTIONS IN THE MALE HIERARCHY OF TROOP II

	Subordinate				Times dominant
Dominant	Alpha	Beta	Gamma	Delta	
Alpha	—	11	5	10	26
Beta	4	—	1	5	10
Gamma	1	0	—	4	5
Delta	2	0	0	—	2
Times subordinate	7	11	6	19	
Times involved in male-male dominance sequences	33	21	11	21	
Times dominant %	79	45	42	9	
Times subordinate %	21	55	58	91	

ing the alpha male of troop II may not be true for Nilgiri langurs as a whole, they may reveal the general nature of the determinants of the alpha male status.

(1) The alpha male played a major role defending troop integrity in intertroop altercations and territorial battles (Poirier, 1968c, 1969a). He did not actively intervene in intratroop disputes, however. Only once did the alpha male protect another troop member during an intertroop encounter; when the beta male attacked the alpha female, the alpha male immediately chased him off. This isolated incident does not suggest that a special relationship existed between the alpha male and alpha female.

(2) The alpha male was involved in more dominance sequences than other males, although such activities were infrequent.

(3) The alpha male was involved in fewer dominance sequences outside the male hierarchy than any other male.

(4) The alpha male had greater freedom of movement and choice than other individuals. He took desired foods, moved over desired progression routes, and sat wherever and with whomever he pleased.

(5) Troop activity was patterned after the alpha male. Although any animal might lead the troop's progression, movement was keyed to the alpha male. When he stopped to rest, the troop did likewise; when he moved, the troop began moving.

(6) A threat by the alpha male almost always elicited subordinate responses which were more frequently directed toward him than dominant gestures issued by him. Subordinate animals immediately responded to the alpha male's presence by subordinating themselves to him. His status was usually recognized without his issuance of a threat.

K. DYNAMICS OF THE DOMINANCE HIERARCHY

The Nilgiri langur dominance hierarchy was maintained with minimal aggression, as most dominance interactions were accomplished using subtle forms of threat behavior. The major system mediating interindividual behavior in dominance situations consisted of visual cues in the form of direct movement, facial expressions, and body positions or attitudes. Displacement was a very common pattern used by dominant animals; the subordinate individual responded by moving away. Physical attack was rare; tactile signals such as biting, slapping, and wrestling seldom occurred.

Because vocalizations indicated high states of arousal, their subordination to nonvocal cues suggests a minimal amount of tension in most dominance encounters. A dominant animal need not vocalize to emphasize its state of agitation or to attract attention of a subordinate. Subordinate monkeys seemed constantly aware of a dominant animal's location, and the troop as a whole was always cognizant of the position of the alpha male. In a stable hierarchy such as that manifested by Nilgiri langurs, recognition of relative dominance status is quickly communicated by a "status indicator" (Altmann, 1962b) such as a stare threat, a head bob, or by presenting or mounting. The actual behaviors involved in dominance interactions and pertinent features of their expression in Nilgiri langurs are described in the Appendix.

L. Stability of Dominance Hierarchy

The dominance hierarchies in the various Nilgiri langur troops studied appeared quite stable. Hierarchies were clearly defined and maintained through subtle postures and gestures; fighting was rare. No shifts in the various hierarchies were recorded. Interestingly, in troop II, which was in rather constant flux and during the fourth month of the study was joined by a male trio (Poirier, 1968b, 1969a), only the alpha male dominated the beta male (Table XVII). The beta male appears to have been the dominant individual of the male trio prior to its joining troop II. Although in the new troop structure he dropped to the beta rank, he appeared to retain his former status with the two other joining males (gamma and delta). No shifting of ranks occurred when the male trio entered a new social environment.

M. Acquired versus Derived Status

The few reports available from long-term studies suggest that a mother's rank is important in helping determine her infant's status. The concept of identification was introduced from psychoanalysis to explain the fact that Japanese macaque infants with dominant mothers tend themselves toward dominance. Infants of higher-ranking mothers identified successfully with troop leaders. Offspring of lower-ranking mothers were unable to identify with troop leaders during childhood, and in the Takasakiyama troop they became peripheral members or deserted the troop (Itani et al., 1963).

Studies of the Koshima (Kawai, 1958) and Minoo B (Kawamura, 1958) troops in Japan show that in paired competition for food successful monkeys were often infants of higher-ranking mothers. Koford's (1963) report on the rhesus of Cayo Santiago indicated that adolescent sons of the highest-ranking females held a high rank in the adult male hierarchy. Koford suggests "apparently because of protection by their mother during youth, sons of high-ranking mothers have attained top status in the band without going through the stage of being peripheral or sub-dominant males" (Koford, 1963, p. 147).

Infants of lower-ranking baboon mothers exhibited considerable insecurity in the form of a greater frequency of alarm cries and more demands on the mother. This led to intensification of the mother-infant bond. Offspring of dominant females, however, acted more secure and exhibited more freedom from their mother (DeVore, 1963).

Dominance was not an important feature in the life of an adult female Nilgiri langur and her status was seldom apparent in her relations with other troop members. Therefore it is unlikely that her status had any measurable affect upon that of her infant(s). Most likely, the dominance status of a mother was less influential on the development of her infant than was her "temperament," which may have affected the total pattern of her maternal behavior. Every infant had free acess to every other infant; females of all ranks had free access to all infants (Poirier, 1968d).

Japanese macaque troops are comprised of central and peripheral portions. Infants born in the central part of the troop identify with troop leaders; in turn they themselves become leaders. On the contrary, the Nilgiri langur troop is divided into semiautonomous age/sex subgroups. There is little contact between the adult male and the female subgroup, especially when infants are present (Poirier, 1968d, 1969a). Unlike some Japanese macaque infants, the Nilgiri langur infant had minimal contact with the adult male(s) of the troop.

N. Dominance as Correlated with Mode of Life

Field studies suggest that dominance behavior, especially as manifested by adult males, is more frequent and intense in terrestrial than arboreal species. The freedom from predation enjoyed by the arboreal langur not only provides a clue in explaining the lack of an elaborate social hierarchy, it also helps to explain the more relaxed, nonaggressive, individual social interactions. DeVore (1963, p. 314) suggested that ". . . intragroup aggressive and agonistic behavior decreases by the degree to which a species is adapted to arboreal life." He further noted a correlation between degree of sexual dimorphism and dominance activity. "The evidence does suggest . . . that increased predation on the ground leads to increased morphological specialization in the male with accompanying changes in the behavior of individuals and the social organization of the troop (DeVore, 1963, p. 315). Although Struhsaker (1969) reports exceptions among African cercopithecines, both of

DeVore's hypotheses tend to be supported by observations of the Nilgiri langur dominance system.

O. COMPARATIVE DATA

1. Nature and Frequency of Dominance Behavior

The relaxed tenor of Nilgiri langur social life is one of the most obvious characteristics of its social order. Dominance interactions were infrequent and usually accomplished without physical aggression. The overwhelming majority of Nilgiri langur dominance interactions involved solely the two interacting animals, were enacted without physical contact, and were of short duration. However, dominance interactions among bonnet macaques (one of the least pugnacious of the macaques), for example, are characterized by physical contact, are noisy, and often last 5 minutes or more (Simonds, 1963; Poirier, 1968b). Jay (1965a, p. 248) noted a similar contrast between North Indian langurs and rhesus macaques: "rhesus are more intense, quicker moving, more easily provoked to threat, more aggressive and more vocal than are the relaxed langurs."

2. Role of the Alpha Male

A major role of the alpha male among macaques (Simonds, 1963) and baboons (Hall and DeVore, 1965), for example, is the repression of intratroop aggression. The Nilgiri langur alpha male, in contrast, rarely aided another animal in intratroop conflict situations. The major function of the Nilgiri langur male (alpha male in multiple-male troops) was protection of the troop against extratroop dangers, especially encroachment by males of neighboring troops (Poirier, 1968b, 1969a, 1970a).

3. Alliance Formation

Macaque and baboon individuals often combine ranks in a common attack against another. Animals often gain and maintain status by coordinating their activities. Simonds (1965, p. 185) noted that one bonnet macaque male ". . . never initiated a threat sequence without the support of a more dominant male, but he would join any threat in which other males threatened those immediately below him." Hall and DeVore (1965, p. 65) noted for the chacma baboon: "threat behavior is often seen between combinations of individuals indicating that an individual may sometimes seek support or 'enlist' the threatening behavior of another individual." In contrast, only one combined attack was noted in 1250 hours of observation of the Nilgiri langur.

XI. GROOMING BEHAVIOR

A. Introduction

Grooming is a highly interdependent activity among nonhuman primates serving a dual biological, social function. Biologically, grooming helps to maintain the hair free of ectoparasites and prevents wound infection. Socially, grooming helps an individual to gain the social acceptance of others. In most nonhuman primates grooming behavior ". . . is highly reciprocal and is a basis of much social conditioning in both natural and captive primate groups and is an important substratum for social integration" (Carpenter, 1964, p. 148). Most monkeys respond immediately to grooming; in addition, they are often strongly motivated to groom as well as to be groomed. Grooming is apparently a pleasurable experience constituting one of the most important expressions of close social ties. Furthermore, grooming is important "from the viewpoint of group integration; not only does it depend, seemingly, upon a previous state of positive conditioning in the participating animals but the behavior further enhances and strengthens the social relationships" (Carpenter, 1940, p. 191).

It seems likely that grooming behavior is learned within the context of the social group. Monkeys raised in laboratory environments restricting social contact groom infrequently. Such animals use ". . . no specific gestures to invite grooming and the attempts of one restricted monkey to groom another were generally met with indifference rather than with the active cooperation seen in the feral group." In no instance did "a pair of restricted monkeys spend as much as ten seconds in continuous grooming" (Mason, 1963, p. 165).

Grooming is the most widespread nonhuman primate pattern involving tactile stimulation (Marler, 1965). Mode of grooming varies; tree shrews use their tongue and teeth for grooming which is facilitated by the procumbant lower incisors which form a "tooth comb." Among various lemurs, the teeth and tongue regularly accompany the hand (Buettner-Janusch and Andrew, 1962). The hands, however, are typically the grooming agents; the mouth is then applied when a foreign object is located. One hand typically parts the hair and the other holds it down; the lips and tongue or one of the hands picks the foreign matter. The matter is then conveyed to the mouth and usually ingested if not disagreeable. Although any part of the body may be groomed, the tendency in social grooming is to groom areas inaccessible to an animal itself, for example, the head, face, neck, back, and anogenital region.

B. INITIATION OF GROOMING

The invitational phase to grooming deserves special attention in connection with communicative interchanges and social coordination. Most Nilgiri langur grooming occurred in situations in which one animal initiated the bout by presenting itself to another for grooming. For example, when the back was to be groomed, the animal presenting for grooming simply moved to another and sat with its back toward it.

Should the animal approached for grooming not respond, the presenting animal often groomed it without invitation. After a short time the presenting animal would lie down and then usually be groomed. Thus the brief initial grooming sequence served as a stimulus for the initiation of grooming by the partner.

> The beta male of troop II moves to the alpha male and presents his back for grooming. The alpha male fails to respond; the beta male then grooms the alpha male's arm in a cursory fashion. He again presents his back, to which the alpha male responds by turning away. The beta male moves in front of him and grooms his arm. He lies in front of the alpha male and is groomed on the back and arm for 45 seconds (March 1, 1966).

Approximately as many presentations for grooming went unanswered as were responded to. It was fairly common for two animals to sit next to each other mutually presenting with neither commencing to groom the other. The following sequence is typical.

> The troop-II delta male approaches the beta male. The delta male presents his back to the beta male as an invitation to grooming. Instead of grooming, the beta male presents in turn to the delta male. The beta male leans back presenting his chest. Both animals subsequently sit facing one another. The delta male presents once again but without result. Both animals just sit next to one another resting (May 8, 1966).

Grooming occasionally followed a mounting sequence. Either the mounting or mounted animal groomed, usually the latter. One monkey rarely approached another and simply began to groom.

C. METHOD OF GROOMING

Grooming was usually accomplished with the hands; one hand, either the palm or back of the hand was used, typically parted the hair by brushing it against the grain and the other picked through the hair.

Extraneous matter was seldom removed; however, when it was, it was smelled and tasted; some was ingested and some immediately spit out.

Because of their very short thumbs, and the difficulty therefore of opposition, Nilgiri langurs were inept at picking material from the hair. Consequently, extraneous material was more often removed with the mouth than is the case among macaques (personal observation) and other terrestrial primates with pronounced, better developed thumbs. Except for occasional intense grooming bouts, such as recorded in Section XIII,E, the groomer rarely used its tongue or teeth in grooming.

D. Reaction to Grooming

Grooming was quite obviously a very pleasurable experience for the animal being groomed. The groomed animal was relaxed, often closed its eyes, and during longer bouts typically lay prone or supine on a branch. Occasional penile erections were witnessed during longer grooming bouts; these occurred in both male-male or female-male sequences.

The relaxed attitude of the groomed animal is marked. The animal being groomed, especially if it is lying down, is in a very vulnerable position but an attack never occurred during a grooming sequence. Also, there were no observations of a grooming animal being attacked because it pulled too hard on the hair of the groomed animal.

E. Length, Climatic, and Diurnal Variations in Grooming

The length of social grooming sequences varied from under 1 minute to approximately 23 minutes. Most sequences (50%) lasted approximately 1 minute (see Fig. 5). The attenuation of grooming bouts was related to the fact that 62% of all the sequences witnessed were nonreciprocal; there was no alternation between respective groomer-groomee roles. The brevity of grooming sequences was also related to the fact that more than two animals were very rarely involved in any one sequence, and when one of the participants departed the sequence terminated.

Most grooming, both social and nonsocial or allogrooming (Sparks, 1968), occurred between 1200 and 1600, which was the time of resting or intermittent leisurely feeding. At this time the animals were in the trees sheltered from the hot afternoon sun. Less social grooming occurred outside this time period, for the animals were either gorging themselves after the evening's sleep or were moving and feeding heavily before retiring for the evening. There was no increase in grooming activity immediately before retiring for the evening. Furthermore, troop dispersal was

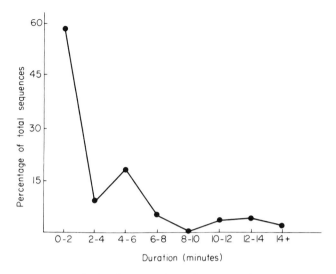

FIG. 5. The duration of social grooming bouts.

quite wide during feeding and periods of movement and there was little maintenance of physical contact. During resting periods, however, the troop was rather tightly knit into the respective subgroups. The hourly frequency of grooming behavior is given in Table XVIII.

Climatic conditions seemed to have some effect upon the incidence and frequency of grooming. Most grooming bouts (55%) occurred on sunny, warm days, 36% occurred on cloudy or partly cloudy days, and only 9% were recorded during rainy periods. This necessarily means that

TABLE XVIII
FREQUENCY OF GROOMING BEHAVIOR

Hour of occurrence	Percentage of occurrence
0500–0600	0
0600–0700	0
0700–0800	1
0800–0900	2
0900–1000	5
1000–1100	14
1100–1200	22
1200–1300	14
1300–1400	13
1400–1500	12
1500–1600	10
1600–1700	6
1700+	1

less grooming was witnessed during the two monsoon periods, November to the beginning of January and May through August, than at other times of the year. The dearth of grooming behavior on rainy days appears to be a result of the fact that an animal commonly remains huddled up to maintain warmth and this minimizes interanimal contact.

F. RELATIONSHIP OF AGE, SEX, AND DOMINANCE TO SOCIAL GROOMING

1. Age

Adult-adult grooming bouts comprised 63% of the total social grooming sequences. At least one adult animal was involved in 81% of the sequences recorded. Time spent in grooming increased sharply with increasing age.

In sharp contrast to many nonhuman primates, adult Nilgiri langurs rarely groomed young animals. Subadults were the youngest age category groomed by adult males. Adult females seldom groomed young animals, including either their own or infants of other females. Just 7% of the total grooming sequences occurred between an adult female and an infant. Female-infant grooming bouts frequently were cursory acts; there was little evidence of the intense mother-infant grooming reported for North Indian langurs, macaques, and baboons, for example. The longest female-infant grooming bout recorded lasted only 5 minutes. This involved two adult females grooming an infant 2. Adult female–infant 2 grooming was slightly less frequent than adult female–infant 1 grooming. Adult female–juvenile grooming was rarely recorded. Infants were never observed to invite grooming from adult females, in fact, they often resisted an adult female's advances.

An infant 2 occasionally groomed an adult male. Such sequences were, however, of short duration. Most began after an adult male had inadvertently frightened a youngster. Similar age class animals groomed one another more frequently than they groomed animals of different age classes, that is, most social grooming occurred within age/sex subgroups. Indeed, 65% of all grooming involved animals of the same sex and 71% involved animals of the same age class.

The percentage of time each class in troops I–IV assumed the groomer or groomee role is given in Table XIX.

2. Sex

There was a distinction in the amount of time adult males and adult females assumed the groomer-groomee roles. Adult males assumed the groomee role slightly more often than adult females. Adult females assumed the groomer role in 50% of the total grooming bouts, however,

TABLE XIX
ASSUMPTION OF ROLES IN GROOMING SEQUENCES: TROOPS I–IV

Age	Role	Percentage of total
Adult male	Groomer	28
	Groomee	36
Adult female	Groomer	50
	Groomee	34
Subadult	Groomer	12
	Groomee	19
Juvenile	Groomer	8
	Groomee	4
Infant	Groomer	1
	Groomee	7

and adult males in only 28%. This differentiation cannot simply be explained by noting the prevalence of females. If grooming partners were selected at random, that is, if each animal had the same opportunity to be groomed as any other, we would expect adult females to be groomed approximately 1.3 times more frequently than the adult males; that is, in proportion to their numerical ratio in the adult population. Since this was not the case, there must be other factors influencing the percentage of time adult males were groomed. Two possibilities exist: either the adult males were capable of forcing the adult females to groom them, or the adult females were attracted to the adult males and took frequent opportunities to groom them. Since adults were normally separated into sex subgroups between which there was minimal interaction, the second possibility seems less likely. Indeed, adult females rarely sought adult males in order to groom them, whereas an adult male could move to an adult female and be groomed almost at will.

Duration was a major differentiating factor in grooming. More instances of female-female grooming lasted for longer time periods than did male-male sequences (Table XX). This suggests that females were either under a stronger stimulus to groom one another or that the latent conflict that may exist in the male dominance hierarchy (see Section IV,C for further discussion) prevented males from being in sustained interanimal contact.

There is also an apparent time differential in heterosexual grooming bouts, however, its significance is not immediately evident. The longest heterosexual grooming bouts consisted of males grooming females.

3. Dominance

Dominant animals more frequently assumed the groomee role and less frequently the groomer role than did subordinates. Dominant animals

TABLE XX
LENGTH OF GROOMING SEQUENCES

Time (minutes)	Percentage of total	
	Homosexual grooming	
	Male-male grooming	Female-female grooming
0–1	53	30
1–2	25	9
2–4	0	18
4–6	13	9
6–8	0	14
8–10	0	14
10–12	9	0
12–14	0	6
	Heterosexual grooming	
	Male-female grooming	Female-male grooming
0–1	59	53
1–2	0	5
2–4	14	12
4–6	13	30
6–8	14	0
8–10	0	0

assumed the groomee role 70% of the time and subordinates 30% of the time, that is, dominant animals were groomed 2.3 times more frequently than were subordinates. Dominance status was also related to the role of initiator and terminator of grooming sequences. Dominant animals initiated the majority of the grooming sequences (67%), illustrating the fact that dominants could more readily move to subordinates and be groomed than vice versa. However, subordinates terminated most sequences (59%). The reason for this is unclear; however, the relaxed attitudes observed during grooming may be misleading. A subordinate animal may actually be tense and may depart as soon as it feels the dominant individual is placated. There is the additional possibility that the grooming animal simply did not derive pleasure from the activity and left as soon as the situation permitted.

A dominant animal almost always commanded a longer period of grooming than did a subordinate. Shorter grooming sequences of adults grooming subadults, for instance, frequently consisted of a cursory picking over of the body. Longer sequences of subadults grooming adult males involved a more careful search over the entire body of the animal being groomed. Two examples are given.

> The troop V adult male approaches a subadult female and sits next to her for approximately 10 seconds. He grooms her back in a cursory manner which involves simply a brief parting of the hair. He sits and she moves away (July 21, 1966).
>
> The beta male of troop III sits in front of the subadult female with the half-tail. He presents his side by reaching to an overhead branch which he grasps with his left hand; he then leans back. She grooms his side for 25 seconds, then moves around to groom his chest for an additional 40 seconds. She works rather intently with her face close to his chest. She stops a moment, then pulls his hair tuft to her face and begins to groom it. She grooms his head for 10 seconds. The male looks at her and she again begins to groom, this time concentrating on his thigh. The thigh is groomed for 15 seconds. She stops and sits a moment and then moves away. A total of approximately 90 seconds was spent grooming the male (July 22, 1966).

G. Functions of Grooming

1. Biological Function

The prime biological function of Nilgiri langur grooming presumably was cleansing the hair of dirt and ectoparasites. Approximately 62% of all social grooming concentrated on areas that an animal could not effectively cleanse by itself. The anatomical focus of grooming is presented in Table XXI.

TABLE XXI
Focus of Social Grooming

Focus	Percentage of total	Could area be effectively self-groomed?
Chest	20	Yes
Back	19	No
Head or hair tuft	17	No
Side	9	No?
Neck	7	No
Thigh	7	Yes
Leg in general	6	Yes
Shoulder	4	No
Stomach	3	Yes
Arm in general	3	Yes
Facial region	2	No
Underarm	2	No
Base of tail	2	No
Foot	0.09	Yes

Social grooming probably served a reduced biological function for Nilgiri langurs when compared to terrestrial species. The arboreal habitat occupied by the Nilgiri langur probably substantially reduced the number and/or concentration of ectoparasites in the environment. For example, since most fecal matter drops to the ground, there is little opportunity for infestation by ectoparasites which eventually reside in fecal material. Struhsaker (1969) notes that forest-dwelling cercopithecines groom less frequently than their savannah-dwelling relatives and correlates this difference with an increase of ectoparasites in a savannah environment.

2. Social Functions

As suggested earlier, grooming is most important from the standpoint of social integration. It depends upon a previous state of positive conditioning in the participants and it enhances and strengthens social relationships, perhaps as a result of the fact that the tactile stimuli show every sign of serving as a reward for the animal being groomed (Marler, 1965; Poirier, 1968a).

In light of its potential function, it is interesting that social grooming was not a prominent activity among the Nilgiri langurs. Only 134 social grooming bouts were recorded, that is, one grooming interaction per 9.3 hours of observation. A listing of the social grooming sequences observed in troops I–IV appears in Table XXII.

TABLE XXII

TOTAL GROOMING SEQUENCES: TROOPS I–IV

Groomer	Animal being groomed[a]								Times did grooming
	AF	AM	SAM	SAF	SA	J	I 2	I 2	
AF	35	17	0	0	0	0	5	4	61
AM	6	20	0	3	4	0	0	0	33
SAM	0	0	3	0	0	0	0	0	3
SAF	0	4	0	0	0	1	0	0	5
SA	0	0	0	0	7	3	0	0	10
J	0	0	0	3	3	0	0	0	6
In 1	0	0	0	0	0	0	0	0	0
I 2	0	4	0	0	0	0	0	0	4
Times received grooming	41	45	3	6	14	4	5	4	122 + 12 sequences of unidentified monkeys

[a] For abbreviations see footnote to Table X.

A prime function of Nilgiri langur social grooming is the reduction of tension between monkeys involved in agonistic encounters. Approximately 45% of all Nilgiri langur grooming bouts were initiated by agonistic situations. The most frequent response to the invitation of grooming was a general relaxation of tension and the adoption by the recipient of a posture inviting further grooming. Grooming indirectly favored a bond by diverting the recipient to an action, that is, invitation to further grooming, which was incompatible with agonistic behavior. One of the most common responses of a subordinate Nilgiri langur was to groom the dominant monkey, which almost immediately terminated a dominance sequence. Similarly, the embrace-grooming combination was common in aggressive situations.

H. SELF- OR ALLOGROOMING

Self-grooming was limited to those body parts where visual inspection could accompany manipulation. The legs and thighs were areas most often in focus. As in social grooming, the hand was typically the grooming agent; the mouth was occasionally employed to free debris. Individuals usually sat while self-grooming, but occasionally they did so while lying down. Monkeys might groom themselves intently, but rarely continued the activity longer than 5 minutes.

Self-grooming sometimes appeared to be a contagious activity; other animals were stimulated to groom while observing nearby monkeys pick through their own hair. Occasionally, an animal watching another groom itself interrupted and assumed the groomer role. The interruption of self-grooming was never rejected.

I. SCRATCHING

Nilgiri langurs frequently scratched themselves. The arms, legs, and sides were usually the focal points. Both the hands and the feet were used to scratch other parts of the body. The hands were used mainly to scratch the arms, legs, head, face, and back. The feet scratched the side and occasionally the head. When scratching with the feet, Nilgiri langurs resembled a dog pursuing the same activity. Both the arms and legs had a high degree of mobility; an animal was capable of reaching the midpoint of its back with its hands.

In vigorous scratching, the monkey moved four bent fingers, excluding the pollex, up and down the selected areas with great force. The middle or index finger was used to scratch delicate areas such as the corner of the eye. Scratching was occasionally effected by rubbing the back, chest, chin, or mouth on a branch. This occurred most often after a rain.

J. Comparative Data

Grooming is the most frequent social contact occurring in rhesus macaque troops (Koford, 1963). Perhaps four-fifths of the grooming sequences recorded on Cayo Santiago occurred between a female and her various young, including mature daughters with their own infants, or among brothers and sisters (also see Sade, 1965). "Even some of the top males occasionally groom immatures" (Koford, 1965, p. 195). Grooming between adult males and females is a prominent pattern of rhesus courtship behavior. Longer grooming bouts are usually reciprocal. Grooming occupies a large part of the bonnet macaque's day (Poirier, personal observation; Simonds, 1965). Grooming commences when the group awakens and continues sporadically until they move to the sleeping trees. Simonds notes that there is no sexual differentiation in grooming; age and personality seem to be the determining factors. Dominant males and females do most of the grooming and are in turn groomed most often.

Among baboons grooming is largely confined to the females (Hall, 1965; Bolwig, 1959). Hall feels that time differences have a more important bearing on the social significance of grooming than frequency. Longer grooming bouts characterize female-female or female-juvenile sequences. Washburn and DeVore (1961) note that baboon grooming is more apparent in the early part of the day and again in the later afternoon. *Anubis* baboons form grooming clusters, the nucleus of which is often a dominant male or a mother with a very young infant. Dominant males are sought out for grooming.

K. Conclusion

Nilgiri langur grooming was infrequent when compared to the amount of grooming reported for common langurs or for most other monkeys. In support of the present findings, Tanaka (1965) reported only five instances of grooming during his 1-month study of five Nilgiri langur troops. Although grooming occurred infrequently, Sugiyama (1965b, p. 237) reported that South Indian common langurs were involved in an average of 9.6 grooming interactions per day. He added that the interactions were ". . . far less than in the case of macaques." However,

Jay (1965a, p. 216) notes that among North Indian common langurs
". . . mutual relaxed grooming . . . may occupy more than five hours
a day." The dichotomy in social behavior occurring among North
Indian langurs on the one hand and South Indian common and Nilgiri
langurs on the other is interesting. This is considered further in the subsequent sections. It is interesting, however, that both South Indian and
Nilgiri langurs have the weakest troop structures among langurs. This
can be correlated, among other things, with lack of grooming behavior.
Grooming functions as a cohesive bond, and since it is primarily lacking
in South Indian *P. entellus* and *P. johnii* langurs there is a reduction in
a powerful centripetal force which binds the troop together.

The lack of Nilgiri langur grooming behavior can also be correlated
with the relatively relaxed quality of daily social behavior. Assumption
of a terrestrial habitat leads to increased terrestrial predation, pronounced sexual dimorphism, and increased pugnacity. Increased pugnacity appears to be related to increased frequency of grooming behavior,
for grooming helps maintain peaceful social relationships. Thus the
dearth of Nilgiri langur grooming behavior is related to the infrequency
of dominance interactions, which seems ultimately related to the arboreal
habitat. Marler (1965, p. 551) states:

> Grooming is particularly prominent in species in which dominance rela-
> tions play a significant role, such as rhesus and baboons. . . . By com-
> parison, mutual grooming is much less frequent in adult gorillas (Schaller,
> 1963), and it is tempting to see a correlation with the subtler and more
> peaceful dominance relationship in this species.

XII. PLAY BEHAVIOR

A. FUNCTIONS OF PLAY BEHAVIOR

Much has been written about nonhuman primate play behavior, although a clear-cut definition of just what constitutes play behavior is
not readily agreed upon. Loizos (1967) has compiled one of the most
comprehensive reviews of play behavior in nonhuman primates and
should be consulted for references of the current status in the attempt
to define the concept. Setting aside the definitional argument of the
precise boundaries of play behavior, it should be noted that play seems
to be a universal characteristic of primates. This does not imply, however,
that there are no species-specific characteristics in the play repertoire.

Beach (1945) has suggested that the amount, duration, and diversity of play behavior is related to a species' hierarchal position on the phylogenetic scale. Both Lorenz (1956) and Morris (1964) propose that a distinction be made between animals whose mode of survival is specialized, either structurally or behaviorally, and those who are opportunists. The latter exhibits a restless curiosity or neophilia—love of the new (Morris, 1964). Behaviorally, neophilic animals share a tendency to play and generally maintain high activity levels. According to Morris, primates are the supreme example of neophilic animals.

Play is one of the first nonmother-directed activities appearing in ontogenetic development. According to Harlow (1963), play behavior begins in the second stage of rhesus macaque infant development. It is generally accepted that involvement in play behavior is fundamental for the development of future skills requisite for survival. The hopping, chasing, and wrestling bouts characteristic of play help increase a youngster's muscular coordination. Key elements of a nonhuman primate's life, the development of social bonds, grooming, components of sexual behavior, and aggression, are to some degree learned and rehearsed in the play group. "Playfulness . . . is rightly regarded as a useful index of the physical and psychological well-being of the young primate. Its prolonged absence raised the suspicion of retardation, illness or distress" (Mason, 1965, p. 530).

Although detailed field studies of the role of play behavior in the socialization process are lacking, most observers agree that the experiences gathered in play behavior facilitate the development of skills upon which adult social relationships are founded. Play may help youngsters find their place in the existent social order (Carpenter, 1934). The basis of the dominance hierarchy, for example, may be formed in the play group where individuals learn each other's aggressive and defensive capabilities. The wrestling bouts characteristic of play behavior give a growing primate practice in behavior that, at least in part, influences its adult social position. As a consequence, subadult males tend to do less chasing and more wrestling as they mature.

In addition to social play, nonsocial play assumes a major proportion of a young primate's daily routine. The nonsocial play that directs an infant's attention to the immediate environment probably helps acquaint it with its surroundings as well as facilitating its muscular coordination and development (Mason, 1965; Washburn and Hamburg, 1965). Furthermore, the motivation to play and explore bring about a diversified sampling of the environment which is probably of great importance in adaptation.

B. Observational Definition of Nilgiri Langur Play

Play behavior was recorded as any unstereotyped relaxed behavior in which an animal was involved in vigorous activity seemingly without directed purpose (Schaller, 1963). Under this definition, whereas hanging from a branch in order to drop to a lower one was not recorded as play, hanging from the same branch and kicking the feet into the air was. Most play behavior, social or nonsocial, involved either one or a combination of running, climbing, hanging, chasing, or wrestling. Individuals sometimes pulled leaves and branches and threw them about. Inanimate objects such as dead twigs and vegetation, such as moss and epiphytes growing on the twigs, were often play objects. Play behavior thus consisted of vigorous activity. Additional cues for defining play behavior included the open-mouth gesture and gamboling, both of which are discussed in Section XII,M.

C. Overall View

During or following the prolonged midmorning and afternoon rest and feeding period, young Nilgiri langurs engaged in playful behavior. From 3 months until approximately 4 years of age for females and perhaps 5 or 6 years for males, they spent some portion of each day in social play behavior. Overall, however, Nilgiri langurs were not playful, and hours frequently passed without witnessing a play sequence. The dearth of play behavior in smaller troops was attributable to the lack of youngsters. Even in larger troops such as III, and IV, however, which contained representatives of all age classes, usually no more than 2 hours per day were spent in play activities. A total of 180 play interactions, or one play sequence per 6.9 hours of observation, was recorded.

A listing of the play sequences recorded in troops I, II, III, and IV appears in Table XXIII. All but three instances of adult-infant play behavior occurred in troop II in which an infant 2 had no peers. Ten of these sequences occurred during the joining of a male trio with the troop (Poirier, 1968a, 1969a). At times it was impossible to distinguish the initiator of a play bout, therefore sequences are listed just once. For example, infant-juvenile or juvenile-infant play is simply listed as infant-juvenile. The vertical column of the table does not indicate the play sequence initiator.

Most Nilgiri langur play behavior occurred in infant, juvenile, and subadult play groups. Adults seldom engaged in play except in troops

TABLE XXIII

PLAY BEHAVIOR IN TROOPS I–IV[a]

	I	J	SA	A	Uniden-tified	Mother	Crow	Squirrel	Observer
Infant	31	12	8	34	0	4	1	2	2
Juvenile	—	10	10	6	0	0	0	0	2
Subadult	—	—	22	—	0	0	0	0	0
Adult	—	—	1	1	0	0	0	0	0
Unidentified	—	—	—	—	41	—	—	—	—
Total involvement in play	89	38	41	42[b]	41	4	1	2	2

[a] For abbreviations see footnote to Table X.

[b] An explanation of the high incidence of adult involvement in play is given in the preceding paragraph and in the citations.

in which there were no peers with whom a youngster could engage itself. Mothers, who bore the brunt of their youngsters' early exuberance, were seldom more than passive participants. Nevertheless, until the infant-2 stage, adults were extremely tolerant of playful infants.

D. Ontogeny of Play Behavior

Harlow (1963) designates five stages in the development of play behavior in laboratory rhesus macaques. The first stage is presocial during which infants explore and manipulate all the objects in the test area, including other infants. They do not engage them in play interactions, however. The second stage is rough-and-tumble play during which infants romp, wrestle, and roll about vigorously. The third stage is approach-withdrawal play which is characterized by pairs of monkeys chasing one another about without necessarily engaging in physical contact. The fourth stage of integrated play involves both rough-and-tumble and approach-withdrawal play. The fifth stage of aggressive play appears at the end of the first year and is characterized by biting and pulling. It results in little or no injury, however. Aggressive play graduates into true aggression.

In the present study specific attention was paid to the ontogeny of play behavior from the tenth day until 112 days of age in four Nilgiri langur infants born in troop I. Following Harlow's scheme, four stages of Nilgiri langur play development were designated which approximated Harlow's interpretation. Two variations did appear in this *very limited* sample, however. (1) The presocial stage did not involve physical contact with other infants. (2) The second and third stages of social play behavior, approach-withdrawal and wrestling, were in reverse order of the macaque situation. Rhesus macaque rough-and-tumble play appeared prior to approach-withdrawal play.

E. Presocial Play, Investigation, and Manipulation—Stage I

The first 20 days of life were presocial, during which Nilgiri langur infants occasionally explored and manipulated objects in the environment. This always occurred in the mother's presence. The infant usually retained tactile contact with the mother and reached for objects in the vicinity. The following is typical.

> The alpha female's infant sits at her chest and reaches for objects surrounding them. He pulls moss from the branch and holds it in his hand.

He looks at it and turns it about in his hands. The moss is discarded and a leaf taken which receives similar treatment (December 16, 1965).

At approximately 24 days of age the infants began to leave their mothers to climb about adjacent branches. They always remained within easy reaching distance, however. The first infant-infant contact occurred on the 25th day; it involved two infants clumsily climbing over one another. There appeared to be no clear motive such as play. The sequence is described as follows.

The alpha female's infant male and the gamma female's infant female approach each other. When they meet the infant male climbs over the female. No social interaction is involved and each infant appears almost oblivious to the other's existence (December 22, 1965).

From the 24th to the 38th day, an infant wandered 1 or 2 ft from its mother to explore the immediate environment. It still generally avoided contact with other infants in the vicinity, however. These excursions often included pulling a leaf from a branch, placing it in the mouth, running a short distance, and then discarding the leaf. Twice a small twig was carried in the hand a short distance and then discarded. These nonsocial play efforts can be visualized as a progressive focusing of the manipulatory play movements upon particular inanimate objects in the environment. The early efforts were primarily random.

By approximately the 45th day, infants began to perform the characteristic playful walk termed "gamboling." This act is characterized by a "bouncy" form of locomotive rather than striding across a branch. The impression is reminiscent of a young colt frolicking in an open field.

F. Approach-Withdrawal Play—Stage II

During the seventh week playful social contacts were rather frequently established and the first instances of chasing were recorded. The earliest examples of approach-withdrawal play were truncated, usually lasting under 90 seconds. Two infants approached one another, mouths open, and randomly slapped and touched briefly. Then they parted. Chase sequences were carried out over distances of 2 or 3 ft; they often terminated prior to the establishment of contact:

The gamma female's infant gambols to the delta female's infant who sits approximately 10 inches from her mother. The gamma female's infant's mouth is open. She slaps at the delta female's infant who slaps back. They exchange a few rapid slaps and then the gamma female's infant gambols off (January 16, 1966).

By the eighth week a consistent feature of play behavior was the transportation of a leaf or twig in the mouth or hand which often became the object of playful chasing.

G. WRESTLING—STAGE III

Wrestling bouts began at approximately 9 or 10 weeks of age. The bouts were usually short as the infants tired rapidly and returned often to rest by their mothers. Wrestling bouts involved rolling and tumbling about on the branches and mouthing the arms, legs, and tail of the opponent. The following example is typical.

> The delta and gamma females sit resting side-by-side. The delta female's infant sits in front of her; the gamma female's infant approaches and slaps her arm. They face each other, mouths open, and begin to wrestle. They roll about the branch and pull one another's arms, head, tail, and legs. After 30 seconds they return to their mothers. They rest approximately 1 minute and then resume vigorous wrestling (February 2, 1966).

H. INTEGRATED PLAY—STAGE IV

By the tenth week playful social contacts were actively sought whenever two or more infants were in proximity. Such contact rarely involved more than two individuals, however. Integrated play was more complicated than earlier manifestations of play behavior and there was a noticeable prolongation of play sequences. An infant often ran past another and was then chased; once the pursued animal was caught, the participants wrestled. When the sequence ended, the infants either returned to their mothers or rested a moment and then resumed the interaction.

I. NONSOCIAL OR INDIVIDUAL PLAY

The distinction between social and nonsocial play is not fundamental. The "characteristics that affect responses to inanimate objects, such as complexity, size, or mobility of the stimulus, also influence reactions to social stimuli" (Mason, 1965, p. 529). An examination of Nilgiri langur behavioral acts directed toward inanimate objects during nonsocial play supports this assumption. Many activities characteristic of Nilgiri langur social play such as threats, swinging, and kicking occurred with objects as well as with individuals. Interestingly, many of these patterns were performed even when playmates were available.

J. Early Stages of Nonsocial Play

The earliest stages of Nilgiri langur nonsocial play were primarily focused on the infant's own body. The hands, feet, and tail were particularly enticing objects during the first 2 months of life. An infant often sat by its mother mouthing its own hands. Occasionally, an infant lay on its back, opened and closed its fists, and randomly kicked its feet into the air. The picture is reminiscent of a young child staring at its opening and closing hands, kicking its feet into the air, and smiling. Sometimes the langur infant lay in a prone position and slapped at its feet. The tail was often pulled or batted about and occasionally chased in a manner similar to a dog after its own tail.

K. Later Stages of Nonsocial Play

As the infant's motor coordination developed, nonsocial play became more strenuous and involved. Hopping from branch to branch, swinging from a branch and kicking into the air, carrying leaves and twigs in the mouth, and various other acts which seemed to function primarily as energy releasers were frequently observed. Later stages of nonsocial play often involved manipulation and tossing about of broken branches and other objects. The following are typical sequences of nonsocial object-oriented play behavior.

> A 3-month-old female breaks off a small twig with her mouth and hands and gambols along a branch with it. She sits close to her mother turning the twig about in her hands and mouth. She manipulates the twig for almost 10 minutes and then drops it and returns to her mother's chest February 26, 1966).
>
> The delta female's 3-month-old infant hangs by her hands from a branch and kicks at an opposing branch. She pursues the activity for approximately 30 seconds then rights herself and gambols off down the branch where she sits and rests (February 24, 1966).

L. Relation of Age to Nonsocial Play

Nilgiri langur nonsocial play behavior was characteristic of the infant-1 developmental stage; it was less important in the infant-2 stage and rare in older individuals. When it occurred in older age groups, it was usually of a more vigorous nature. The nonsocial play of older youngsters involved hanging and swinging from branches as compared to object-carrying by infant 1's. This probably primarily reflects maturation of the motor skills. In troops such as troop II in which no nonadult peer groups existed, the youngster engaged in nonsocial play until a later age than did a youngster in troop I which had peer orientation. The troop-II youngster directed its activities either toward the adults, who seldom responded, or toward the inanimate environment. Occasionally a crow or squirrel was chased.

M. INITIATION OF SOCIAL PLAY

Social play provided the Nilgiri langur infant the first opportunity to associate and interact closely with other youngsters in the troop. Any one or a combination of the following behavioral cues were taken as indicative of a playful mood: branch shaking, dragging an object past another animal, grabbing and pulling another's tail, gamboling, hanging and kicking, lying on the back and kicking, slapping, and the open-mouth approach. Gamboling and the open-mouth approach were the signals most often employed to indicate playfulness. A playful animal approached another with its mouth open and head tilted slightly to one side. If the animal being approached wished to play, it reciprocated similarly. The cue was distinguished from the open-mouth gesture characteristic of dominance interactions as the tension wrinkles commonly produced at the mouth corners were absent and the teeth were not prominently displayed. Altmann (1962b) refers to such communicatory actions as "metacommunication," that is, messages involving communication about communication. "Primates include in their repertoire a set of social messages that serve to affect the way in which other social messages are interpreted" (Altmann, 1962a, p. 279).

Other behaviors commonly associated with play initiation served to attract the attention of prospective playmates. A playful langur might jump up and down on a branch, inviting others to engage it in a playful interaction. A playful youngster often dragged a piece of branch, leaf, or other object past another. This usually attracted the other animal's attention who chased and/or wrestled with the transporter in an attempt to gain the displayed object.

> An infant female in troop I drags a small twig in her hands past the infant male. The male immediately rises and gambols off after the female. The male catches her, pulls the stick away and runs to his mother with it in his mouth. The female chases him for a moment but then returns to her mother (February 25, 1966).

N. FORMS OF SOCIAL PLAY

1. Wrestling

Chasing and wrestling were the most typical forms of Nilgiri langur social play. Subadults engaged in wrestling bouts far more frequently than they engaged in chasing. The reverse was true of infant play sequences. Wrestling bouts were largely restricted to age-mates; however, subadults sometimes grappled with juveniles. Despite the fact that a subadult often outweighed by as much as three times the juvenile with whom it grappled, such a weight disparity never resulted in injury.

Wrestling was characterized by a number of behavioral patterns occurring singly or as constellations. Wrestling animals frequently gently bit or mouthed each other in the angle made by the shoulder and neck or along the back, although the skin was never broken. Occasionally, two youngsters approached each other bipedally with their mouths open and their arms swinging down in front of them. Upon contact they wrestled. Mounting was occasionally witnessed during wrestling bouts; however, many of the patterns outlined in the Appendix (Section D) for mounting behavior were absent.

Wrestling usually occurred in the trees; the animals rarely descended to the ground to play. It was not uncommon for animals to wrestle or slap and kick at each other while hanging from a branch with only one hand or foot. Falls were relatively frequent but the individuals never seemed to be injured as a result. Most often, however, wrestling occurred in a sitting or prone position. The participants soon became a mass of tangled arms, legs, and tails.

2. Chasing

Chasing was the major form of presubadult play behavior. More than two animals were rarely involved in any single chase sequence. Play chasing often resulted from an attempt to obtain some desired object which another animal dragged or carried about. The chase sequence terminated when one animal tired or when one caught the other, in which case they grappled. Young animals following each other frequently made repeated descending leaps from one branch to another several times in rapid succession.

O. General Characteristics of Social Play

1. Formation of Play Groups

Nilgiri langur social play was rarely pursued in clusters of more than four animals. In larger troops, however, as many as five or six animals joined and left a wrestling spree before it terminated. Play groups were fluid aggregations with monkeys coming and going. There appeared to be no preferred play combinations. Animals did not seem to seek particular individuals with whom they chose to interact in a playful fashion.

In larger troops containing infants, juveniles, and subadults play was not restricted to age-mates. The play behavior of older monkeys frequently assumed the form of vigorous wrestling, however, which although not punishing to the youngsters often became too rough and strenuous. This tended to segregate youngsters into separate play groups of their own.

2. Length and Diurnal Variation in Play

Most play bouts were short; sequences involving infants usually lasted under 5 minutes. Sequences involving juveniles and subadults, however, might last 10–15 minutes. Play encounters involving four or more animals lasted longer than sequences involving only two participants.

The incidence of play behavior varied with the time of the day. Play was primarily concentrated in the following four periods: (1) an early morning play period occurring before the troop moved to the top of the sleeping trees for sunning and early morning intermittent feeding, (2) a period occurring when the older monkeys settled for the midmorning rest interval, (3) a period that occurred during the midafternoon rest period, and (4) a period of play when the troop settled into the sleeping trees for the evening. At any time the troop remained stationary, however, youngsters might be seen engaged in brief play sequences.

Although it appeared that less play occurred during the monsoon periods, it was not positively determined whether or not weather had a marked effect upon the incidence of play behavior.

3. Vocalizations in Play and Adult Interference

Play bouts were usually silent. Vocalizations were emitted only when play became too rough, for example, if an older subadult wrestled too vigorously and scared an infant or juvenile. Adults never came to the aid of youngsters who were frightened during the course of a play bout. One instance was recorded in which a subadult female attempted to play with an infant 2 that was running to its mother. The infant protested the abrupt approach by screeching loudly. The infant's mother who sat close by made no attempt to chase the female away. The infant finally freed itself from the grasp of the playful female and ran to sit by its mother.

P. RELATION OF PLAY TO AGE

1. Late Subadult Play

In terms of the life cycle, the amount of time a Nilgiri langur was engaged in play behavior increased until the juvenile or subadult developmental stage. Beyond the early subadult stage (3–4 years for males), play behavior markedly decreased with age. In addition to the physiological changes associated with the onset of sexual behavior, play-fighting seemed to reduce the amount of social play and of general activity. Quite possibly the motivation eliciting play became inadequate for submerging agonistic behavior. As the severity of play-fighting increased with consequent pain and frustration, involvement in play behavior decreased.

Play-fighting is a special category of play behavior characteristic of late subadult male play sequences. As male langurs matured, play became more and more vigorous. Sometimes the observer found it practically impossible to distinguish between play wrestling and aggressive wrestling sequences. During play-fighting older animals vigorously slapped, kicked, and hit each other. If younger animals were in the play group, they usually left and gave the subadults a wide berth. Vigorous play-fighting normally lasted 2 or 3 minutes before terminating.

Female langurs manifested a sharp decrease in play activities upon attainment of subadult status. This seems to strengthen the argument that play serves a more important function in the male maturation process. Subadult females were more interested in adult female activities, especially the care of the young, than they were in vigorous play behavior. Furthermore, where presubadult play often occurred in heterosexual play combinations, subadult play occurred in male and female play groups.

2. Adult-Adult Play

Only one instance of adult-adult play was recorded. It involved a very brief wrestling bout between an adult male and female. The interaction occurred at a time of increased sexual activity in troop IV, and there is the possibility that it had some sexual connotation. This could not be validated, however. The sequence follows.

> The alpha male and an adult female rest sitting next to each other. The male faces the female, his mouth open. The female responds with an open-mouth cue. They wrestle from a sitting position; they kick at each other and mouth each other's necks. The sequence terminates in approximately 30 seconds after which they sit quietly again (June 9, 1966).

3. Adult-Youngster Play

Adult-infant and adult-juvenile play was sometimes witnessed in troops in which a youngster lacked peers. Adult-infant play was most frequent in troop II in which the infant 2 and delta male were rather frequently involved in short wrestling and chasing bouts. Overall, however, adult-infant play was very rare and the troop-II situation seems to be an anomaly.

Q. INTERSPECIFIC PLAY

In addition to play with inanimate objects and with their group members, Nilgiri langur youngsters occasionally directed their play behavior toward the giant squirrel and common jungle crow with whom they shared the forest canopy. Such play was seemingly nonreciprocal. It involved sneaking up on and chasing the squirrel or crow through the trees. The sequences were recorded as play and not as interspecific aggression, for the youngsters exhibited the open-mouth or gamboling behavior. Two of the three instances of interspecific play follow.

> The troop-II infant 2 locates a squirrel which it chases through the tree for a distance of 20 ft. The infant shortly terminates the chase and sits disinterestedly 10 ft from the squirrel (March 1, 1966).
> The troop-II infant 2 chases a crow through the trees. The youngster sneaks up behind the crow and jumps, landing on the branch where the crow perches. The crow flies off cawing loudly with the infant gamboling behind in hot pursuit. The infant again moves above the crow and jumps to the branch where it rests. The crow is startled and flies away with the infant in close pursuit. The infant repeats the pattern of sneaking up on the crow and jumping at it. The total sequence lasts 9 minutes (March 22, 1966).

Among other things, these sequences illustrate the fact that when conspecifics are absent, especially peers, other objects in the environment are the focus of playful animals. After acceptance by a troop, the observer also was occasionally the focus of playful youngsters. Infants and juveniles in troops I, II, and III sometimes jumped above my head and then gamboled away. The following is exemplary.

> As I stand observing troop II, the infant 2 jumps above my head showering me with moss and other debris. He looks at me; when I look back, he gambols off to sit by the delta male. The infant repeats this pattern three times consecutively before tiring and moving away (March 11, 1966).

R. DISCUSSION AND CONCLUSION

The amount of play behavior and the animals participating in play activity differ from species to species. In contrast to the dearth of play behavior recorded by the present study and substantiated by Tanaka (1965) for the Nilgiri langur troops he studied, North and South Indian langur youngsters spent considerable time in vigorous play behavior. Macaques and baboons also indulge in a good deal of play activity. In baboons, for instance, play ". . . is the first behavior seen at the sleeping places before dawn and the last behavior to cease at night" (Hall and DeVore, 1965, p. 87).

Since play behavior helps a young monkey find its place in the social order, it is likely, although unproven, that the limited number of playmates accessible in many Nilgiri langur troops influenced the socialization process. A maturing youngster in an average Nilgiri langur troop unquestionably had fewer opportunities for social interaction with its peers than the macaque or baboon infant that matured in a troop containing 15 or 20 young playmates. Play behavior probably serves a less important function in the social system of Nilgiri langurs than it does in macaque and baboon societies. The consequences of this are still to be determined, however.

XIII. PATTERNS OF SEXUAL BEHAVIOR

A. Female Estrous Cycle

Physical or behavioral signs of estrous were never recorded. Nilgiri langur females appear to lack the perineal swelling common, for example, to macaque (except bonnet macaques), baboon, and chimpanzee estrous females. Therefore a feral female's physiological state could be determined only when it was expressed behaviorally.

Nilgiri langur females possess an everted clitoris, a highly unusual trait in Old World Monkeys. Occasionally, the clitoris underwent a color change, becoming somewhat darker pink in appearance. Furthermore, the clitoris sometimes appeared more pronounced than usual. Possibly these periods of pronounced eversion and color change coincide with some physiological change such as the onset of estrous. They were too infrequently observed, however, for any conclusions to be drawn.

B. Mating Behavior

Only one attempted mating, which occurred between the beta male and alpha female in troop II, was observed. The sequence was almost immediately disrupted by the interference of the female's young offspring.

> The beta male moves behind the alpha female. The female remains seated as the male attempts to mount. The female refuses to rise so he thrusts slowly against her back. The infant 2 runs over and pushes the male aside. The male's only response is to sit and manipulate his penis. The

alpha male moves over and grunts, displacing the beta male (August 3, 1966).

Although one cannot draw conclusions from only one instance, the interference of the infant 2 during this attempted mating sequence is interesting. Such interference could be an expression of what Jay (1965a) characterizes as harassment behavior in North Indian common langurs. North Indian langur consort pairs are harassed by less dominant males who run about barking, threatening, or slapping at the consort pair. Harassment was directed almost exclusively at the male and not the estrous female.

C. Copulatory Patterns

From observations of mounting behavior in contexts other than copulation, the copulatory posture of the Nilgiri langur appears similar to that described for most monkeys. The basic copulatory patterns, for example, mounting, thrusting, and placing the feet about the ankles or lower leg of the mounted animal, were all evident in Nilgiri langur play or dominance mountings.

D. Lack of Mating Behavior

Although the absence of mating behavior cannot be adequately explained, several possibilities are offered. Nilgiri langurs were observed during all months of the year, during their waking and part of their sleeping hours, and in all types of weather. Thus the dearth of information on mating behavior is quite likely not the result of failure of the observational method. A second possibility might be that all the females in the study area were either pregnant or nursing during the observational period. This, however, was not the case. A third possibility, suggested by Haddow (1952) for African red-tailed monkeys, is that mating occurred late in the evening when the observer was generally not present. This may have been true; however, as this pattern has not yet been clearly demonstrated for any of the nonhuman primates, this explanation seems of minimal likelihood.

A feasible, albeit tentative, explanation is that the observer's presence upset the study populations. In most primates sexual behavior is the activity most vulnerable to disruption by outside influences. It has been suggested, for example, that the presence of an observer may help account for the dearth of mating behavior observed in gorillas and chimpanzees.

E. MASTURBATION

Sixteen instances of male self-manipulation were recorded. Six positively terminated in ejaculation and two may have. One instance of ejaculation followed an intense grooming sequence between the alpha and delta male in troop II. The sequence is described as follows.

> *1230:* The alpha male moves to another tree; the delta male follows. The alpha male sits with his back to the delta male as a grooming presentation. The delta male grooms his thigh and back. He stops momentarily and both animals sit quietly. The delta male again begins to groom; he picks through the hair and licks one spot continuously. He also grooms by pulling the hair through his teeth. *1240:* The sequence ends and both males sit side by side. *1243:* The delta male again begins to lick the alpha male's thigh but stops after 20 seconds and turns his back. *1253:* The alpha male yawns three long, drawn-out yawns. He then lies on a branch to rest. While resting, he picks ejaculate from his penis and eats it. Ejaculation must have occurred as a result of the intense grooming for no masturbation was witnessed (March 9, 1966).

The remaining instances of ejaculation were preceded by masturbation. The penis was stimulated by handling, by rubbing between the palms of the hands, and then by working it back and forth between the thumb and index finger. Masturbating males assumed a characteristic pose; the legs were thrust rigidly forward, one foot was placed atop the other; and the toes of the top foot were wrapped about the bottom foot. In all instances the ejaculate was eaten. The following is typical.

> The adult male of troop I sits high in a gum tree approximately 40 yd from the females. His legs are thrust out in front, left foot overlapping the right. He stimulates his fully erected penis by working it between the thumb and forefinger and rubbing it between the palms. After a minute of continuous stimulation, he stops and yawns. He begins again almost immediately. Forty seconds later he ejaculates. He picks the ejaculate from his penis and eats it. He then falls asleep (June 5, 1966).

Instances of male sexual arousal, as manifested by self-manipulation, occurred from February through August; the peak was April and May, when 8 of the 16 occurrences were observed.

Most recorded acts of male sexual arousal occurred during the morning hours between 0900 and 1200. This was the coolest part of the day. The morning hours were also times of intense feeding, intermittent resting, and the least amount of movement.

F. COMPARATIVE DATA

1. Indication of Sexual Receptivity and Incidence of Mating

Jay (1965a, p. 240) notes that a North Indian langur female ". . . must indicate to the male that she is receptive by shaking her head rapidly from side to side, presenting, and dropping her tail on the ground." South Indian common langurs employ a similar mechanism (Sugiyama, 1965a). *Presbytis cristatus* females solicit the male in a similar manner, for example, by making lateral head-shaking movements and then presenting the hindquarters (Bernstein, 1968). Head shaking and presentation occurred in combination among Nilgiri langur females and among males. Rather than indicating sexual receptivity, however, it was interpreted as a submissive gesture.

The dearth of witnessed mating behavior among Nilgiri langurs again emphasizes the wide range of frequency of sexual behavior among nonhuman primates. Frequency of observed mating behavior is not constant in cross-specific comparison; heterosexual behavior certainly plays a much larger role in the social life of some species than in others. Bernstein, for example, reports the incidence of mating among the Malaysian colobid *P. cristatus* at 18 copulations in over 1000 hours of observation. The incidence of mating is higher among common langurs. In contrast to Nilgiri langurs, however, *P. cristatus* males were observed masturbating only once. At this time it appears that the low incidence of Nilgiri langur heterosexual behavior may be an anomaly among colobids.

There seems to be some correlation among monkeys between an increase in arboreality and a decrease in sexuality. Roughly, the more terrestrial colobids, for example, the South and North Indian common langur, and presumably *P. cristatus*, copulated more than their more arboreal counterparts. This may in part be associated with the smaller troop sizes common to arboreal representatives of the genera and to the reduction of aggression, which is related to degree of arboreality. It would be interesting to test this hypothesis for *Macaca* and *Cercopithecus*. Is there a decrease in mating when one compares the terrestrial rhesus to the arboreal lion-tailed macaque? Sugiyama's recent report (1968) resulting from 2½ months of observation of the lion-tailed macaque suggests this may not be true, however, more complete information is needed.

2. Social Stimulation of Mating

Among South Indian langurs peak sexual activity followed troop reconstruction, that is, the joining of a male group with a bisexual troop.

"These cases show that social stimuli, such as severe fighting among males, change of troop leader, or a severe attack with killing of infants activate the sexual activity of females; in other words social stimuli play a direct part in activating sexual activity" (Sugiyama, 1967, p. 233). Goodall (1965) reports that when the chimpanzees in the Gombe Stream Reserve aggregate there is an increase in mating. "It is not known whether these groups are formed as a result of a need for social contact stimulated by sexual excitement, or whether the increase in activity, due to the aggregation of a large number of chimpanzees, has a direct effect upon sexual activity" (Goodall, 1965, p. 451). In contrast, none of the troop reconstructions observed in the present study stimulated sexual activity, although this was attended to specifically.

XIV. THE TROOP: ITS STRUCTURE, FLUIDITY, FUNCTION, AND ADAPTABILITY

The final section of this chapter highlights some of the aspects of the social structure of the Nilgiri langur. It is an attempt to show that the Nilgiri langur social structure results, at least in part, from processes linked to the socialization process, social interaction, and environmental pressures.

A. THE FLUIDITY AND STRUCTURE OF THE TROOP

Internal change (other than that resulting from deaths and births) occurred in three of the five troops under constant surveillance. In a 7-month period, troop II underwent six changes. When the study terminated, the troop was not yet stable. A schema of the changes in troop II is presented in Fig. 6. In addition to changes occurring in troop II, troop IV lost one adult male and a juvenile female, and three females of troop V joined troop II. A fourth troop (I) may have been recently formed. This is suggested by the fact that when first contacted troop I contained only adults, while all troops surrounding it had some infant, juvenile, or subadult members. In addition, the females of troop I bore the first infants in the fall birth season.

Troop II originally contained seven members: one adult male, three adult females, one juvenile male, one infant male, and one infant of undetermined sex. Troop II was contacted during the first month of study. It was observed for approximately 12 hours prior to the first change; subsequently, it was under observation for 353 hours.

On January 11, 1966, an adult male trio joined troop II. The interactions occurring during the first 2 weeks of merging were limited and nonagonistic. Aggressive physical contact was not witnessed. The most aggressive sequence was a chase involving the resident male and a male group member which terminated prior to contact. The usual mode of aggression was some form of threat behavior, frequently a stare threat or displacement.

During the initial period of merging, the male group usually remained somewhat peripheral to the resident troop. Resident animals made little attempt to contact the male group; likewise, the male group made little effort to interact with them. When the resident male moved to a different area of the home range, the original troop members followed as before. Resident troop members remained keyed to the alpha male; when he ate, they ate, when he rested, they rested. No male group member attempted to lead the troop's progression; there was no outward conflict between the resident male and the male group members for the status of troop leader (Fig. 6).

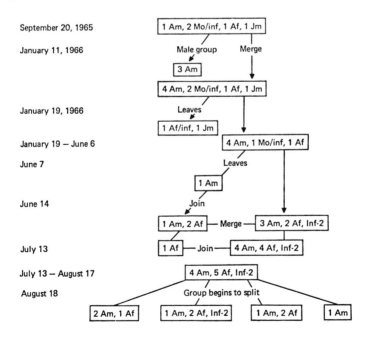

Fig. 6. Schema of social change in Group II.

1. Role of Play Behavior in Troop Change

The incidence of play behavior during the merging process was unexpected and deserves further comment. Play seemed to assume a major role facilitating the integration of the male group into troop II, accounting for 31% of the interactions between the male group and a resident troop member during the first 2 weeks of merging. All the play behavior occurred between the male group and the infant 2. The play was particularly striking in view of the fact that the dominant male of the trio played with the infant during this period but ignored the infant once the trio merged with the troop.

The infant 2 appeared to play a major role in the integration of the male trio with the troop. Infant 2–male group interactions accounted for 42% of the social contact. The infant initiated most of the interaction. Two hypotheses may be suggested concerning the joining male–infant 2 interactions: (1) The males, as new objects in the environment, may have aroused the infant, which resulted in play contacts (Mason, 1965). (2) The males' acceptance of the infant 2 helped facilitate their entrance into the troop structure, for by tolerating the infant 2 they may have been demonstrating submissiveness.

2. Role of the Females

The three adult females played a passive role in the integration of the male group; they were involved in 8% of the mutual interactions. For the most part they ignored the males and seemed to adjust readily to the fact that the troop was enlarged. This adjustment probably resulted from the fact that the males and females were commonly separated into distinct subgroups between which there was minimal social contact.

3. Subsequent Changes in Troop II

On January 18, troop II experienced the second change in a week; an adult female, her infant 2, and a juvenile (perhaps her previous offspring) deserted the troop. They moved across the road, subsequently joining another troop (probably troop XI). This occurred without advance warning. No further changes were noted between January 18 and January 24. On January 24, troop II began to shift from its home range; by January 27, they had relocated on the opposite side of the road. January 29 to June 7 passed without further change. Perhaps the movement to a new home produced a period of troop adjustment during which the troop remained closer together.

On June 7, the beta male (a member of the male trio) left troop II and established contact with troop V, with whom they overlapped and with whom they maintained an agonistic relationship (Poirier, 1968b).

Even so, he established a peaceful relationship with them by feeding in proximity. The adult male of troop V made no attempt to drive him off. Subsequently, troop V, including the beta male, moved out of range of observation; for 1 week they could not be located. On June 14, the beta male rejoined troop II accompanied by two adult females of troop V. The three were not subjected to any greeting behavior, nor did they seem to be the focus of undue attention. Initially, the two females interacted primarily with each other and only rarely with resident members of troop II. When they did interact with troop II, it was primarily through brief grooming bouts, both with other females and with the males. In addition, they initially remained on the troop's periphery during resting and sleeping periods. On July 13, a third adult female joined the troop. Her entrance was unrecorded; she was already present when contact was established at 0750. It was assumed that she also came from troop V, for she made immediate contact with the two females of troop V who had joined previously.

These three females joined troop II without incident. In all cases the first resident member approaching them was the infant 2. In contrast to the nonaggressive reception received from the male group, the females chased the infant away. Neither its mother nor any resident troop members responded to mild threats directed at the infant. The five females subsequently formed a subgroup distinct from the four adult males.

Male/male tolerance dissipated shortly after the third female entered troop II, and brief chasing and threat encounters between the males became commonplace. By August 18, the troop appeared destined for a major split. Three distinct subgroups had formed, moving in divergent directions. One subgroup contained the beta male and a joining female; another the gamma male, the mother, and her infant 2, and a second resident female; the third subgroup contained the delta male and two joining females. The alpha male remained unattached and continually moved unimpeded back and forth between the different subgroups. The alpha male appeared highly nervous; he vocalized, urinated, and defecated much more than usual. The male's movement between the subgroups seemed to be an attempt, apparently abortive, to have them follow him. The following example from our field notes is typical:

> 1040: The alpha male begins to grunt (a coordinating vocalization issued during progression) and moves toward two joining females. The beta male, who sits next to him, remains. He grunts as the alpha male moves off. . . . 1045: The alpha male sits near two of the females and occasionally looks in their direction, then he moves 15 ft away to sit alone. . . . 1050: He returns to his starting point . . . the beta male moves adjacent to him. A subgroup under the gamma male's leadership moves in the direction of the Toda hut. . . . 1120: The alpha male moves along the ground

in the direction of the joining females. *1130:* He returns to his starting point. Two females cross the path and sit at the lower end. *1137:* The alpha male crosses on the ground to sit on the same side of the path as the females. *1140:* He recrosses the path to his starting point. The beta male's subgroup moves in a northeasterly direction. *1150:* The alpha male sits quietly, occasionally picking at his feet (August 18, 1966).

Unfortunately, the observational period for troop II ended here and the results of the alpha male's efforts can not be reported.

B. Factors Involved in Troop Change

1. Unbalanced Sex Ratio

The unstable conditions characterizing troop II are not clearly understood. The influences prompting the male group's attachment to, as well as the desertion of the family group from, troop II are unclear. The unbalanced sex ratio of troop II may have facilitated the adult females' movement from troop V to troop II, however. Prior to their addition, troop II contained four adult males and two adult females. Troops II and IV were the only ones in which adult males numerically predominated. Troop IV also experienced a composition change in which an adult male deserted the troop, resulting in a composition of two adult males and one adult female.

The addition of three adult females to troop II rebalanced the sex ratio in favor of the females. Carpenter (1964) has suggested that primate troops operate about "homeostatic" or "sociostatic" equilibrations. When the sex ratio or troop size deviates too far from the norm for a species, important regulatory processes operate to rebalance the equilibrium.

2. Behavior of Joining Animals

The relative ease with which the six individuals joined troop II probably resulted largely from their assuming a nonagonistic posture. Laboratory studies suggest that the resistance of newcomers to aggression increases the duration and intensity of group attack which continues until they submit (Bernstein, 1964; Sugiyama, 1960). Nonresistant, submissive individuals are more likely to be positively received by a troop than animals manifesting opposite traits.

The relatively fluid nature characterizing troop II resulted from two factors. (1) Nilgiri langur females appeared to adjust easily to the presence of strange animals. This is especially important because females usually constituted the adult majority in a troop. (2) Autonomous subgroup behavior, which resulted in the weakening of troop cohesion, precluded the formation of a united front to face an intruder. The individual action of a male against an intruder while other troop members (including other males) ate or performed "nonrelevant" acts, suggests a lack of troop solidarity in the wake of intrusion by outside animals.

Carpenter (1940) has suggested that there are two types of primate groupings. Some species, such as baboons, howlers, and macaques, are characterized by a high degree of troop closure or lack of permeability, and animals seldom enter from outside. Other species, such as chimpanzees and gorillas, have a high degree of permeability and relatively fluid troop organization. Langurs seem to be polarized at both extremes; the North Indian common langur has a stable troop structure (Jay, 1965a), while the South Indian (Sugiyama, 1967) and Nilgiri langur troop is more fluid. Among South Indian langurs, "many instances of change in langur society have been observed" (Sugiyama, 1967, p. 227). Sugiyama concluded that a troop was struck by change ". . . about every three years on the average" (Sugiyama, 1967, p. 232).

C. SUGGESTED EXPLANATIONS FOR THE WEAK TROOP STRUCTURE

The integration of primate groups is presumably a complex process relying on mutually reciprocal patterns of behavior modified and made specific by learning and conditioning (Washburn and Hamburg, 1965). Almost every phase of behavior of which an animal is capable enters, to some degree, into the determination of its gregariousness and the qualities of its complex social organization. It is therefore suggested that the social organization characteristic of the Nilgiri langur troops under observation resulted not only from a lack of predation and correlated factors but also from processes linked to a dearth of social interaction, the existence of subgroups, and the mother-infant relationship. The dearth of grooming and play behavior, both of which tend to establish and reinforce social bonds in other species, may have militated against the formation of strong troop ties. The laxness characterizing a mother's attitude toward her infant and the early onset of weaning must also, in some as yet undetermined way, affect the Nilgiri langur socialization process.

1. Grooming and Play

How the processes outlined above operate can only begin to be understood with the limited knowledge gained in 1 year. In lieu of controlled laboratory testing of these behaviors, we must rely on a comparative approach. We find, for example, that both the macaques and baboons, which have the most cohesive social troops, also spend much of the day engaged in grooming and play behavior. A similar correlation exists among Indian langurs between an increase in grooming and play behavior and troop cohesion. North Indian langurs, who groom and play more frequently than either South Indian or Nilgiri langurs, also have the strongest troop structure. The North Indian langur, baboon, and macaque are more terrestrial however, and as indicated earlier this also influences troop structure and composition.

2. Mother-Infant Relationship

The earliest influences exerted upon Nilgiri langur infants originate from within the social group formed by the mother and other adult females. The mother-infant relationship is a locus for the transference of knowledge and experience. Harlow's (1962, 1963; Harlow and Harlow, 1961) studies illustrated the extreme importance of a mother's influence upon the social development and subsequent behavior of her infant. By providing support, nourishment, and protection, the mother makes survival possible. In addition, the mother-infant relationship prepares the infant for later social adjustment. A mother's behavior toward her infant is considered as representative of the larger social aggregate (Mason, 1965). The maternal relationship may be the prototype of all later social bonds (Jensen *et al.*, 1967). Within the mother-infant framework, there are impressive species-specific differences which may contribute to broad and persistent differences in future behavior (Jay, 1962; Tinklepaugh and Hartman, 1932; Tomilin and Yerkes, 1935).

In most primates the intense female interest in her young assures it the protective environment and care needed until it becomes an independent troop member. The relative casualness of Nilgiri and South Indian langur mothers toward their infants, as compared to the very protective attitude assumed by North Indian langur, baboon, or macaque mothers, must have some effect upon the social development of the infant, hence upon troop structure. A continuum can be usefully applied when discussing the protective behavior of the langur mother. The North Indian langur mother occupies one end of the scale, the Nilgiri and South Indian langur mother the other. A North Indian langur mother was constantly aware of her infant (Jay, 1962); however, a South Indian langur mother ". . . does not always keep constant and close watch over it (her infant) . . ." (Sugiyama, 1967, p. 233). Interestingly, South Indian and Nilgiri langurs, among whom mothers show the least protective behavior, are characterized by the weakest social organizations. Of course, here again, the degree of terrestrialness may be a factor.

The quality of the mother-infant bond influences the infant's socialization experience; hence it affects the social order. The nature of the mother-infant relationship becomes extremely important if one considers every phase of behavior of which an animal is capable to enter into the determination of its gregariousness and the quality of its social order. This is developed further by Poirier (1968a,c,d).

D. Functions of a Nilgiri Langur Troop

1. Troop Protection and Survival

Troop life is not as important for the Nilgiri langur as it is, for example, for the savannah-dwelling baboon and the nonforest-dwelling cercopithecines (Struhsaker, 1969) for whom life in an organized troop is a prerequisite for survival. Nilgiri langur social relations were not oriented primarily toward protection of the individual by cooperative troop action. Instead, an individual protected itself by dashing through the nearest trees. Individual monkeys were not dependent upon the protection which in other species is afforded by a large adult male with impressive canines. The fact that some Nilgiri langurs lived apart from a bisexual troop for varying lengths of time is witness to the decreasing importance of troop life. In contrast, no healthy baboon was recorded living apart from a troop (Washburn and DeVore, 1961).

2. Learning and the Prolonging of Youth

Given the previous information, we are disposed to inquire into what holds the Nilgiri langur troop together. What are the functions of the troop? Why does it exist? If Nilgiri langur troop members do not take full advantage of the opportunities that troop living normally offers in the form of grooming, play, and protective behavior, why choose a troop mode of existence? Washburn and Hamburg (1965) suggest that a primary raison d'etre for troop existence is learning. The troop is the center of knowledge and experience far exceeding that of its individual constituents. "It is in the group that experience is pooled and generations linked" (Washburn and Hamburg, 1965, p. 613). From watching other troop members, an animal learns what foods to eat, what behavioral patterns to employ, who the existing predators might be and the correct usage of the communication repertoire. An animal learns the mode of survival by living in a troop where it benefits from the shared knowledge of the species. Laboratory studies by Mason (1963, 1965) and Harlow (1963), among others, suggest the dire consequences resulting from the lack of a proper context for social learning.

Troop existence is the sociological response to the primate biological adaptation of prolonged immaturity. "Since the prolongation of pre-adult life is biologically expensive for the species, there must be major, compensatory advantages in the young's remaining relatively helpless for so long" (Washburn and Hamburg, 1965, p. 620). The compensation is learning. Despite the restraints it imposes upon the social order, the long period of infancy has a selective advantage. ". . . it provides the species with the capacity to learn the behavioral requirements for adapting to a wide variety of environmental conditions" (Washburn and Hamburg, 1965, p. 620).

3. Learning Maternal Behavior and the Mother-Infant Relationship

Social traditions are passed, in the first instance, through the mother-infant relationship. The mother-infant dyad is the context in which the infant learns its social responses and how to fulfill its biological requirements, for example, which foods to eat. To be a skilled or even adequate mother, a female must be reared within a social group. She must have had ample opportunity to observe the interactions between other mothers and infants.

Harlow (1963) demonstrated the importance of social contact for the proper ontogeny of maternal behavior. Rhesus mothers raised in laboratories and deprived of normal socialization experiences were deficient in their maternal behavior (Harlow and Harlow, 1961). Such mothers manifested gross abnormalities, ranging from marked indifference to active rejection of the infant. Females raised in isolation or in other abnormal conditions failed to nurse effectively; their infants consequently required hand rearing.

Harlow (1963) further demonstrated the consequent importance of maternal behavior upon the socialization of an infant. A female reared by a poor mother (i.e., one raised outside the normal social context in isolation) was herself woefully inadequate. "We only know that these monkeys without normal mothering and without peer affectional relationships have behaved toward their infants in a manner completely outside the range of even the least adequate of normal mothers" (Harlow, 1963, p. 184). Continuation of this trend would have a dysgenic effect upon the species, upsetting the infant's fundamental social relationships with the mother which are the primary context for the earliest life experiences.

4. Socialization

At this point one must ask in what way orderly social relations depend upon prior experience. Must a young monkey or ape undergo a period of socialization (must it live at least part of its life in close contact with others) before it can perform effectively as an adult member of the species? Mason's (1960, 1961a,b, 1963) studies on rhesus macaques aptly demonstrated the effect of environmental and social restriction on the maturation process. Animals with restricted social experiences, for example, those raised in isolation, showed strikingly abnormal patterns of sexual, grooming, and aggressive behavior. "The restricted monkeys, on the one hand, groom infrequently. They used no specific gestures to invite grooming and the attempts of one restricted monkey to groom another were generally met with indifference, rather than with the active cooperation seen in the feral group" (Mason, 1963, p. 165). Animals deprived of normal social experiences were more aggressive than feral animals participating in social interactions. Restricted monkeys fought more often and the fights were longer and more severe than those of feral animals.

Restricted monkeys were not attracted to other members of the species, nor were others attracted to them. Feral monkeys were either indifferent to restricted monkeys or actively avoided them. "The entire series of tests indicates that restricted monkeys are not attracted to other members of their group and were either indifferent to them or actively avoiding contact with them. . . . The results indicate . . . that our socially deprived monkeys were attracted to some feral animals, but that they were not attractive companions to each other, nor to monkeys born in the field" (Mason, 1963, pp. 168–169). Mason's studies emphasize the fundamental importance of the socialization process in the primate's life. The full development of an animal's biological potentialities seems to require the stimulus and direction of social forces that are usually provided in the social troop.

5. Further Cohesive Bonds

The importance of group life for the fulfillment of one's biological potentialities and for passing on pooled knowledge having been established as almost universal, what binds a troop together once these conditions are satisfied? What factors bind the Nilgiri langur troop together when the most important cohesive bonds formed in grooming and play relationships and the protective role of the male are lacking? Two possibilities exist:

(1) The ability to meet nutritional requirements. The expressed antagonism between various Nilgiri langur troops and the defense of definite areas against encroachment by others leads individuals to cohere into a social aggregate in order to take full advantage of available food supplies. If an animal could feed at random wherever it pleased, there would be no detrimental effects if it deserted the troop. Because only

troop members were allowed within the favorite areas of the home range, however, monkeys remaining within the troop could best meet their dietary needs. Although some Nilgiri langurs lived outside bisexual troops, they occupied areas not ordinarily overlapping with bisexual troops, suggesting that only troop members took advantage of food sources within the home range.

The above conclusion seems especially applicable in areas of high population densities. In such circumstances, nongroup animals living interstitially between two bisexual troops had a reduced chance to satisfy their nutritional requirements. They were forced to live in less desirable surroundings because they were continually harried by bisexual troop males. Consequently, areas with high population densities, such as area C containing troops II, V, XIII, and XIV, had no nongroup animals. All the individuals encountered in high-population-density environments belonged to bisexual troops. [A number of authors, for example, Davis *et al.* (1948) for rats, Hediger (1953) for numerous captive animals, and Howard (1920) for birds, likewise suggest that a prime function of territories is to increase the accessibility and availability of food sources for animals occupying the area.]

(2) The attraction of the females. The second cohesive force was the females. Although some females existed outside bisexual troops, they seemed to do so for only a very short time. Therefore males attached to bisexual troops were more likely to have ready access to females than males outside troops. Although the instance of the male trio joining troop II suggests that some males living outside bisexual troops could satisfy their sexual needs by joining a troop, it seems more likely that males with whom resident females associated throughout the year had a greater chance of doing so.

The preceding argument has three weak points. (a) No Nilgiri langur mating was recorded. Therefore further data is needed before one can conclude that the adult females constituted an attractive force holding or drawing the males to the troop. (b) The one instance in which a male trio joined a bisexual troop did not show an increase in sexual activity as might be expected [especially in light of Sugiyama's (1965b) data on South Indian langurs]. (c) A sexually aroused male could always release his tensions through masturbation, which was recorded. This does not contend that sex behavior is the prime binding force in the nonhuman primate society, rather, it suggests that sex is perhaps one of many binding fibers.

Jolly (1966) postulated an interesting hypothesis concerning the cohesive forces in primate societies. Noting that the socialization process began at an early age, she remarked that ". . . adult friendly behavior originates directly from the contact, grooming and play behavior of the infant" (Jolly, 1966, p. 159). Developing this point further, Jolly noted that if the friendly relations were in fact derived from the mother-infant relationship it would suggest that social primates retained many infantile characteristics. "The *original* cohesive force in primate social evolution would then be an infantile or juvenile attraction to others that was retained in the adult" (Jolly, 1966, p. 163).

E. THE IMPORTANCE OF TROOP LIFE FOR FEMALES

The foregoing leads one to the conclusion that the troop way of life is more important for some individuals than others. Adult females were the cornerstone of Nilgiri langur society. In addition to being numerically predominant, they were the focus of infant and juvenile social interaction. Most important, the successful transference of the species behavioral repertoire and social traditions was primarily the task of the adult female within the mother-infant relationship.

Nilgiri langur society is female focal. Primarily because of the females' presence the troop existed; the major role of the Nilgiri langur male was to father the succeeding generations. The male played a minor part in the protection of the troop and in the social development of the infant. In contrast, in most terrestrial species, such as baboons, macaques, and gorillas, the dependence of other troop members on the adult males creates male-focal social organizations.

Hall's (1965) studies on patas monkeys provide an interesting correlate to the Nilgiri langur situation. Patas groups are one-adult-male multiple-female organizations. The sole function of the adult male in addition to paternity is that of lookout for extragroup disturbances. The patas, similar to the Nilgiri langur, group is organized about the adult females. Hall (1965, p. 275) concluded ". . . the one-male unit of the wild is highly organized around the adult female, the adult male having a clearly defined role as breeder and watcher for external threat, with all the major initiative *within* the group coming from the adult female." The patas, similar to the Nilgiri langur, group is female focal. A further correlate is found among South Indian langurs. Sugiyama (1965b, p. 411) notes that ". . . the foundation of the troop is an aggregation of females and their young. Males may be regarded as those who joined this fundamental aggregation afterwards."

F. SUMMARY

In summary, the Nilgiri langur troop organization is best understood if one considers the forms of social interaction. The Nilgiri langur troop functions less as a vehicle providing its members a context in which to gain social fulfillment, for example, in grooming and play, than as a context in which to learn the behaviors necessary for survival. A prime concern of the Nilgiri langur troop is the initiation of youngsters, by their mother and other females, into the lifeways of the species. Females shoulder the primary burden of socialization, and thus the Nilgiri langur troop is considered female focal.

G. The Adaptability and Flexibility of the Troop*

Although somewhat neglected until recently, an intriguing topic of investigation is the ongoing adaptation of nonhuman primates to changes in niche. Many of the study areas during the current research underwent rapid change as a result of human destruction. Adverse affects upon Nilgiri langur populations notwithstanding, this rapid change provided an unusually fertile situation for studying the processes of behavioral adaptability and intertroop variability. Since some of this material has been presented in detail elsewhere (Poirier, 1969b), only highlights are considered here.

1. Dietary and Communicatory Variations

Both dietary and communicatory variations have been previously discussed in the appropriate subsections.

2. Terrestrial Behavior

There is a strong correlation between the amount of time spent terrestrially and human alteration of the habitat. Although classified arboreal, a term needing greater clarification, Nilgiri langurs did, when forced, move rather rapidly and for extended periods of time along the ground. Troop II, for example, spent considerable time terrestrially while its original home range was being totally destroyed by the forest department. Troop IV inhabiting an area surrounded by cultivated fields, was very often terrestrial in order to take advantage of the food the gardens offered. Man's intrusion into the plateau drastically affected the amount of time a troop spent on the ground. Troops inhabiting home ranges recently destroyed, or in which cultivated tracts existed, were more terrestrial than adjacent troops in untouched areas. This is illustrated in Table XXIV.

* Behavioral flexibility is used here in a general sense to indicate adjustments of the organism to ecological conditions and change.

TABLE XXIV

TERRESTRIAL BEHAVIOR

Troop	Occurrence per hour of observation of terrestrialness	Comments
I	Once per 187 hours of observation	Stable home range
II	Once per 35 hours of observation	Home range destroyed
III	Once per 16 hours of observation	Home range destroyed
		Many *Acacia* seedlings present
IV	Once per 7 hours of observation	Home range included
		Large cultivated tracts

3. Flexibility According to Age and Sex

All members of the Nilgiri langur population did not adjust to conditions of change similarly. Variations in levels of adaptability seem to occur along age/sex boundaries. For example the infants of troops I and IV appeared to be in the forefront in their willingness to accept the new dietary item *E. globulus*. The acceptance of this dietary item probably follows the pattern: level 1: infant; level 2: mother, other infants, siblings; level 3: males.

In contrast to the fact that troop I and IV males are likely to be the last to accept the gum leaves as food, observations made during the shift of troop II's home range suggest that adult males are less conservative in their behavioral patterns (and therefore perhaps more adaptable) than adult females. The selective advantage of this has been previously discussed.

4. Discussion

Learned behavior becomes highly significant as one ascends the phylogenetic ladder. Since each Nilgiri langur troop has learning experiences different from any other, any one Nilgiri langur troop cannot represent the species. This is a basic characteristic of all higher primate social organization.

> One of the relics of the theory that most primate behavior is instinctive was the notion that the behavior of a whole species might be meaningfully described on the basis of a small sample from a single location, and even that the behavior of a genus might be inferred from the behavior of such a sample of the behavior of the species. Today it is clear that learning is important in social behavior and ecological adaptation, and that the behavior of even a species cannot be described on the basis of limited samples. (Jay, 1968, p. 173)

Behavioral incompetence leads as rapidly to extinction as does failure in the morphological framework or an organic deficiency, and thus selection acts rigorously on behavior (Nissen, 1958). The ability of the Nilgiri langur to adapt to the changing ecology in the plateau area suggests considerable behavioral plasticity on its part. Selective forces may be operating that favor the more adaptable, behaviorally plastic members of the local langur population. Were it not for their ability to adjust rather rapidly to changing conditions, we would expect the extinction of the Nilgiri langur in many plateau localities in the very near future.

Nonhuman primates are generally endowed with the ability to meet many of the challenges of a changing environment. Behavioral flexibility, such as discussed in this chapter and instances reported by others (i.e., Jay, 1968; Maples, 1969; Singh, 1969), was an essential trait facilitating the shift from a pongid to a hominid condition during the Pliocene. Frisch (1968) supports this conclusion. Washburn and Hamburg (1965, p. 615), noting the ability of primates generally to live a life adapted to anticipate not only daily needs but occasional crises, perhaps best sum up this point.

> A system that could meet only day-to-day problems would not survive for long, and evolution, through natural selection, builds a substantial margin of safety into the individual animals and into their way of life. The group moves more, is more exploratory, is more playful than there is any need for on the average day, but by so doing it is preparing for crises. The individual animals appear stronger and more intelligent than necessary for normal activity, but survival requires coping with the rare event. (Washburn and Hamburg, 1965, p. 615)

XV. CONCLUSION

In this chapter we have attempted to highlight some of the more significant aspects of Nilgiri langur ecology and social behavior. Although some of the points were developed in previously published accounts, much of the current material results from ongoing analyses. It should be evident that although the Nilgiri langur is similar in many respects to other colobids there are important differences. Where possible, correlations were established between ecology and social behavior and between ecology, social behavior, and troop structure. In order to understand Nilgiri langur troop structure and social behavior, one must consider two somewhat opposing requirements. The first is the need to adapt to the ecological niche; the second concerns the need to provide a social context in which youngsters can be initiated into the lifeways of the troop.

Nilgiri langurs responded to ecological pressures in various ways. The relatively small troop sizes, troop spacing, and the high degree of flexibility that allowed adaptation to new food plants and movement into new ecological niches are important examples. The dearth of grooming behavior in their arboreal habitat, the loud intertroop vocalizations, a lessening of gestural communicatory patterns, and perhaps the dearth of observed dominance behavior were interpreted as behavioral responses to ecological conditions (see Rowell, 1966, for an interesting discussion of baboon adaptations to forest living). The maintenance of a troop structure, however fluid, provided the context for infant socialization and a medium for the conveyance of troop traditions.

It must be emphasized that a good portion of the data analyzed for this paper was collected on troops currently experiencing some form of human harassment, the extreme being the home range destruction discussed earlier. It is strongly felt, however, that except for certain aspects of the behavioral system, that is, those discussed in the section on flexibility, the picture presented is a fairly representative sample of Nilgiri langur social structure and behavior. Rather intensive survey work convinced this author that the daily routine, methods of feeding, and the relative expression of dominance, grooming, and sexual behavior vary little from troop to troop. The patterns of the mother-infant dyad, home range utilization, and territorial behavior appear to be the same in all the troops observed. Clearly, the need now is for a study of undisturbed Nilgiri langur populations, but there are relatively few accessible areas where this can be accomplished.

APPENDIX: THE DOMINANCE REPERTOIRE

A. ATTACK BEHAVIOR

Any behavior involving physical contact or whose end product was likely to be physical contact, was labeled "attack behavior."

The intensity and duration of attack and threat behavior (Section D) varied with the social situation and the animals involved. More complex attack patterns occurred between participants of relatively equal status, such as the alpha and beta males. The closer the individuals in dominance status, the less likely an immediate submissive response by the subordinate individual. Quantification of attack sequences, where physical contact was or was not established, indicated that aggression by adult males was more frequent than aggression by others. The incidence of attack behaviors in 63 encounters in which the ages of both participants were determined is given in Table A.

TABLE A
QUANTIFICATION OF TOTAL ATTACK SEQUENCES

Attacker	Victim	Frequency
Adult male	Adult male	20
Adult male	Adult female	13
Adult male	Subadult	11
Subadult	Subadult	8
Adult female	Adult female	4
Adult male	Infant 2	4
Subadult	Juvenile	2
Infant	Infant	1

B. REPERTOIRE OF ATTACK BEHAVIORS

1. Slap or Lunge

This was the least aggressive attack cue. Slapping or lunging usually occurred at a distance several feet from the recipient, but during a wrestling bout could come from close quarters. If the sequence involved animals of close dominance status, such as encounters between low-ranking females, the subordinate occasionally returned the slap of the dominant individual. Slapping or lunging often preceded chasing if the recipient did not respond with a sequence-terminating subordinate gesture.

2. Chase

The vast majority of chase sequences (92%) occurred in adult-adult encounters. Dominant males sometimes initiated their chase by rushing forward and lunging at their objective. Adult females, particularly, chased their victims without obvious forewarning. Chase sequences were usually of short duration. Attacking animals seldom vocalized during the chase, but subordinates called loudly. The attacker frequently interrupted its charge prior to reaching its victim; the animals then sat apart and vocalized. The victim often continued to utter subordinate sounds long after the dominant animal terminated the chase.

Following a chase sequence, the attacker frequently attempted to appease its victim. This was especially true in adult female or adult male–infant and juvenile encounters. Appeasement was initiated when the dominant animal began to emit what was classified as the *subordinate segmented vocalization,* a high-pitched sound produced with the mouth one-quarter open and with a slight grin. This was usually followed by an embrace. As tension subsided, the subordinate animal frequently groomed the dominant one, further reducing tension, before departing.

3. Wrestle and Bite

Attack sequences rarely ended with wrestling and/or biting. The wrestling that occurred was most common in adult-adult encounters and was characterized by short slapping bouts and tumbling about the branches. Sequences were usually brief (1 minute being an upper limit) and terminated when one participant ran away screeching. Biting was recorded just once during a wrestling bout; it occurred in a female-female encounter. The bite was aimed at the nape of the neck or shoulder, an area covered with a mane of hair.

C. ATTACK INTERRUPTION

Attack sequences may be interrupted by either the dominant or subordinate participant prior to aggressive physical contact. Chase sequences often ended when the subordinate animal presented, shook its head, or shook its head and presented to the dominant animal. [Head shaking indicates sexual receptivity on the part of North Indian langur females (Jay, 1965a), however, among Nilgiri langurs although rare it is a subordinate gesture utilized by both males and females.] The attacking animal frequently halted its charge prior to reaching the victim and sat facing it vocalizing. In less intense situations, the attacking monkey might face its victim and bite air at it. In still milder situations, after a brief chase the attacker might sit and bob its head vertically.

D. THREAT BEHAVIOR

Behavioral components that could precede the infliction of pain or punishment by one animal on another were subsumed under the category of "threat behavior."

Threat behaviors were usually issued at a distance from the recipient. Except for assertive touching, mounting, and embracing, physical contact was not established, nor was it the likely end product of threat behavior. Threat gestures included facial expressions such as staring, the open-mouth threat, vertical head bobbing, and grinning (see Fig. 3a). Threat postures included limb or body movements such as advancing toward the subordinate animal and occasionally touching as in mounting or the assertive touch. A category of rarely observed threat behaviors included manipulation of inanimate objects, for example, branch shaking.

The Nilgiri langur threat repertoire, especially as manifested by facial gestures, was less extensive than that described for rhesus macaques and baboons. Nilgiri langurs lacked such threat and submissive expressions common to the head and face as lip-smacking, ear flattening, eyebrow raising, yawning, and blinking the eyelids. The absence of these gestures may be attributable to a lack of facial musculature specialization. Furuya (1961–1962) noted a similar dearth of facial expressions among Malaysian silvered leaf monkeys.

Displacement by the dominant animal and the observance of dominance distance by subordinates were the most frequently observed components of less intense threat behavior. Dominance distance refers essentially to what Hediger (1953) termed "individual distance." Spacing behavior of this type, unlike that implied by the term "territory," does not have reference to specific and fixed points within the physical environment. Rather, the area of space surrounding the dominant monkey, which is dominance distance, is a mobile phenomenon encircling the animal as it moves from place to place. The area of space maintained fluctuates with an animal's social position. The maintenance of dominance distances, however, is ignored during grooming. Wynne-Edwards (1962) views the maintenance of consistent individual distances as a result of counterposed synagonistic and agonistic tendencies.

Dominance distances were evident in the normal course of daily activities and they became exaggerated when dominant animals were tense or irritated. At that time, very dominant individuals could be surrounded by an area whose outside perimeter was approximately 3 ft. Both adult males and females maintained areas into which less dominant individuals did not enter without first submitting. The most common submissive movement was presentation.

Slight postural shifts and visual cues were extremely important patterns com-
municating potentially hostile situations. The chief visual cues were looking or *staring*
at a subordinate animal. The effectiveness of stare threatening was obvious when
individuals of approximately equal dominance stature were in proximity. Such animals often went to extremes, sometimes appearing ridiculous to the observer, to avoid
looking at each other. Their heads constantly moved about looking everywhere yet
nowhere, staring into space. The following example is from our field notes.

> The beta and gamma males of troop II sit back-to-back within 6 inches
> of each other. The beta male turns to face the gamma male who looks
> away. The beta male turns away; the gamma male looks to his immediate
> right. The beta male faces left, then both turn away. The beta male faces
> left; the gamma male begins to turn then looks skyward. The beta male
> looks away from the gamma. The gamma male looks to his right but turns
> away as soon as the beta male shifts positions. The gamma male finally
> departs (April 8, 1966).

Turning the back to a dominant animal was a variant of looking away. Both behaviors exhibited areas opposite those used in threat gestures.

Presenting, being mounted, and shaking the head and presenting, were frequently
issued submissive gestures. Presentation behavior appears to represent an incipient
flight response; the presenting animal exhibits the remnants of its original flight
response in a stressful situation. The act typifying subjugation was the movement of
the hindquarters toward the fear stimulus. Perhaps the hindquarters of the animal
being the antithesis of the front end, which was used in threat behavior, reduced the
probability of attack (Marler, 1965).

Presenting was elicited in various situations. Most often presenting was issued
from within a few feet of the recipient, usually within reaching distance. Presenting
had three major functions. (1) It stemmed the possibility of attack by a more
dominant individual. (2) It allowed a subordinate animal to pass close to a dominant individual by presenting to the latter to placate it. It was a signal that no
usurping of status was intended. (3) Presentation was occasionally issued to win
the recipient's favor, for example, to gain permission to sit close by, to eat from the
same food source, or perhaps to groom a more dominant individual.

A less intense form of presentation, designated *rear end flirtation,* was given by
a subordinate when passing a dominant animal. The animal issuing the "flirt" walked
somewhat more slowly than usual, with a slight flexion of the fore and hind limbs,
and turned its hindquarters toward the dominant animal as it passed by. If the dominant individual ignored the act, the subordinate animal continued on by. If for any
reason, however, the dominant animal shifted positions, the performer of the "flirt"
walked by almost sideways.

Freezing was a submissive posture employed in a manner similar to presentation.
If a dominant animal moved toward a subordinate, the latter often stopped all activity and assumed a tense upright sitting posture. Although the dominant animal
seldom threatened, its mere presence was sufficient to elicit freezing from the subordinate individual. Freezing was witnessed most often during feeding bouts. Upon
approach of a dominant animal, a subordinate often stopped eating and remained
perfectly still. A rapid glance from the dominant monkey often sufficed to reduce
tension. The subordinate thereupon had the option to continue eating or move away.
If the subordinate animal ate or left prior to a *releasing gesture* from the dominant
animal, it was attacked.

The most usual response to presentation was mounting by the dominant individual. Mounting was a rather consistent indicator of dominance status when it occurred in adult and subadult dominance sequences. In infant and juvenile interactions in which dominance relationships were not certainly fixed, however, it had an ambiguous meaning. Infants and juveniles often mounted in play situations. In addition, in infant and juvenile interactions, mounting was often reciprocal; the mounted animal often mounting the mounter in turn.

There were two types of dominance mounting behavior, a "full" mount in which one animal covered the back of another and a "symbolic" mount. In the latter, the "mounting" animal merely touched the rear of the presenting monkey. The mounting animal frequently remained seated during a symbolic mount. Symbolic mounting sequences usually involved an adult male and juvenile. Because a juvenile could not support an adult's full weight, the adult occasionally stood bipedally on a branch and simply touched the youngster's back. Sometimes, however, an adult stood with one foot on a branch, wrapped the other about the juvenile's calf and leaned slightly forward. Adult-juvenile mounts were usually brief. The full mount was the end product of a number of conjoined acts, all or a combination of which may be witnessed (Poirier, 1968a).

Twice a more dominant animal stood directly over an individual who was lying on a branch. The dominant animal paused a moment and then moved on. This appeared to be a variation of mounting behavior.

E. Establishment of Physical Contact in Threat Sequences

During threat sequences subordinates often sought rather than avoided physical contact with dominant monkeys. This was especially true of adult female, adult male–juvenile and adult male–infant 2 encounters. Typically, the contact pattern was the embrace or embrace-grooming complex. The subordinate ran to and embraced the dominant animal. The subordinate frequently groomed the dominant monkey before moving away. Juveniles and infants quite often chased adult males and forced themselves to the adult male's chest before their tension subsided. One instance was recorded in which a juvenile chased an adult male for 5 minutes screeching loudly. The juvenile's tension visually subsided only after the male embraced it. A typical example follows:

> Troop II's alpha male moves past the infant 2 during a whoop display and almost knocks the infant off the branch. The infant screeches loudly and chases the male. After running approximately 25 ft, the alpha male sits facing the infant and begins to emit the segmented vocalization. The infant 2 immediately runs to the male and sits in front of him screeching violently for 10 seconds. The male then stretches out his arm, the infant runs to him, and they sit hugging each other. The male hugs the infant to him. The infant grooms momentarily before departing (August 2, 1966).

The Nilgiri langur embrace pattern seemed to mimic the mother-infant embrace-clinging posture (Poirier, 1968a). Corresponding displays in other species have likewise been interpreted as regressions to infantile behavior (Andrew, 1964; Bolwig, 1957; Kummer, 1967a,b). Harlow and Harlow (1961) demonstrated the effect of "contact comfort" on the behavior of laboratory rhesus macaque infants. This original function of the maternal behavior, to protect the infant which is embraced or carried, may well be present in the dominance embrace behavior. Embracing reduced the probability of attack by a dominant animal and relieved the distress of the subordinate. The addition of grooming to embracing behavior was especially interesting since grooming also helped alleviate tension. Thus the recourse to embrace-grooming in aggressive situations had a definite stress-reducing effect. Occasionally, rather than embrace a subordinate monkey, a dominant animal simply placed its hand upon the lower status animal (the "comforting touch") whereupon tension almost immediately decreased. The physical contact which was sought in dominance encounters appeared to reestablish a peaceful relationship between the animals involved.

Grooming was another form of contact very frequently observed during threat sequences. Grooming served to reduce tension between monkeys involved in agonistic encounters. Approximately 45% of all grooming bouts witnessed followed agonistic sequences (Poirier, 1968b). The most frequent response to the initiation of grooming was a general relaxation of tension and the adoption by the recipient of a posture inviting further grooming. Grooming indirectly favored a bond by diverting the recipient to an action, invitation to further grooming, which was incompatible with agonistic behavior (Marler, 1965).

F. ATTACK AND THREAT VOCALIZATIONS

Vocalizations were less frequently issued in the attack-threat repertoire than nonvocal cues; vocalizations functioned primarily as submissive cues. Dominant animals emitted what were basically "calling-attention-to" vocalizations which attracted the attention of other troop members to the interaction and who, therefore, avoided the participants. A vocalization designated the grunt was most frequently employed. Occasionally two males in a very high state of tension began to grind their canines. Canine grinding, usually recorded in intertroop encounters, indicated extreme tension.

G. APPEASEMENT AND SUBMISSIVE VOCALIZATIONS

An important appeasement vocalization emitted by the dominant animal was designated, for lack of precise descriptive characteristics, the "suborindate segmented sound." This signaled that a dominance interaction had terminated. The subordinate

TABLE B

TOTAL INCIDENCE OF EMBRACING OR EMBRACE-GROOMING

Dominant	Subordinate	Frequency
Adult male	Adult female	25
Adult male	Infant 2	9
Adult male	Adult female	7
Subadult male	Juvenile	4
Adult male	Adult male	3
Subadult	Subadult	3
Infant	Infant	2
Adult male	Juvenile	1
Adult male	Subadult male	1
Subadult	Infant 2	1

TABLE C
TOTAL VOCALIZATIONS IN DOMINANCE INTERACTIONS

Vocalization	Frequency given by dominant animal	Frequency given by subordinate animal
Canine grind	9	7
Gruff bark	1	0
Grunt	96	19
Hiccup	7	0
Hoho	7	1
Hollow subordinate sound	0	5
Pant	3	0
Screech	0	19
Subordinate segmented sound	19	54
Squeak	5	61
Squeal	5	106
Whoop	1	1

animal usually responded by moving to the dominant animal whereupon they embraced.

The most prevalent submissive vocalizations were *squeaks, squeals, and screeches.* Squeaks were primarily emitted by infants and juveniles; squealing was strictly a subadult and adult female vocalization. These three major subordinate sounds were probably all related, but there were tonal and pitch variations which appeared to be correlated with age and sex differences. The screech, the most intense submissive vocalization, was emitted by animals of either sex (most commonly females, however) and by all age groups except infant 1's. Infant-1's produced a vocalization designated a *scream* which was not recorded for other individuals.

Nilgiri langur vocalizations could not be a strictly dichotomized into dominant and submissive vocalizations. Table C illustrates the fact that only three vocalizations were exclusively emitted by dominant individuals and two vocalizations were emitted only by subordinates. Individuals therefore could not be consistently labeled dominant or subordinate simply by recording the vocalizations issued during an interaction. The respective dominance statuses of the animals involved had first to be recognized.

REFERENCES

Altmann, S. A. (1962a). The social behavior of anthropoid apes: An analysis of some recent concepts. *In* "Roots of Behavior" (E. L. Bliss, ed.), pp. 277–284. Harper & Row, New York.

Altmann, S. A. (1962b). A field study of the sociobiology of rhesus monkeys, *Macaca mulatta*. *In* "The Relatives of Man" (J. Buettner-Janusch, ed.), pp. 338–345. N. Y. Acad. Sci., New York.

Andrew, R. J. (1964). The displays of primates. *In* "Evolutionary and Genetic Biology of Primates" (J. Buettner-Janusch, ed.), pp. 227–309. Academic Press, New York.

Beach, A. F. (1945). Current concepts of play in animals. *Amer. Natur.* **79,** 523–541.

Bernstein, I. (1964). The integration of rhesus monkeys introduced to a group *Folia Primatol.* **2,** 50–63.

Bernstein, I. (1968). The lutong of Kuala Selangor. *Behaviour* **14,** 136–163.

Bolwig, N. (1957). Some observations on the habits of the chacma baboon. *Papio ursinus.* S. *Afr. J. Sci.* **54,** 255–260.

Bolwig, N. (1959). A study of the chacma baboon. *Papio ursinus. Behaviour* **14,** 136–163.

Booth, A. H. (1957). Observations on the natural history of the olive colobus monkey, *Procolobus verus* (von Beneden). *Proc. Zool. Soc. London* **129,** 421–430.

Booth, C. (1962). Some observations on behavior of *Cercopithecus* monkeys. *In* i"The Relatives of Man" (J. Buettner-Janusch, ed.), pp. 477–487. N. Y. Acad. Sci., New York.

Boulière, F. (1962). "Natural History of Mammals." Knopf, New York.

Buettner-Janusch, J., and Andrew, R. J. (1962). The use of the incisors by primates in grooming. *Amer. J. Phys. Anthropol.* **20,** 127–129.

Burt, W. H. (1943). Territoriality and home range concepts as applied to mammals. *J. Mammal.* **24,** 346–352.

Carpenter, C. R. (1934). A field study of the behavior and social relations of the howling monkeys (*Alouatta palliata.*) *Comp. Psychol. Monogr.* **10,** 1–168.

Carpenter, C. R. (1940). A field study of the behavior and social relations of the gibbon (*Hylobates lar*). *Comp. Psychol. Monogr.* **15,** 1–212.

Carpenter, C. R. (1963). Societies of monkeys and apes. *In* "Primate Social Behavior" (C. F. Southwick, ed.), pp. 24–52. Van Nostrand, Princeton, New Jersey.

Carpenter, C. R. (1964). "Naturalistic Behavior of Nonhuman Primates." Pennsylvania State Univ. Press, University Park, Pennsylvania.

Darling, F. F. (1952). Social life in ungulates. *Struct. Physiol. Soc. Animaux* **134,** 221–226.

Davis, D. E., Emlen, J. T., and Stokes, A. W. (1948). Studies on home range in the brown rat. *J. Mammal.* **29,** 207–225.

DeVore, I. (1963). A comparison of the ecology and behavior of monkeys and apes. *In* "Classification and Human Evolution" (S. L. Washburn, ed.), pp. 301–316. Viking Fund, New York.

DeVore, I., and Hall, K. R. L. (1965). Baboon ecology. *In* "Primate Behavior" (I. DeVore, ed.), pp. 20–52. Holt, Rinehart & Winston, New York.

Ellefson, J. O. (1968). Territorial behavior in the common white-handed gibbon, *Hylobates lar* Linn. *In* "Primates: Studies in Adaptation and Variability" (P. Jay, ed.), pp. 180–200. Holt, Rinehart & Winston, New York.

Emlen, J. T. (1960). Introduction. *In* "Animal Sounds and Communication" (W. E. Lanyon, ed.), pp. ix–xiii. Intelligence Printing Co., Washington, D. C.

Francis, W. (1906). "Madras District Gazetteers: Madura." Govt. Press, Madras.

Frisch, J. (1968). Individual behavior and intertroop variability in Japanese macaques. *In* "Primates: Studies in Adaptation and Variability" (P. Jay, ed.), pp. 243–253. Holt, Rinehart & Winston, New York.

Furuya, Y. (1961–1962). The social life of silvered leaf monkeys. *Trachypithecus cristatus. Primates* **3,** 41–60.

Furuya, Y. (1969). On the fission of troops of Japanese monkeys, II. General view of troop fission of Japanese monkeys. *Primates* **10,** 47–71.

Goodall, J. (1965). Chimpanzees of the Combe Stream reserve. *In* "Primate Behavior" (I. DeVore, ed.), pp. 425–474. Holt, Rinehart & Winston, New York.

Grigg, B. (1880). "A Manual of the Nilgiri District." Govt. Press, Madras.

Gumperz, J. J. (1962). Types of linguistic communities. *Anthropol. Linguistics* pp. 28–49.

Haddow, A. J. (1952). Field and laboratory studies on the African monkey, *Cercopithecus ascanius schmidti*. *Proc. Zool. Soc. London* **122**, 297–394.

Hall, K. R. L. (1960). Social vigilance behavior in the chacma baboon. *Papio ursinus*. *Behaviour* **16**, 261–294.

Hall, K. R. L. (1965). Behavior and ecology of the wild patas monkey, *Erythrocebus patas*, in Uganda. *J. Zool.* **140**, 15–87.

Hall, K. R. L., and DeVore, I. (1965). Baboon social behavior. *In* "Primate Behavior" (I. DeVore, ed.), pp. 53–111. Holt, Rinehart & Winston, New York.

Harlow, H. F. (1962). Development of affection in primates. *In* "The Roots of Behavior" (E. L. Bliss, ed.), pp. 157–166. Harper & Row, New York.

Harlow, H. F. (1963). Basic social capacity of primates. *In* "Private Social Behavior" (C. F. Southwick, ed.), pp. 174–185. Van Nostrand, Princeton, New Jersey.

Harlow, H. F., and Harlow, M. K. (1961). A study of animal affection. *Natur. Hist.* **70**, 48–55.

Haugen, I. (1953). "The Norwegian Language in America." Univ. of Pennsylvania Press, Philadelphia, Pennsylvania.

Hediger, H. (1953). "Studies of the Psychology and Behavior of Animals in Zoos and Circuses." Butterworth, London and Washington, D. C.

Howard, H. E. (1920). "Territory in Bird Life." John Murray, London.

Hutton, A. F. (1951). *J. Bombay Natur. Hist. Soc.* **48**, 681–695.

Itani, J. (1959). Paternal care in the wild Japanese monkey. *Macaca fuscata fuscata*. *Primates* **2**, 84–98.

Itani, J. (1963). Vocal communication of the wild Japanese monkey. *Primates* **4**, 11–67.

Itani, J., Tokuda, K., Furuya, Y., Kano, K., and Shin, Y. (1963). Social construction of natural troops of Japanese monkeys in Takasakiyama. *Primates* **4**, 2–42.

Jay, P. (1962). Aspects of maternal behavior among langurs. *In* "Relatives of Man" (J. Buettner-Janusch, ed.), pp. 468–477. N. Y. Acad. Sci., New York.

Jay, P. (1963a). The Indian langur monkey (*Presbytis entellus*). *In* "Primate Social Behavior" (C. F. Southwick, ed.), pp. 114–124. Van Nostrand, Princeton, New Jersey.

Jay, P. (1963b). Mother-infant relations in langurs. *In* "Maternal Behavior in Mammals" (H. R. Rheingold, ed.), pp. 282–304. Holt, Rinehart & Winston, New York.

Jay, P. (1965a). The common langur of North India. *In* "Primate Behavior" (I. DeVore, ed.), pp. 197–250. Holt, Rinehart & Winston, New York.

Jay, P. (1965b). Field studies of monkeys and apes. *In* "Behavior of Nonhuman Primates" (A. M. Schrier, H. F. Harlow, and F. Stollnitz, eds.), pp. 525–591. Academic Press, New York.

Jensen, G. D., Bobbitt, R. H., and Gordon, B. N. (1967). *In* "Social Communication Among Primates" (S. A. Altmann, ed.), pp. 43–53. Univ. of Chicago Press, Chicago, Illinois.

Jolly, A. (1966). "Lemur Behavior: A Madagascan Field Study." Univ. of Chicago Press, Chicago, Illinois.

Kaufman, I. C., and Rosenblum, L. A. (1966). A behavioral taxonomy for *Macaca nemestrina* and *Macaca radiata:* Based on longitudinal observation of family groups in the laboratory. *Primates* **7**, 206–258.

Kaufman, I. C., and Rosenblum, L. A. (1969). The waning of the mother-infant bond in two species of macaque. *In* "Determinants of Infant Behavior IV" (B. M. Foss, ed.), pp. 41–59. Methuen, London.

Kaufmann, J. H. (1962). "Ecology and Social Behavior of the coati, *Nasua narica* on Barro Colorado Island." Univ. of California Press, Berkeley, California.

Kawai, M. (1958). On the system of social ranks in a natural troop of Japanese monkeys I: Basic and dependent rank. *Primates* **1**, 111–130.

Kawamura, S. (1958). Matriarchal social ranks in the Minoo-B troop: A study of the rank system of Japanese monkeys. *Primates* **2**, 181–252.

Kawamura, S. (1963). The process of sub-cultural propagation among Japanese macaques. *In* "Primate Social Behavior" (C. F. Southwick, ed.), pp. 82–91. Van Nostrand, Princeton, New Jersey.

Koford, C. B. (1957). "The Vicuna and the Puna." *Ecol. Monogr.* **27**, 153–219.

Koford, C. B. (1963). Rank of mothers and sons in bands of rhesus monkeys. *Science* **141**, 356–357.

Krishnamurti, S. (1958). "Horticultural and Economic Plants of the Nilgiris." Co-operative Printing Press, Coimbatore, India.

Kummer, H. (1967a). Dimensions of a comparative biology of primate groups. *Amer. J. Phys. Anthropol.* **27**, 357–366.

Kummer, H. (1967b). Tripartite relations in hamadryas baboons. *In* "Social Communication among Primates" (S. A. Altmann, ed.), pp. 63–73. Univ. of Chicago Press, Chicago, Illinois.

Kummer, H. (1968). "Social Organization of Hamadryas Baboons." Univ. of Chicago Press, Chicago, Illinois.

Kummer, H., and Kurt, F. (1963). Social units of a free-living population of hamadryas baboons. *Folia Primatol.* **1**, 11–28.

Lancaster, J. B., and Lee, R. (1965). The annual reproductive cycle in monkeys and apes. *In* "Primate Behavior" (I. DeVore, ed.), pp. 486–514. Holt, Rinehart & Winston, New York.

Leigh, C. (1926a). Weights and measurements of the Nilgiri langur (*Presbytis johnii*). *J. Bombay Natur. Hist. Soc.* **30**, 223.

Leigh, C. (1926b). Breeding seasons of the Nilgiri langur. *J. Bombay Natur. Hist. Soc.* **30**, 691.

Loizos, C. (1967). Play behavior in higher primates: A review. *In* "Primate Ethology" (D. Morris, ed.), pp. 179–219. Aldine, Chicago, Illinois.

Lorenz, K. (1956). Play and vacuum activities. *In* "L'instinct dans le comportement des Animaux et de L'Homme" (S. Autori, ed.). Fondation Singer-Polignac, Paris.

McCann, C. (1933). Observations on some of the Indian langurs. *J. Bombay Natur. Hist. Soc.* **36**, 618–628.

Maples, W. (1969). Adaptive behavior of baboons. *Amer. J. Phys. Anthropol.* **31**, 107–109.

Marler, P. (1965). Communication in monkeys and apes. *In* "Primate Behavior" (I. DeVore, ed.), pp. 544–584. Holt, Rinehart & Winston, New York.

Marler, P. (1968). Aggregation and dispersal: Two functions in primate communication. *In* "Primates: Studies in Adaptation and Variability" (P. Jay, ed.), pp. 420–439. Holt, Rinehart & Winston, New York.

Marler, P. (1969). *Colobus. guereza:* Territoriality and group composition. *Science* **163,** 93–95.

Mason, W. A. (1960). The effects of social restriction on the behavior of rhesus monkeys. I. Free social behavior. *J. Comp. Physiol. Psychol.* **54,** 582–589.

Mason, W. A. (1961a). The effects of social restriction on the behavior of rhesus monkeys. II. Tests of gregariousness. *J. Comp. Physiol. Psychol.* **54,** 287–290.

Mason, W. A. (1961b). The effects of social restriction on the behavior of rhesus monkeys. III. Dominance tests. *J. Comp. Physiol. Psychol.* **54,** 694–699.

Mason, W. A. (1963). The effects of environmental restriction on the social development of rhesus monkeys. *In* "Primate Social Behavior" (C. F. Southwick, ed.), pp. 161–174. Van Nostrand, Princeton, New Jersey.

Mason, W. A. (1965). The social development of monkeys and apes. *In* "Primate Behavior" (I. DeVore, ed.), pp. 514–543. Holt, Rinehart & Winston, New York.

Mason, W. A. (1966). Social organization of the South American monkey *Callicebus moloch:* A preliminary report. *Tulane Stud. Zool.* **13,** 23–28.

Mason, W. A. (1968). Use of space by *Callicebus* groups. *In* "Primates: Studies in Adaptation and Variability" (P. Jay, ed.), pp. 200–217. Holt, Rinehart & Winston, New York.

Morris, C. (1946). "Signs, Language, and Behavior." Prentice-Hall, Englewood Cliffs, New York.

Morris, D. (1964). The response of animals to a restricted environment. *Symp. Zool. Soc. London* **13,** 99–118.

Morris, R. C. (1927). Elephants eating dirt. *J. Bombay Natur. Hist. Soc.* **36,** 96.

Nissen, H. (1958). Axes of behavioral comparison. *In* "Behavior and Evolution" (A. Roe and G. G. Simpson, eds.), pp. 183–206. Yale Univ. Press, New Haven, Connecticut.

Osman Hill, W. C. (1934). A monograph on the purple-faced leaf monkeys (*Pithecus vetulus*). *Ceylon J. Sci.* **19,** 23–89.

Pocock, R. O. (1928). The langurs or leaf-monkeys of British India. *J. Bombay Natur. Hist. Soc.* **32,** 660–678.

Poirier, F. E. (1964). The Communication Matrix of the Celebes Ape (*Cynopithecus niger*): A Study of Sixteen Male Laboratory Animals. Univ. Microfilms, Ann Arbor, Michigan.

Poirier, F. E. (1966). The Nilgiri langur (*Presbytis johnii*) mother-infant dyad. 65th Annual Amer. Anthropol. Ass. Meetings, Pittsburgh, Pennsylvania.

Poirier, F. E. (1968a). Tactile communication among Nilgiri langurs: With reflections on its role among humans. 3rd Annual Southern Anthropol. Ass. Meetings, Gainesville, Florida.

Poirier, F. E. (1968b). "'The Ecology and Social Behavior of the Nilgiri langur (*Presbytis johnii*) of South India." Univ. Microfilms, Ann Arbor, Michigan.

Poirier, F. E. (1968c). Analysis of a Nilgiri langur (*Presbytis johnii*) home range change. *Primates* **9,** 29–43.

Poirier, F. E. (1968d). The Nilgiri langur (*Presbytis johnii*) mother-infant dyad. *Primates* **9,** 45–68.

Poirier, F. E. (1968e). Nilgiri langur (*Presbytis johnii*) territorial behavior. *Primates* **9,** 351–364.

Poirier, F. E. (1969a). The Nilgiri langur troop: Its composition, structure, function and change. *Folia Primatol.* **10,** 20–47.

Poirier, F. E. (1969b). Behavioral flexibility and intertroop variability among Nilgiri langurs (*Presbytis johnii*) of South India. *Folia Primatol.* 11, 119–133.

Poirier, F. E. (1970). Characteristics of the Nilgiri langur (*Presbytis johnii*) dominance structure. *Folia Primatol.* 12, 161–187.

Poirier, F. E. The ecology and social behavior of the St. Kitts green monkey (*C. aethiops sabaeus*) (in prep.).

Prater, S. H. (1965). "The Book of Indian Animals." Diocesan Press, Madras.

Ripley, S. (1967). Intertroop encounters among Ceylon gray langurs (*Presbytis entellus*). *In* "Social Communication Among Primates" (S. A. Altmann, ed.), pp. 237–253. Univ. of Chicago Press, Chicago, Illinois.

Rosenblum, L. A., and Kaufman, I. C. (1967). Laboratory observations of early mother-infant relations on pigtail and bonnet macaques. *In* "Social Communication Among Primates" (S. A. Altmann, ed.), pp. 33–41. Univ. of Chicago Press, Chicago, Illinois.

Rowell, T. (1966). Forest living baboons in Uganda, *J. Zool.* 149, 344–364.

Ryley, V. K. (1913). The Bombay Natural History Society mammal survey of India. *J. Bombay Natur. Hist. Soc.* 22, 13–26.

Sade, D. S. (1965). Some aspects of parent-offspring and sibling relations in a group of monkeys, with a discussion of grooming. *Amer. J. Phys. Anthropol.* 23, 1–17.

Schaller, G. (1963). "The Mountain Gorilla." Univ. of Chicago Press, Chicago, Illinois.

Simonds, P. E. (1963). Ecology of Bonnet Macaques. Unpublished doctoral dissertation, Univ. of California, Berkeley, California.

Simonds, P. E. (1965). The bonnet macaque of South India. *In* "Primate Behavior" (I. DeVore, ed.), pp. 175–197. Holt, Rinehart & Winston, New York.

Singh, S. (1969). Urban monkeys. *Sci. Amer.* 108–115.

Southwick, C. H. (1962). Patterns of intergroup social behavior in primates with special reference to rhesus and howling monkeys. *In* "The Relatives of Man" (J. Buettner-Janusch, ed.), pp. 436–455. N. Y. Acad. Sci., New York.

Southwick, C. H., Beg, M. A., and Siddiqi, M. R. (1965). Rhesus monkeys of North India. *In* "Primate Behavior" (I. DeVore, ed.), pp. 111–160. Holt, Rinehart & Winston, New York.

Sparks, J. (1968). Allogrooming in primates: A review. *In* "Primate Ethology" (D. Morris, ed.), pp. 148–176. Aldine, Chicago, Illinois.

Spate, O. H. K. (1954). "India and Pakistan: A General and Regional Geography." Methuen, London.

Sterndale, R. A. (1884). "Natural History of the Mammals of India and Ceylon." Thacker, Spink, Calcutta.

Struhsaker, T. T. (1967). Auditory communication among vervet monkeys (*Cercopithecus aethiops*). *In* "Social Communication among Primates" (S. A. Altmann, ed.), pp. 281–323. Univ. of Chicago Press, Chicago, Illinois.

Struhsaker, T. T. (1969). Correlates of ecology and social organization among African cercopithecines. *Folia Primatol.* 1, 80–119.

Sugiyama, Y. (1960). On the division of a natural troop of Japanese monkeys at Takasakiyama. *Primates* 2, 109–148.

Sugiyama, Y. (1965a). Behavioral development and social structure in two troops of hanuman langurs (*Presbytis entellus*). *Primates* 6, 73–106.

Sugiyama, Y. (1965b). On the social change of hanuman langurs (*Presbytis entellus*) in their natural condition. *Primates* 6, 381–418.

Sugiyama, Y. (1967). Social organization of hanuman langurs. *In* "Social Communication Among Primates" (S. A. Altmann, eds.), pp. 221–236. Univ. of Chicago Press, Chicago, Illinois.

Sugiyama, Y. (1968). The ecology of the lion-tailed macaque [*Macaca silenus* (Linnaeus)]—A pilot study. *J. Bombay Natur. Hist. Soc.* **65**, 283–292.

Tanaka, J. (1965). Social structure of Nilgiri langurs. *Primates* **6**, 107–122.

Tinbergen, N. (1951). "The Study of Instinct." Oxford Univ. Press (Clarendon), London and New York.

Tinklepaugh, O. L., and Hartman, O. G. (1932). Behavior and maternal care of the newborn monkey (*M. mulatta, M. rhesus*). *J. Genet. Psychol.* **40**, 257–286.

Tomilin, M. I., and Yerkes, R. M. (1935). Chimpanzee twins: Behavioral relations and development. *J. Genet. Psychol.* **46**, 239–263.

Tsumori, A. (1967). Newly acquired behavior and social interactions of Japanese monkeys. *In* "Social Communication Among Primates" (S. A. Altmann, ed.), pp. 207–221. Univ. of Chicago Press, Chicago, Illinois.

Washburn, S. L., and DeVore, I. (1961). The social life of baboons. *Sci. Amer.* **204**, 62–71.

Washburn, S. L., and Hamburg, D. (1965). The implications of primate research. *In* "Primate Behavior" (I. DeVore, ed.), pp. 607–622. Holt, Rinehart & Winston, New York.

Wynne-Edwards, W. C. (1962). "Animal Dispersion in Relation to Social Behavior." Oliver & Boyd, Edinburgh and London.

Yamada, M. (1963). A study of blood relationships in the natural society of the Japanese macaque. *Primates* **4**, 43–67.

AUTHOR INDEX

Numbers in italics refer to the pages on which the complete references are listed.

SUBJECT INDEX

A

Abnormal behavior, 195–249, *see also* Social isolation
  birth and, 206–207
  defined, 197–198, 242
  maternal, 238–242, 243
  peer deprivation and, 209–210
  peer-only rearing and, 210–211
  sexual, 235–238
  social isolation and, 219–238
Activity level, 204, 205, 207, 215
  tree shrew, 152–153, 156–157, 161, 185
Adaptability, 64–65
  troop, 368–370
African red-tailed monkey, 300, 353
Age, *see also* Maturation
  abnormal behavior for, 204–205, 233, 242
  dominance and, 74, 104, 319, 320
  grooming and, 332
  infant transference and, 309
  learning set and, 18
  motherless mother aggressiveness and, 241–242
  play and
    nonsocial, 346–347
    social, 349–350
  problem-solving skills and, 20
  ratio in Nilgiri langur troop, 264, 266
  sex preferences and, 121–125
  social isolation and, *see* Social isolation
  species preferences and, 118, 119, 137
  vocalizations and, 285
Aggression, 129, 210, 355, *see also* Agonistic behavior, Attack
  captivity and, 202, 203
  dominance and, 73, 75–77, 79, 86, 103, 267, 320, 326, 327, 371–373
  mother-infant separation and, 212, 217

motherless mothers and, 239–242
Nilgiri langur, 321, 322, 324–326, 357
  intertroop, 296–299
play and, 340, 343
sex and, 205
social isolation and, 120, 220, 222, 225–227, 233–234, 236, 238–242, 243
tree shrew, 160, 161, 164, 167–171, 185
  vocalization during, 171, 173
troop change and, 357, 360
Agonistic behavior, 90, *see also* Aggression
  captivity and, 201, 202
  communication and, 278–279, 291
  dominance and, 73–79, 82, 84–87, 89–93, 95, 100–104, 320, 326, 376
  grooming and, 337
  space and, 202
Alarm call, 285, 286
Alliance, 77, 322, 327
Allogrooming, 176, 330, 337
Alternation tasks, 34
Approach, 75, 113
  conditioned, 7, 8, 10
Arousal, emergence from isolation and, 231–232, 235
Asphyxia neonatorum, 206–207, 242
Association theory, 11–12, 14–15, 18, 57
Attack, dominance and, 75, 371–373, 374
Attention, 12, 14, 24, 126
  experience and, 32
Avoidance, 75
  conditioned, 7, 8, 10, 230
  social isolation and, 120

B

Baboon
  captivity of, 202
  dominance in, 72, 75–78, 80, 82–84, 88–96, 98–102, 267, 326, 327

392